Water for Life

Water is a precious resource essential for all forms of life. It can be thought of as the blood of the earth. Although there is plenty of water to meet the demand for the present population and even for a projected population of about 9 billion, there is significant spatial and temporal variation in the global distribution of this precious resource. As a result, there are water-rich and water-poor countries with the latter facing water stress and water scarcity which in extreme situations can lead to water-related conflicts and even 'water-wars.' The World Health Organization (WHO) has identified unsafe drinking water as a major killer in the world. The motivation for writing *Water for Life: Drinking Water, Health, Food, Energy Nexus* is primarily to throw light on the multi-faceted uses and importance of water in life, in particular to highlight the water, health, food, and energy nexus. It is hoped that the contents will help students in civil engineering, geography, and earth and social sciences to perceive the big picture of water management for all human and biotic populations without causing negative effects on the environment.

Water for Life
Drinking Water, Health, Food, Energy Nexus

A. W. Jayawardena

Routledge
Taylor & Francis Group

LONDON AND NEW YORK

First published 2023
by Routledge
4 Park Square, Milton Park, Abingdon, Oxon OX14 4RN

and by Routledge
605 Third Avenue, New York, NY 10158

Routledge is an imprint of the Taylor & Francis Group, an informa business

© 2023 A. W. Jayawardena

British Library Cataloguing-in-Publication Data
A catalogue record for this book is available from the British Library

Library of Congress Cataloging-in-Publication Data
Names: Jayawardena, A. W., author.
Title: Water for life : drinking water, health, food, energy nexus / A. W. Jayawardena.
Description: Abingdon, Oxon ; New York, NY: Routledge, 2023. |
Includes bibliographical references and index.
Identifiers: LCCN 2022026831 | ISBN 9781032358888 (paperback) |
ISBN 9781032360010 (hardback) | ISBN 9781003329206 (ebook)
Subjects: LCSH: Water resources development. |
Water-supply. | Hydrology. | Public health.
Classification: LCC TC405 .J39 2023 |
DDC 363.6/1–dc23/eng/20221012
LC record available at https://lccn.loc.gov/2022026831

ISBN: 978-1-032-36001-0 (hbk)
ISBN: 978-1-032-35888-8 (pbk)
ISBN: 978-1-003-32920-6 (ebk)

DOI: 10.1201/9781003329206

Typeset in Sabon
by Newgen Publishing UK

Contents

7   Water and health   129

11 Water and energy   231

Preface

Water is a precious resource essential for all forms of life. Despite its importance in life, it does not receive the attention it deserves compared to that received for some other resources the world needs. Although there is plenty of water to meet the demand for the present population and even for a projected population of about 9 billion, there is significant spatial and temporal variation in the global distribution of this precious resource. As such, it is not always available when and where needed. As a result, there are water-rich countries and water-poor countries with the latter facing water stress and water scarcity, which in extreme situations can lead to water-related conflicts and even 'water-wars.'

The main use of water is for drinking followed by the needs for growing and processing of food the human and other biotic populations need. It has been highlighted that approximately 29% of the world's population does not have access to safe drinking water. The World Health Organization (WHO) has identified unsafe drinking water as a major killer in the world. There are also similar outcomes caused by the lack of access to basic food by a significant part of the global population leading to malnutrition and sometimes preventable deaths. In addition to these two basic needs, there are other uses of water such as for energy, industry, transportation and recreation. On the negative side, too much water and/or too little water can lead to water-related disasters and conflicts.

The motivation for writing this book is primarily to shed light on the multifaceted uses and importance of water in life, in particular to highlight the water, health, food and energy nexus. Although it is not meant to be a textbook, it includes some engineering contents that can help undergraduate and graduate students in civil engineering, geography and earth and social sciences to perceive the big picture of water management for all human and biotic populations without causing negative effects on the environment. It can also be a source of reference for high school teachers who can impart the importance and value of water in daily life to students.

In writing this book, the author benefitted from the published work of others. Appropriate acknowledgements have been made in citing such work in relevant parts of the text. Last but not least, the author expresses his gratitude to Tony Moore, Senior Editor for Taylor and Francis, who gave continuous encouragement.

Despite the care taken to ensure the correctness of the material presented, it is still possible that there may be typographical errors and/or omissions due to oversight. The author would be grateful if the readers would kindly bring to his attention if they find any such errors and/or omissions. After all, to err is human and to forgive is divine.

A. W. Jayawardena May 15, 2022

Author bio

A. W. Jayawardena is currently an adjunct professor in the Department of Civil Engineering at the University of Hong Kong where he has been teaching for many years. He has also been a visiting professor, Chu Hai College of Higher Education, Hong Kong; a technical advisor to the Research and Development Centre, Nippon Koei Co. Ltd, Japan; Research and Training Advisor to the International Centre for Water Hazard and Risk Management (ICHARM) under the auspices of UNESCO and concurrently a professor in the National Graduate Institute for Policy Studies (GRIPS), Japan; a guest /visiting professor of Beijing Normal, Hohai and Tsinghua Universities, China; Honorary Professor in the Department of Statistics and Actuarial Sciences of the University of Hong Kong; an adjunct professor of Vellore Institute of Technology, India; Senior Engineer, Howard Humphrey & Sons, Consulting Engineers, UK; and an engineer in the Irrigation Department of Sri Lanka. His awards include Visiting International Fellowship of the Environmental and Water Resources Institute (EWRI), American Society of Civil Engineers; International Award, Japan Society of Hydrology and Water Resources, and Invitational Fellowship, Japan Society for the Promotion of Science (JSPS). He has authored the books *Environmental and Hydrological Systems Modelling, Fluid Mechanics, Hydraulics, Hydrology and Water Resources for Civil Engineers*, and authored/co-authored over 200 publications.

Chapter 1

Introduction and scope

The water–food–energy nexus is central to sustainable development with water driving the food and energy sectors. Demand for all three is increasing, driven by a rising global population, rapid urbanization, changing dietary habits and economic growth. Water is essential for all forms of life, but the value it deserves is not recognized compared to the values attached to some other resources the world needs. Fresh water is abundant in nature and is sufficient to meet the demand for the present population and even for a future projected population of about 9 billion. However, there is significant spatial and temporal variations in the distribution of this precious resource across the globe, making some countries water-rich and some others water-poor. Water-rich countries in terms of high annual per capita availability of water include Iceland, Gabon, Papua New Guinea, Canada and New Zealand, in that order. At the other end of the spectrum, in some countries in North Africa and Arabian Peninsula, the per capita availability runs into negative terrain, implying that water comes from trans-boundary sources. The internationally recognized water scarcity level is 500 m^3 per capita per year. Water-poor countries suffer from water stress and water scarcity, which in uncontrollable situations can lead to water conflicts at inter-community and intra-community levels. Water conflicts in extreme cases can lead to 'water-wars'. Virtual water plays an important role in attempts to minimize the disparities between water-rich and water-poor countries.

Access to safe drinking water is recognized as a basic human right, but the unfortunate inconvenient truth is that it is not available for about 2.1 billion people (29% of the world population) on earth and that the people who are deprived of having access to safely managed drinking water live in low-income countries. Safely managed drinking water in this context implies water available on premises and free from contamination. To share this precious resource among all living things in an equitable manner while protecting the environment requires effective water resources management, good water governance and sometimes changing the way water is used. Failure to do so can result in water-related conflicts, which in extreme

DOI: 10.1201/9781003329206-1

situations can lead to 'water-wars'. This book aims to highlight the multi-faceted uses of water as well the water-related disasters resulting from too much or too little water and the conflicts that can arise when the ownership of water is contested.

1.1 SCOPE AND LAYOUT

The contents of this book are organized into 16 chapters with each chapter dealing with water and different aspects of life. The second chapter introduces earth's water resources followed by the chemistry of water, global perspective of drinking water, supply and demand of domestic water, drinking water from source to tap, water and health, water and food, water and irrigation, water and soil, water and energy, water for transport, water for industry, water for recreation, water and disasters and water-related conflicts in each succeeding chapter. Where relevant, lists of references for follow-up are also given at the end of each chapter. Brief descriptions of coverage of each chapter are given in the following paragraphs.

The total freshwater in land areas of earth is about 2.5% of the total water resources on earth. Of this, the extractable percentage is about 0.5% since much of the freshwater is trapped in ice caps and glaciers (about 1.925%). The most dynamic part of water is the surface waters contained in rivers, freshwater lakes and inland seas that has a residence time of about 2.3 years compared to the residence times of about 3,000 and 10,000 years, respectively, for water in the oceans, and ice caps and glaciers. Although there is plenty of water available worldwide, the per capita share of water is decreasing with time as the global population is increasing unabatedly. This is an issue that leads to water stresses, water scarcity and water conflicts. Under the auspices of the United Nations, several international initiatives are being actively pursued to meet the Sustainable Development Goals (SDG 6 in particular) by 2030. These and other related issues are presented in Chapter 2.

Water is a chemical compound made of two of the universe's most abundant elements, hydrogen and oxygen. The chemical formula of water, H_2O, is better known than that of any other chemical compound, but from the point of predicting its chemical properties, it is one of the least known substances on earth. Water is colourless, odourless, tasteless and transparent and is a powerful universal solvent as well as a means of transporting people, goods and services. In the human body, water flushes body waste (mainly urine), helps digestion of food and regulates body temperature by sweating and respiration. Water exists in all three phases – solid, liquid and gas – and one of its unique properties is that its density in solid form is less than its density in liquid form with the highest density at 4°C. All other substances that exist in the three phases have the following general relationship:

$$\rho_{solid} > \rho_{liquid} > \rho_{gas}$$

Water also has the highest surface tension among all liquids except mercury that allows objects with a higher density to float on the surface rather than to sink, which is due to the strong hydrogen bond in the water molecule. Other unique properties include high capillarity, high boiling and melting points. Water can generally be classified as blue water, green water, white water, grey water and black water depending on the occurrence, function and the quality. These and other related properties are presented in Chapter 3.

Chapter 4 is about water and health. The health of a nation depends upon the level of cleanliness of the domestic water supply. The World Health Organization (WHO) has concluded that water-borne diseases is the leading killer in the world (Berman, 2009). According to some data sources, the deaths due to unsafe water as a percentage of the total deaths range from a high of 14.5% in poor and developing countries to near zero in developed countries. In this context, safely managed water means water available on premises when needed and free from contamination. It is a problem that is often ignored or sidelined by the developed countries as it is only a problem of the poor and the developing countries. With the emergence of COVID-19, health authorities all over the world are advising and promoting hand washing as a precaution against transmission of novel corona virus, but the big question is where is the water for 29% of the world population. This is an issue that needs attention at all levels.

Water has contributed to the elimination of hunger through irrigation and diseases through public water supplies. Unlike in the ancient times when people used to live near sources of water, the present-day living expects the water to be delivered to the place of living. This requires the necessary infrastructure, which includes finding a source of water, harnessing it and storing where necessary, treating and conveying potable water to the place of living. Such tasks require inputs from several disciplines and civil engineering is a major one. Sources include rainwater, surface water, groundwater, sea water, imported water, soil moisture and recycled water. The demand for this precious resource comes from the sectors such as domestic, agricultural, industrial, recreational, transportation and energy. From the point of view of optimal management of available water resources, the engineers' task is to balance the supply and demand. Some engineering aspects of collection and storage, reservoirs and reservoir sedimentation, water intakes, wells etc. are also presented in Chapter 5.

Once finding a source of water, harnessing, collecting and storing has been completed, the most important remaining task is to treat the water. The treatment processes involved consist of sedimentation, coagulation, flocculation, filtration, fluoridation, disinfection and aeration. These tasks need inputs from biologists, chemists and civil engineers. The treated water needs to satisfy the standards set by the local water authorities who generally follow the standards set by the World Health Organization. Finally, a distribution system to convey the treated water to consumers' ends needs to be designed and constructed. For consumers who do not depend on public water supplies, some aspects of treatment at individual household level are also presented in Chapter 6.

Water contamination can take place in different forms. They include debris and sediment contamination, microbial contamination due to the presence of pathogens such as viruses, bacteria, protozoa and helminths, chemical contamination and radiological contamination, which is very rare. Microbial contamination can lead to several types of water-borne diseases, some of which can be fatal. Some details of the water-borne diseases including their causes, modes of transmission, diagnosis, prevention and treatment are given in Chapter 7.

Food comes next to water for the survival of living things. Globally, nearly one in nine people, or about 820 million people, are hungry or undernourished, and about 132 million people live with acute hunger that approaches starvation (McCarthy and Sanchez, 2020). In sub-Saharan Africa, the percentage of population undernourished is much higher. Even as hunger rises around the world, more people are becoming overweight or obese, with nearly a third of the global population falling into this category, according to the report. Growing and production of food takes about 1,000 times more water than the drinking water requirement. The main sources of food are cereals that include rice, corn, wheat and potatoes. The main source of water for agriculture is rainfall supplemented with irrigation. The Food and Agricultural Organization (FAO) estimates that the water withdrawal for agriculture would increase by 11% in 2030 compared to that in 2000. Water is also needed for livestock and fisheries. The United Nations Sustainable Development Goal 2 (SDG2) aims at ending hunger, achieving food security and improved nutrition, and promoting sustainable agriculture by 2030. Chapter 8 gives some details of evapo-transpiration and crop water requirements as well as other related issues of water and food.

Irrigation played an important role in human civilization. Hydraulic civilization flourished along the banks of rivers in Egypt, Mesopotamia (land between the Tigris and Euphrates Rivers, which at present is Iraq), India and China. Irrigation has a positive impact on alleviating poverty, which is the root cause of malnutrition. It has been reported that in India, 69% of people in un-irrigated districts are poor whereas only 26% are poor in irrigated districts (World Bank, 1991). Irrigation requirements estimated using a simple water balance equation depends on crop water evapo-transpiration and other sources of water such as rainfall. The total irrigated area in the world as of 2012 has been approximately 2% of the world's land area with China taking the top place followed by India, United States, Pakistan, Iran and Indonesia, in that order. Micro-irrigation has been playing an increasingly important role in China. These and other related issues are presented in Chapter 9.

Soil is a medium that receives water from rainfall and irrigation, stores it together with nutrients and transmits it to plants, which in turn produce the food via photosynthesis that humans and animals need for their lives. The growth and yield of cultivated plants depend heavily on soil properties and on oxygen in soil water. Chapter 10 gives details of basic soil properties, soil fertility, soil erosion, soil stability and soil failure, as well as the basic geotechnical properties of soil. It also describes the soil water system, including the mechanics of moisture movement in soils, the soil hydraulic parameters, the governing equations of moisture movement in partially saturated soils and the approaches for their solution.

Chapter 11 deals with water and energy, which has a symbiotic relationship. Energy is required for providing water services and water resources are required for the production of energy. Hydro-electricity is the largest renewable source of energy in the world, meeting about 16% of global electricity needs. Approximately 90% of global energy production is water-intensive. The amount of energy needed to provide potable water varies from source to source. Desalination is the most expensive method of providing potable water. The provision of water for agriculture, which needs no treatment, requires energy for delivering the water from the source to the demand area, which depends upon the quantity of water delivered and the elevation difference between the source and the destination. Energy is also required for wastewater treatment before discharging into receiving waters as well as for industrial processes either as a cooling agent or as inputs to the manufacturing process. A relatively new type of demand for water is in the area of hydraulic fracturing (or fracking) for extracting oil and/or gas from subterranean rocky formations. The United Nations Sustainable Development Goals (SDG 6 and SDG 7) aim to ensure availability and sustainable management of water and sanitation for all and access to affordable, reliable, sustainable and modern energy for all by 2030. Other forms of water-related energy include tidal energy, wave energy, blue energy and shale energy, which requires hydraulic fracturing.

Water, apart from being the most important ingredient of life, is also a medium of transport. It carries nutrients necessary for the metabolism of all living things and also carries away the waste generated in living beings. Plants have an ingenious hydraulic system of transporting water and nutrients from the soil to the leaves in the canopy against gravity, which in tall trees can exceed 100 m in height. Water is also a mode of transport for the movement of people, goods and services over oceans, lakes, canals, rivers and other waterways. World trade heavily depends on ocean transport. This topic is discussed in Chapter 12.

Industrial demand for water is relatively new compared with the agricultural and domestic demands. The major industries that use water include manufacturing processes, electricity-generating industries, iron and steel industry, textile industry, paper and pulp manufacturing industry, beverage industry, automotive industry and for fire-fighting, which is essential in urban living where the rates of flow need to be very high but lasts for a short time. In developed countries, industries are the major users of water accounting for approximately 55% of total water withdrawals whereas in less developed countries, it is about 9%. Every manufactured product uses water during some part of the production process including water used for fabricating, processing, washing and diluting, cooling and transporting, incorporating water into a product or for sanitation needs within the manufacturing facility. Industries such as food, textile, pulp and paper, iron and steel, chemicals, petroleum etc. use large quantities of water. This topic is discussed in Chapter 13.

Recreation, discussed in Chapter 14, is a popular activity among people of all ages. Recreation in water can be with active participation such as swimming and diving or passive participation such as observing a recreational activity by others. It is also a major contributor to tourism. Water-related tourism includes ocean tourism, canal tourism, beach tourism, river tourism, water park tourism, ski tourism and hot spring tourism. Active water-related recreation can lead to health hazards if the quality of water is below safety levels.

Too much water or too little water can lead to water hazards, which may become a disaster if the vulnerable community lacks the coping capacity. Water-related disasters include floods, droughts, storms (monsoons, cyclones, hurricanes, thunderstorms, tornados, tropical depressions and tropical storms), landslides, avalanches, tsunami and water-related biological disasters. In terms of the cost and damage induced by various types of natural disasters, 'water-related disasters' by far exceed those by any other natural disaster.

International initiatives to reduce disaster risks include International Decade for Natural Disaster Reduction (IDNDR), Yokohama strategy and related plan of action, United Nations International Strategy for Disaster Reduction (UNISDR), United Nations Office for Disaster Risk Reduction (UNDRR), Hyogo Framework for Action (HFA) 2005–2015, Sendai Framework for disaster Risk Reduction 2015–2030, international day for disaster risk reduction and International Flood Initiative (IFI). Such initiatives aim to reduce disaster risks by information and experience sharing, knowledge exchange and training. These, including some statistics of major water-related disasters, are presented in Chapter 15.

With the unabated increase in human population, the competition for water, which is a finite resource, is rapidly increasing, resulting in conflicts related to the use and availability of water. Such conflicts may occur regionally and/or across countries when the ownership of a source of water is contested. With over 286 international rivers and 592 trans-boundary aquifers shared by 153 countries (UN, 2018), it is natural to expect some conflicts about the way the water has to be shared. Water conflicts can arise because of territorial disputes, competition over resources and/or political reasons. Internal conflicts within countries can occur when there are disparities in the use and accessibility to water from different communities. Conflicts are most likely to affect the downstream countries when upstream countries take unilateral control over shared water. Such conflicts at times can lead to 'water-wars'. With the per capita share of water availability decreasing with time, such conflicts could also be expected to increase in the future. Causes, types and strategies to reduce water-related conflicts as well as some major water-related conflicts in the world are highlighted in Chapter 16, the last chapter. The Appendix gives a list of names for water in some other languages.

REFERENCES

Berman, J. (2009): WHO: Waterborne Disease Is World's Leading Killer, October 29, 2009. www.voanews.com/archive/who-waterborne-disease-worlds-leading-killer

McCarthy, Joe and Sanchez, Erica (2020): Malnutrition Is the Leading Cause of Death Globally: Report 2020 (globalcitizen.org) May 14, 2020.

United Nations (2018): Sustainable Development Goal 6: Synthesis Report 2018 on Water and Sanitation, United Nations.

World Bank (1991): Report No. 9518-IN India Irrigation Sector Review (in Two Volumes), Volume 1: Main Report December 20, 1991 (World Bank Document).

Chapter 2

Earth's water resources

2.1 INTRODUCTION

Water is a precious resource that is essential for all forms of life. It is abundant in nature but has significant temporal and spatial variability. With increasing population, the per capita share of water on earth is decreasing, and in some regions, it has reached levels to the extent that communities face water stress and water scarcity. Whereas lack of safe drinking water is a major problem for over a billion inhabitants of the earth, too much water also brings about misery, agony and destruction to many people, places and infrastructure. The former may be attributed to the physical lack of water, pollution or unaffordability, and the latter is attributed mainly to urbanization and livelihood issues.

The total amount of freshwater easily accessible, which consists of surface waters and groundwater, is about 14,000 km^3. Of this, only about 5,000 km^3 are being used by humans. Thus, there is plenty of freshwater available to meet the demands of the present population of over 7 billion and even to meet the needs of future populations up to about 9 billion. However, there is significant spatial and temporal variation in the distribution of this global resource, which can and will lead to water stresses and water scarcities in some places in the globe during certain times. To ensure that all living things in the world share this precious resource in an equitable manner requires effective water resources management, good water governance and sometimes changing the way water is used.

2.2 DISTRIBUTION OF WATER ON EARTH

Water is abundant in nature but not often found in places when and where needed. Table 2.1 gives the approximate distribution of earth's water resources from which it can be seen that temporary storage of water within the hydrological cycle occurs at four main places. They are:

- Oceans 1,350,400 (10^3 km^3)
- Ice caps and glaciers 26,000 (10^3 km^3)

Table 2.1 Approximate distribution of earth's water resources

Item	Area (km² × 10³)	Volume (km³×10³)	% of total water
Atmospheric vapour (water equivalent)	510,000 (at sea level)	13	0.0001
World ocean	362,033	1,350,400	97.6
Water in land areas:	148,067	(124,000)	–
Rivers (average channel storage)		1.7	0.0001
Freshwater lakes	825	125	0.0094
Saline lakes; inland seas	700	105	0.0076
Soil moisture;			
vadose water	131,000	150	0.0108
Biological water	131,000	(Negligible)	–
Groundwater	131,000	7,000	0.5060
Ice-caps and glaciers	17,000	26,000	0.9250
Total in land areas (rounded)		33,900	2.4590
Total water, all realms (rounded)		1,384,000	100
Cyclic water:			
Annual evaporation			
From world ocean		445	0.0320
From land areas		71	0.0050
Total		516	0.0370
Annual precipitation			
On world oceans		412	0.0291
On land areas		104	0.0075
Total		516	0.0370
Annual outflow from land to sea			
River outflow		29.5	0.0021
Calving, melting, and deflation from ice-caps		2.5	0.0002
Groundwater outflow		1.5	0.0001
Total		33.5	0.0024

- Groundwater 7,000 (10^3 km³)
- Lakes and inland seas 230 (10^3 km³)

The four main processes of water transfer between these storages are:

- Precipitation 516 (10^3 km³/year)
- Evaporation 516 (10^3 km³/year)
- Surface runoff 29.5 (10^3 km³/year)
- Groundwater flow 1.5 (10^3 km³/year)

The estimates of annual renewable water resources and access to renewable water resources globally have been reported to be 39.6 and 29.7 km³, respectively (WWDR3, 2009; Table 10.1). The latter works out to be about 75% of the total renewable water resources. Regionally they are 9.8 , 4.0,

13.2, 0.25, 4.4 and 8.1 km^3, respectively, for Asia, Eastern Europe and Central Asia, Latin America, Middle East and North Africa, sub-Saharan Africa and Organisation for Economic Co-operation and Development (OECD) for total renewable resources and 9.3, 1.8, 8.7, 0.24, 4.1 and 5.6 km^3, respectively, for Asia, Eastern Europe and Central Asia, Latin America, Middle East and North Africa, sub-Saharan Africa and OECD for access to renewable water resources (WWDR3, 2009; Table 10.1).

The total water in the land areas works out to about 2.459% of the earth's entire water resources. The extractable percentage of this is of the order of about 0.5% (from lakes, rivers and groundwater). The largest volume of freshwater is found in the ice caps and glaciers (26,000 × 10^3 km^3, or, 0.925%), which is sufficient to keep the world's rivers flowing for nearly 1,000 years (annual surface runoff from rivers is about 29.5 × 10^3 km^3). The rest of the freshwater is stored in inland lakes, soil water, groundwater, atmospheric water and rivers and streams. Of this fresh water, the extractable part is only of the order of about 1%. On a long-term basis, the total water resources on earth are in a stable equilibrium state through the processes of the hydrological cycle. The temporal and spatial variability and the changing lifestyles lead to water stress and water scarcity.

The human population in the world has always been increasing except for short-term falls in the 14th and 17th centuries. These falls were mainly due to pandemics caused by 'black death'[1] and plague. It is also projected that the population will continue to increase until about 2050. The positive trends can be attributed to a number of factors such as improved medical facilities, low infant mortality rates, increased life expectancy and increased food production. The population which in 1750 was 791 million has surpassed 7.9 billion as of November 2021, (Figure 2.1) (en.wikipedia.org/wiki/World_population), with the highest rate of growth of 2.2% per year recorded in 1963. It has taken over 200,000 years for the world's population to reach 1 billion and only 200 years more to reach 7 billion.

The world population surpassed the 7 billion mark at the end of October 2011. Regionally, Asia is the home to 60% of the world population followed by Africa with 16%, Europe with 10%, Latin America and Caribbean with 9%, North America with 5% and Oceania with 0.5%. China, India, the United States, Indonesia, Pakistan, Brazil, Nigeria, Bangladesh, Russia and Mexico rank as the 10 most populous countries whereas Singapore (with over 7,940 persons/km^2), Bangladesh, Taiwan, Lebanon, South Korea, Rwanda, the Netherlands, Haiti, India and Israel rank as the 10 most densely populated countries in the world in that order (en.wikipedia.org/wiki/World_population). Future projections of world population vary

---

1 Black death was the most devastating pandemic in human history which is reported to have killed about 30–60% of European population during 1348–1350. It is said to be caused by a bacterium carried by rat fleas living on 'black rats' travelling in merchant ships.

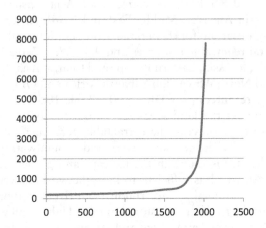

Figure 2.1 World population (vertical axis population in millions; horizontal axis time in years).

significantly from source to source depending upon the scenario assumed and range from a low of about 3.2 billion to a high of about 24.8 billion by the year 2050.

Despite the past trends, it is also well known that continued increase is not sustainable as the competition for resources will begin to dominate the growth rate sooner or later. An important resource that could dominate the growth rate is water.

The total global runoff to the oceans (exorheic) is estimated to be about 37,200 km³ annually and the corresponding runoff to inland receiving waters (endoheic) is about 940 km³ (WWDR3, 2009). In terms of the runoff discharging into the oceans, Amazon River takes the first place with its discharge more than the combined discharge of the next few large rivers.

Water needs is another parameter, which is constantly on the increase. In Asia and Africa, the needs are likely to exceed the supplies in the foreseeable future and the alternative then is the re-use of water. In the not so distant future, it is conceivable that water will play the same or even a more important role than oil in geopolitical conflicts.

A country or region may be considered as 'water-rich' or 'water-poor' by the per capita amount of water available. Water-rich countries include Iceland (highest), Gabon, Papua New Guinea, Canada and New Zealand with respective annual per capita water availabilities of 294,340, 176,370, 154,610, 84,510 and 79,810 m³ whereas water-poor countries include Botswana (lowest), Chad, Namibia and United Arab Emirates with respective annual per capita water availabilities of –7,460, –3,280, –1,940 and –910 m³ (Countries Compared by Environment > Water > Availability.

International Statistics at NationMaster.com). The world average is about 14,000 m³. The negative availability implies that water has to be imported from trans-boundary sources.

2.3 WATER AVAILABILITY

2.3.1 Some indicators of water availability

< 1,000 m³/year	Catastrophically low
1,100–2,000 m³/year	Very low
2,100–5,000 m³/year	Low
5,100–10,000 m³/year	Average
10,000–20,000 m³/year	High
> 20,000 m³/year	Very high

The highest (in 1995) per capita water availability was in Canada with 170,000–180,000 m³/year whereas North Africa and Arabian Peninsula had only about 200–300 m³/year. China in 1990 had 2,427 m³/year, which is slightly less than the threshold for being classified as a water-stressed country. However, by 2025, when the population is projected to exceed the 1.5-billion mark, China will have only 1,818 m³/year. In Northern China, the situation is even worse.

Oil-rich countries such as Kuwait, Qatar, Bahrain, Saudi Arabia and the United Arab Emirates (UAE) are among the countries with least water per capita.

Israel and Jordan are also among the water-scarce countries. The scarcity has also led to conflicts. Israel controls Palestinian use of water in the occupied 5,890 km² West Bank. The per capita water availability in Jordan in 1990 had been only 327 m³/year.

2.4 RESIDENCE TIME

From the rates of transfer (Table 2.1), it can be seen that the average length of time water resides in any one storage (residence time) varies considerably. For example, the residence time in the oceans is given by

$$\frac{\text{Volume in storage}}{\text{Rate of transfer}} = \frac{1,350,400}{445} \approx 3000 \text{ years}$$

Similarly, the residence times in the ice caps and glaciers is about 10,000 years, in groundwater is about 4,700 years and in rivers, lakes and inland seas is about 2.34 years.

Table 2.2 Potential water resources in different parts of the world

Continent	Total runoff km³/annum	Stable runoff km³/annum	Stable runoff percentage
Africa	4,225	1,903	45
Asia (except USSR)	9,544	2,900	30
Australia	1,965	495	25
Europe (except USSR)	2,362	1,020	43
North America	5,960	2,380	40
South America	10,380	3,900	38
USSR	3,484	1,410	32
All countries	38,820	14,010	36

Source: After Lvovich (1979).

Thus, it can be seen that the most dynamic part of the hydrological cycle involves surface water on the continents, which has a residence time of some 2.3 years. It should, however, be noted that these figures are not exact. More recent figures compiled in the WWDR3 show slightly different residence times.

Table 2.2 gives the distribution of water resources in different parts of the world. Africa has the highest percentage (45%) of stable runoff whereas Australia has the lowest (25%). The world average is about 36%.

2.5 SOURCES OF WATER

2.5.1 Icecaps and glaciers

Icecaps and glaciers hold the largest volume of freshwater on earth, amounting to approximately 26×10^6 km³ (or about 0.925% of total water resources on earth). The distinction between icecaps and glaciers based on the size is that an extent of less than 50,000 km² is considered as an ice cap whereas an extent larger is considered as a glacier. Both types contain frozen water. Ice caps generally tend to have a flat topography and spread laterally whereas glaciers can have a rougher topography and tend to flow like a frozen river. Larger versions of ice caps are called ice sheets, which can spread to over 50,000 km² in extent. Ice caps and ice sheets contain extremely old ice. The oldest ice drilled in Antarctica is said to be about 2.7 million years old!

Icecaps (Figure 2.2) are found in high-latitude high-altitude places. Those in high-latitude countries are called polar ice caps. Large ice caps in the world include Vatnajokull in Iceland and Austfonna in Norway. They are also found in high-altitude places such as in the Himalayas, Rockies, Andes and the Southern Alps of New Zealand. The mobility of ice caps is limited to a specific area and extent.

Figure 2.2 Icecaps (*Source*: www.nationalgeographic.org/encyclopedia/ice-cap/).

Glaciers, on the other hand, are like rivers of ice that flow across landscapes, usually found in high latitudes and high altitudes having snow in the winter that does not melt completely during summer. Due to high overburden weight, the ice at the bottom layers melt and become slippery, causing glacier movement. They move slowly but over long periods of time, very often causing destruction and deformation of the landscape.

Despite the large amounts of freshwater trapped in glaciers and icecaps, they are not of much use for human use except in places where the melted water feeds large rivers such as the Ganges River that flows across India and Bangladesh. Other benefits of ice caps and glaciers include keeping the earth surface cool by reflecting the solar heat energy due to the high albedo of ice and snow. On the negative side is the effect of melting of ice caps and glaciers due to warming climate that can have far-reaching consequences such as sea level rises.

2.5.2 Atmospheric water

The atmosphere can be sub-divided into a number of vertical regions depending upon their physical characteristics. Up to about 80 km, the constituent composition of gases is quite homogeneous whereas above 80 km

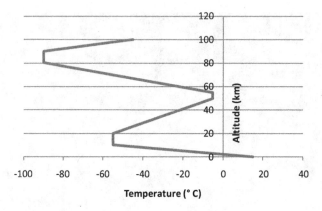

Figure 2.3 Average atmospheric temperature variation with altitude.

the gases are stratified. Therefore, the lower 80 km or so is referred to as the homosphere while that above 80 km is referred to as the heterosphere.

2.5.2.1 Troposphere

Troposphere is the lowest layer of the atmosphere and it contains about 75% of mass and almost all of moisture and dust of the atmosphere. All weather phenomena take place in this region. It extends to about 10–20 km. Temperature decreases in the troposphere (Figure 2.3). The top of the troposphere is known as the tropopause. Its height varies from about 15–20 km in the tropics to about 10 km in the poles.

2.5.2.2 Stratosphere

Stratosphere is the atmospheric shell above the top of the troposphere and below the stratopause. It is calm and with more or less constant temperature in the lower layers (Figure 2.3) and extends to about 10–50 km. The stratosphere is very stable and has very low moisture content. However, there are horizontal winds particularly near the bottom and at higher levels in the polar regions during the winter (winds of about 300 km/h). The stratosphere is very important for life because it contains the ozone, which absorbs the ultraviolet radiation from the sun. The maximum ozone zone lies between 20 and 32 km. Although the thickness of the layer is high, the actual quantity of ozone is very small. If the quantity of ozone were taken down to the ground level, it would compress to a thin layer of about 1 cm by the weight of the atmosphere. Ultraviolet rays are harmful and can lead to skin cancer. The top of the stratosphere is referred to as the stratopause.

2.5.2.3 Mesosphere

Mesosphere refers to the atmospheric shell between the top of the strato-sphere and the mesopause and is about 50 km high. The top of the meso-sphere is referred to as the mesopause and is at a height of about 80 km.

2.5.2.4 Thermosphere

Thermosphere refers to the atmospheric shell extending from the top of the mesosphere at about 80 km to outer space. It includes the ionosphere, which is the atmospheric shell characterized by high ion density extending from about 70 km upwards, and the exosphere, which is the outermost portion of the atmosphere whose lower boundary is at a height of about 500 km. Temperature increases in the ionosphere and the exosphere.

The composition of the atmosphere by volume is given in Table 2.3.

The atmosphere holds about 13,000 km^3 (or about 0.0001%) of earth's total water resources, or about 0.04% of total global freshwater (Gleick, 1996). Water is in the form of clouds (Figure 2.4), which are made of tiny water droplets or tiny ice crystals. The amount of water vapour contained in the atmosphere (Figure 2.4) is a function of several factors such as the availability of a source of moisture, place, temperature and elevation. In any location, the water vapour in the atmosphere is restricted to the lower layers despite the low density relative to dry air. The amount of water vapour contained in the atmosphere is measured by relative humidity, which is

Table 2.3 Principal constituents of the atmosphere

Constituent	% by volume of dry air	Concentration in ppm of air
Permanent composition		
Nitrogen	78.084	
Oxygen	20.946	
Argon	0.934	
Carbon dioxide	0.033	
Neon	0.00182	
Helium	0.000524	
Methane	0.00015	
Krypton	0.000114	
Hydrogen	0.00005	
Important variable gases		
Water vapour	0–5	
Carbon dioxide	0.034	340
Carbon monoxide		< 100
Sulphur dioxide		0–1
Nitrogen dioxide		0–0.2
Ozone		0–10

Figure 2.4 Water vapour in the atmosphere.

defined as the ratio of the amount of moisture in the air to the amount needed to saturate the air at the same temperature. The saturation vapour pressure (SVP) ranges from about 5 mb (at 0°C) to about 50 mb (at about 32°C). Water vapour is the most important greenhouse gas that can cause an increase in climate warming.

The water vapour in the atmosphere is returned back to the surface of the earth in the form of precipitation, which can be in the form of rainfall and/or snowfall. The stages of precipitation consist of lifting, expansion and cooling, nucleation, condensation and raindrop formation. The two important processes in the hydrological cycle, precipitation and evaporation, maintain the cyclic nature of water movement on earth from the surface of the earth to the atmosphere and back.

Cloud seeding is a method of weather modification. It is done by introducing a seeding agent to a cloud. It may be carried out by deploying the seeding agent from above or inside the cloud, from below the cloud and allowing it to disperse naturally by updrafts or thermals, or by ground-based generators for low-hanging cold clouds in mountainous areas. The most widely used purpose is to increase precipitation.

The common seeding agents are silver iodide, potassium iodide, dry ice (solid carbon dioxide) and hygroscopic particles such as common salt. Introduction of a seeding agent, which normally has a crystalline structure similar to that of ice, will induce nucleation and freezing. The results of cloud seeding are mixed. There are no statistically significant results to prove that cloud seeding will always enhance precipitation although there have been observations made in some places where precipitation has been produced as a result of cloud seeding.

Cloud seeding is practiced in many countries in Asia, Europe, the United States, UAE, Russia, Canada, Australia and Africa. Attempts to pre-empt rainfall by cloud seeding were carried out in Beijing prior to the 2008 Olympics to prevent rainfall during the games. One of the better-known examples of cloud seeding in Asia is the Royal Rainmaking Project in Thailand in late 1950s. It has become so popular in Thailand and in neighbouring countries to the extent that in 2005 the European Patent Office granted King Bhumibol Adulyadej the patent on Weather Modification by Royal Rainmaking Technology.

2.5.2.5 Extracting water from fog – occult precipitation

Occult precipitation, also known as horizontal precipitation, is caused by the interception of the fog by the vegetation. The small droplets of water existing in the fog, which normally would remain suspended in the atmosphere, would precipitate after intercepting the vegetation and coalescing and becoming raindrops. Fog can be considered as any cloud that intercepts a topographic surface. The presence of an obstacle such as vegetation can enhance the interception of such small droplets that will coalesce and become raindrops that fall upon the ground. The factors that affect occult precipitation include the type, size, density and homogeneity of the vegetation as well as the exposure to winds. Occult precipitation can occur only when there is air movement, fog, and vegetation simul-taneously. It can occur in places where the humid air is forced to rise due to orographic features or by wind. The humid air cools adiabatically due to lowering of pressure and condenses in tiny droplets that form clouds and fog.

Based on studies carried out in Madeira island, it has been reported that in addition to the contribution to replenish the soil water, occult precipita-tion is also contributing to nutrient cycling and ecosystem bio-geochemistry (https://aprenderamadeira.net/article/precipitacao-oculta).

The fog is usually richer than rain in nitrogen, which is essential for plant growth. Occult precipitation is not normally recorded by a rain gauge but is reported that it contributes 7–28% of the total rainfall in some places (Fernanda Castro, 2017).

2.5.3 Surface water

Surface water includes water stored in natural and/or artificial reservoirs as well as water in rivers and streams, which can be harnessed by direct tapping or by constructing impounding reservoirs.

2.5.3.1 Major freshwater lakes in the world

Lakes are natural reservoirs and the water in lakes are generally clearer than in rivers because of the settling of silt and other suspended matter. However, lakes are not found everywhere. Their use is limited to areas where there are freshwater lakes. Some of the world's large freshwater lakes are

- Lake Superior in United States (surface area 84,650 km^2)
- Lake Victoria in East Central Africa (surface area 69,700 km^2)
- Lake Huron in United States (surface area 61,200 km^2)
- Lake Michigan in United States (surface area 59,600 km^2)
- Lake Nyasa in East Africa (surface area 37,800 km^2)

Other smaller lakes are found in Scotland, Finland, Canada, Denmark, Germany, Switzerland, China, Cambodia etc. Lakes rarely dry up in temperate climates. But in tropics and sub-tropics where the rate of evaporation can exceed the rate of inflow, freshwater lakes may dry up. It is also possible for such lakes to become highly saline like the Dead Sea. The science that deals with the study of lakes is called limnology.

The quality of water in lakes is dependent on factors such as the

- inflow water, catchment and/or rivers,
- effluent discharges,
- location of intake,
- degree of eutrophication (if any),
- depth of intake and
- season.

Abstraction from lakes is done by pumping or by gravity flow if the elevation is high.

2.5.3.2 Major rivers in the world

A river can be classified as large or small using several criteria such as the length, the catchment area, the discharge, and the population served. In terms of the discharge, Amazon River takes the top place with a discharge larger than the combined discharge of the next few large rivers followed by Ganges, Congo, Orinoco, Yangtze, La Plata, Yenisei, Lena, Mississippi, Mekong, Irrawaddy and Ob, in that order. In terms of the length, Nile River takes the top place with a length of some 6,650 km. Some statistics of the 15 longest rivers in the world are shown in Table 2.4. As can be seen, ten of the 15 longest rivers cross national boundaries. Most of them run through countries in Africa, South America and Asia. More details of some

Table 2.4 Fifteen longest rivers in the world

Ranking	Name of river	Length (km)	Catchment area (10^6 km²)	Average discharge (m³/s)	Outflow	Riparian countries
I	Nile	6,650	3.35	2,800	Mediterranean Sea	Ethiopia, Eritrea, Sudan, Uganda, Tanzania, Kenya, Rwanda, Burundi, Egypt, Democratic Republic of the Congo, South Sudan
2	Amazon	6,436	7.05	209,000	Atlantic Ocean	Brazil, Peru, Bolivia, Colombia, Ecuador, Venezuela, Guyana
3	Yangtze River	6,378	1.80	31,900	East China Sea	China
4	Mississippi/ Missouri	5,970	3.22	16,200	Gulf of Mexico	USA
5	Yenisei/ Angara	5,539	2.60	19,600	Kara Sea	Russia Mongolia
6	Ob-Itysh	5,410	2.98	12,800	Gulf of Ob	Russia, Kazakhstan, China, Mongolia
7	Yellow River	5,400	0.75	2,110	Bohai Sea	China
8	Rio de la Plata	4,880	2.58	18,000	Rio de la Plata	Brazil, Argentina, Paraguay, Bolivia, Uruguay
9	Congo	4,700	3.48	41,800	Atlantic Ocean	Democratic Republic of the Congo, Central African Republic, Angola, Republic of the Congo, Tanzania, Cameroon, Zambia, Burundi, Rwanda

(continued)

Table 2.4 Cont.

Ranking	Name of river	Length (km)	Catchment area (10⁶ km²)	Average discharge (m³/s)	Outflow	Riparian countries
10	Amur (Heilong Jiang)	4,440	1.86	11,400	Sea of Okhotsk	Russia, China, Mongolia
11	Lena	4,400	2.49	17,100	Laptev Sea	Russia
12	Mekong	4,350	0.81	16,000	South China Sea	China, Myanmar, Laos, Thailand, Cambodia, Vietnam
13	Mackenzie	4,241	1.79	10,300	Beaufort Sea	Canada
14	Niger	4,200	2.09	9,570	Gulf of Guinea	Nigeria, Mali, Niger, Algeria, Guinea, Cameroon, Burkina Faso, Côte 'Ivoire, Benin, Chad
15	Brahmaputra	3,848	0.71	19,800	Ganges	India, China, Nepal, Bangladesh, Disputed India/China, Bhutan

of the major rivers in the world can be found in the *Handbook of Applied Hydrology* (Ed. Singh, 2016).

Of particular interest in the Asia-Pacific region is the Mekong River, which runs through six countries. It has several rapids, waterfalls and uneven gradients, thus making it difficult for navigation in the entire river. However, it is an important trading route between the Yunnan Province of China and neighbouring Myanmar, Lao PDR and Thailand. The French colonialists in the late 19th century attempted to make the river navigable up to China, but were not successful in taming the Khone Phapheng falls, which are the largest falls of its kind in Southeast Asia. Khone Phapheng falls are located in Champasak Province in southern Lao PDR near the border with Cambodia (13°56′53″N; 105°56′26″E). It is a succession of rapids extending to about 9.7 km along the river with the highest fall reaching 21 m. The Falls are in the area called Si Phan Don, which means 4,000 islands. This section of the river is about 14 km wide during the monsoon season, making the Falls the widest in the world. The main river branches off into many streams, thereby forming many islands some of which are perennial and inhabited by fisherman and the majority appearing only during the dry season, and hence the name 4,000 islands. The river reach containing the 4,000 islands is about 50 km long. The Mekong River is also connected to Tonle-sap (meaning great lake), a shallow lake in Cambodia via the Tonle-sap River, which has a unique feature of flow reversal. It is the largest freshwater lake in Southeast Asia and was designated as a UNESCO ecological hotspot (known as Biosphere) in 1997. During the dry season (November–May), flow takes place from the Tonle-sap Lake to the Mekong River, and vice-versa during the wet season (monsoon season from June to October).

The Ganges River, which flows through India and Bangladesh, originates in the Indian state of Uttarakhand at the confluence of Bhagirathi and Alaknanda rivers at Devprayaghas in western Himalayas, has a length of 2,525 km and empties into the Bay of Bengal. The Ganges River system, considered as the holy river in India, joins with Yamuna River, which has a flow larger than that of Ganges, and after joining with several other rivers in India enters Bangladesh. After entering Bangladesh, the main branch of the Ganges is known as the Padma. The Padma is joined by the Jamuna River, the largest distributary of the Brahmaputra River, which has its headwaters in Nepal. Further downstream, the Padma joins the Meghna River, the second largest distributary of the Brahmaputra. Thereafter it is called Meghna River as it enters the Meghna Estuary, which empties into the Bay of Bengal.

There also exists a large number of relatively small rivers in the Asia-Pacific Region. Details of some of these rivers are given in the UNESCO-IHP publication 'Catalogue of Rivers for Southeast Asia and the Pacific – vols 1–6' (Eds. Takeuchi et al., 1995; Jayawardena et al., 1997; Pawitan et al., 2000; Ibbitt et al., 2002; Tachikawa et al., 2004; Chikamori, et al., 2012). In these publications, details of 121 rivers from Australia, Cambodia, China, Democratic People's Republic of Korea, Indonesia, Japan, Laos, Malaysia, Mongolia, Myanmar, New Zealand, Papua New Guinea, Philippines, Republic of Korea, Thailand and Vietnam can be found.

2.5.4 Groundwater

Groundwater accounts for about 7×10^6 km^3 of freshwater (about 0.5% of total water resources on earth, or about 30% of all freshwater on earth). In terms of freshwater resources, the volume of groundwater storage is second only to the water stored in ice caps and glaciers. It represents an enormous supply of invisible freshwater. The importance of groundwater as a resource is highlighted in the recent issue of the World Water Development Report with the title 'Groundwater – Invisible Made Visible' (WWDR 2022). Although large in quantity, it is not evenly distributed across the earth.

Regionally, the largest users of groundwater are North America and South Asia whereas it is underutilized in Northern Africa and sub-Saharan Africa. Underutilization is not due to lack of the resource but due to lack of investment most notably in infrastructure, institutions, trained professionals and knowledge of the resource. In the Asia-Pacific region, the countries that use groundwater in large quantities, mainly for irrigation, include Bangladesh, China, India, Indonesia, Iran, Pakistan and Turkey.

Groundwater is also the feasible and affordable resource for providing domestic water to rural populations around the world who are not served by public water supplies. This is particularly relevant in sub-Saharan Africa and South Asia where the rural population is large but dispersed. It is also the only resource during droughts.

Extraction of groundwater in large quantities is energy intensive as the water has to be pumped from deep below. However, the high energy cost of pumping is somewhat compensated by the relatively short distances the water has to be conveyed on the surface of the ground. Compared to surface water, groundwater is relatively clean and require less treatment.

With the unabated increase in population and the changing lifestyles, it is estimated that the global food production should increase by 50% by the year 2050 compared to the year 2012 (FAO, 2017). This in turn requires a substantial increase in water. It is therefore important and critical to increase the agricultural productivity and sustainable extraction of groundwater.

Groundwater is stored in aquifers, which are water-bearing geological formations. Most aquifers are found in sand and gravel formations, limestone formations and sedimentary rock formations. They may be confined or unconfined. A confined aquifer holds water under pressure between two impermeable zones. An unconfined aquifer holds water at atmospheric pressure with a lower confining impermeable zone and a free surface at the top. The different types of water in aquifers are

- Connate water – water that has not been in contact with the atmosphere for an appreciable part of the geologic time scale
- Juvenile water – new water of cosmic or magmatic origin
- Meteoric water – water from the atmosphere
- Metamorphic water – water associated with rocks

Figure 2.5 shows the different types of aquifer formations. In confined aquifers, the height to which the water level rises in a bore bole is called the piezometric head and if it is above the ground surface, water will flow out of the bore hole like a spring. In unconfined aquifers, the water is held at hydrostatic pressure between a lower confining impermeable boundary and

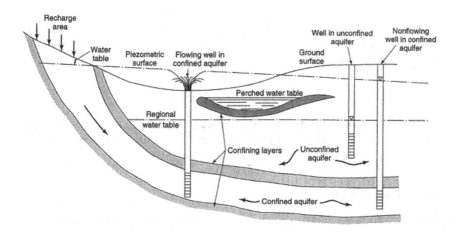

Figure 2.5 Types of aquifers.

an upper free surface boundary. There are also perched aquifers which are unconfined and above the water table with relatively low capacities, leaky aquifers where the confining zones are not well defined and where transfer of water can take place vertically between aquifers and aquitards with poor permeable formations. The hydraulics of groundwater flow, which is assumed to obey Darcy' law for flow through saturated porous media, is described in several textbooks on groundwater hydrology (e.g. Todd, 1980; Cherry and Freeze, 1979; among others) as well as in a book by the author (Jayawardena, 2021).

2.5.4.1 Methods of recharging aquifers

Aquifers can be recharged by a number of methods for a number of reasons. Recharge may be by natural means such as by rainfall or by artificial means such as the following.

(i) Basin method

Basins may be natural or artificially constructed using levees to contain the water, which may be natural from rainfall or artificially diverted from streams. The soil in the basin needs to be silt free to prevent clogging of pores, which will retard and/or prevent infiltration. They include infiltration basins using the soil aquifer treatment (SAT) system in which wastewater is allowed to infiltrate into the subsoil, which undergoes natural physical, biological and chemical purification.

(ii) Stream channel method

This method allows water to spread in streams and channels with slow-moving water. They may be natural or artificial by installing check dams to slow down the flow velocities.

(iii) Ditch method

Ditch and furrows may be long and winding type as well as a tree-like type. The velocities should be sufficient to carry the sediments, which would otherwise clog the pores and retard infiltration.

(iv) Irrigation method

This method uses excess irrigation water during non-irrigating seasons.

(v) Recharge pit method

This method uses pits, which may occur naturally or artificially made to store water, which will infiltrate with time.

(vi) Recharge well method

In this method, the aquifer is recharged using recharge wells.

(vii) Wastewater disposal (soil aquifer treatment (SAT) method)

In this method, wastewater is allowed to seep into the sub-soil naturally or pumped into an aquifer via a recharge well. The wastewater gets filtered during the process of passage through the porous medium, thereby achieving two objectives – treating the wastewater and recharging the aquifer.

2.5.4.2 Problems associated with groundwater extraction

Problems associated with groundwater extraction include mining, land subsidence, saltwater intrusion, violation of water rights and pollution. Groundwater mining is over extraction exceeding the rates of recharge.

(i) Land subsidence

Land subsidence is an irreversible process in which the ground is sinking over large areas. Sometimes, such occurrences can also be seen locally as sinkholes. The primary cause of land subsidence is excessive groundwater pumping, which causes compaction of soils in aquifer systems. Many cities in the world, which pump excessive quantities of groundwater, have been subjected to land subsidence. Examples include the San Joaquin Valley and New Orleans in the United States, some parts of Tokyo metropolitan area in Japan, Bangkok in Thailand, Jakarta in Indonesia, the Hague in the Netherlands, Venice in Italy, Shanghai in China and Mexico City in Mexico. In some cities, the subsidence has been controlled by legal means, restricting the quantities of groundwater pumping. For example, in Tokyo Metropolitan area, which has experienced land subsidence since the early 1900s and accelerated since the end of Second World War due to expansion of economic activities, have brought the rate of land subsidence from a maximum of about 24 cm/year in some areas to about 2 cm/year in recent years with the introduction of the Industrial Water Law, which limits the quantities of groundwater pumping. (Sato et al., 2006). International initiatives to address this problem was the formation of the UNESCO Land Subsidence International Initiative (LASII) in 2018, which was originally called the Working Group, with missions that include improving the scientific and technical knowledge needed to identify and characterize threats related to natural and anthropogenic land subsidence, and to stimulate and enable international exchange of information for sustainable groundwater resources development in areas susceptible to land subsidence. Several International Symposia on Land Subsidence have been held in different countries from Asia, Europe and North America, with the first one in Tokyo in 1969 (Tison, 1969).

(ii) Seawater intrusion

Saltwater intrusion is a common problem in coastal aquifers. The fresh groundwater in coastal aquifers is usually replenished by rainfall. When the rate of abstraction by pumping exceeds the replenishment rate, the freshwater–saltwater interface will move landwards due to the reduction of the hydraulic head in the freshwater aquifer. The saltwater has a higher density and has a higher mineral content, resulting in a higher pressure that pushes the interface landward. The aquifer water then becomes contaminated and any pumping then yields freshwater mixed with saltwater, which is not suitable for domestic and/or agricultural purposes. Further pumping, if continued, will aggravate the problem as a result of reducing the hydraulic head on the freshwater side. The interface occurs not at sea level but at a depth below the sea level of about 40 times the height of freshwater above sea level. This relationship is known as the Ghyben–Herzberg principle, named after its originators. It can be demonstrated using a U-tube filled with seawater and freshwater that will create an interface. The Ghyben–Herzberg equation then is

$$\rho_s g z = \rho_f g (z + h_f) \Rightarrow z = \frac{\rho_f}{\rho_s - \rho_f} h_f \qquad (2.1)$$

where z is the height of sea water, h_f is the height of freshwater, ρ_f and ρ_s the densities of freshwater and sea water, respectively.

There can also be saltwater intrusion due to tidal effects, storm surges, hurricanes and in land areas where there are seawater-conveyancing canals. Under normal conditions, the intrusion of saltwater inland is limited because of the high ground level of aquifer areas, which offer a high hydraulic head. Saltwater intrusion is a problem in many regions in the world where the main source of drinking water is groundwater. Examples of places with problems associated with saltwater intrusion include several coastal regions in the United States including Florida, Louisiana, California and Washington, and in Pakistan, Cyprus, Morocco and the Mekong Delta.

(iii) Water rights

Water rights is a complex issue. In general, water rights give access to water bodies adjacent to a property, which means the owner of the property will have access to adjoining water bodies that include atmospheric, surface and ground water. Two other closely associated rights are the riparian rights, which allow access and usage of flowing bodies of waters like rivers and streams, and littoral rights, which allow access to lakes, seas, and oceans. Property owners can have the right to use the water but they do not own the water. Water rights is a legal entitlement that depends upon the legal system in the relevant region. For example, in the United States, each state has its own laws that govern water rights.

In common law, a property owner has the right to use the water beneath his/her property. However, a problem arises when the property owner extracts excessive water from wells drilled in his/her property but draws water not only from the groundwater beneath his/her property but also from adjoining properties as well since there are no property boundaries for groundwater. It then becomes a legal issue.

There is no unified 'water law' in the world. Even in a particular jurisdiction area, water laws have not reached maturity and are still evolving as new environmental and other challenges come in to play. The oldest water law may perhaps be the 'Roman Law' for water administration, which provided free access to 'aqua publica' but allowed right to request private access to public water with permission from relevant authorities.

Another issue of concern is the contamination of groundwater resulting from disposal of industrial and other harmful wastes. Once a pollutant enters the groundwater, it gets dispersed by diffusion and dispersion and the consequences can be far reaching. An incident that attracted attention and publicity was the case of well water contamination in the city of Woburn in Massachusetts in the United States during the period from mid to late 1970s, which led to the suspicion by the community that the unusually high incidence of leukaemia, cancer, and a wide variety of other health problems were linked to the possible exposure to volatile organic chemicals in the groundwater pumped from two wells. The incident culminated in a court case which also received wide publicity as well as a book titled *A Civil Action* (Harr, 1995) written about the case, which later in 1998 turned into a movie by the same title.

2.5.5 Soil moisture

Soil moisture, sometimes referred to as vadose water, is the water contained in the soil between the ground surface and the water table. Quantitatively, it is about 150×10^3 km^3 (or, 0.01% of total water resources). It is in partially saturated form whereas the groundwater below the water table is in saturated form. All vascular plants absorb water and nutrients from the soil to produce their 'food' together with carbon dioxide absorbed from the atmosphere via the biological process called photosynthesis. This 'food' then becomes the food for all other living beings. It is therefore an important resource for life. More details about the water in the soil is given in Chapter 10.

2.5.6 Seawater – desalination

This is the most abundant source of water on earth (approximately 97%). Although very large in quantity, the quality of seawater is not good enough for domestic or agricultural use (except for flushing and cooling purposes). However, by an expensive process of purification, it is possible to convert seawater into freshwater.

In the hydrological cycle, natural desalination is taking place in large quantities by solar distillation. The distilled (evaporated) water is returned to the surface of the earth as rainwater, which is the primary source of all other sources of freshwater. However, it is uneconomical and unrealistic as a stable source except in regions where there is plenty of sunlight and no other alternative. Other methods of desalination require an external source of energy.

Desalination is the process of removal of dissolved solids from water. It may be used to convert seawater to freshwater or polluted freshwater to freshwater. The salinity present in seawater is normally expressed in mg per litre (or, ppm) of dissolved solids, Cl$^-$ ion or NaCl. Typical solids concentrations in certain types of waters are as follows:

- Brackish water – 1,000–5,000 mg/litre
- Moderately saline water – inland waters with 2,000–10,000 mg/litre
- Severely saline water – coastal and inland water 10,000–30,000 mg/litre
- Sea water > 30,000 mg/litre ($\cong$ 35,000)
- Potable water – max 500 mg/litre (not more than 250 ppm of NaCl)

The different methods of desalting seawater can be classified as follows and are briefly described in the sections that follow.

- Distillation (Heat consuming)
  - Multiple effect
  - Multiple – stage flash – > 85% of all desalting in the world
  - Solar
  - Vapour compression
  - Vapour compression with forced circulation
- Electrodialysis (Power consuming)
- Ion exchange
- Solvent extraction
- Freeze separation (power consuming)
- Reverse Osmosis (Power consuming)

The main consideration for a desalting plant is the cost. It is a variable factor, which depends upon the

- size of plant (unit price decreases with increasing size),
- quality of sea water,
- fuel cost,
- location of plant,
- efficiency (thermal) of the plant – running cost is a substantial part of the total cost. Therefore, high efficiency is very essential, that is, for each unit of energy input, the weight of water processed must be a maximum and
- sharing of fuel costs, that is, desalting is done in conjunction with another process where the rejected heat from the desalting plants can be utilized, or vice versa.

2.5.6.1 Multiple-effect distillation

It is the oldest from of distillation. The source of heat energy is high-temperature and high-pressure steam from an external source. The basic component of the plant is a heat exchanger, which is usually of the shell and tube type. This type permits a large amount of heat transfer surface (area) within a given volume. The purpose of the heat exchanger is to cool or to heat. The hot fluid can be either in the bundle of tubes or in the shell itself. In the case of the hot fluid in the bundle of tubes, condensation takes place within the tubes while evaporation takes place outside of the tubes, that is, inside of the shell. The inlet (high pressure and temperature) steam condenses while transferring the heat to the seawater. This heat evaporates some of the seawater at a slightly reduced pressure and temperature.

This increases the temperature of the seawater, which is fed into the second heat exchanger. It receives the vapour generated from the previous exchanger. In order to cause the flow of heat from the condensing steam to the water inside the tubes, it is necessary to maintain the water temperature less than that of the condensing steam. The vapour formed will be at the same temperature as the water so that the vapour in the second effect will have a slightly lower temperature and pressure than the first. This process can be repeated any number of times until the vapour temperature approaches the temperature of cooling water, that is, when there is no further evaporation. The heat in condensation is absorbed into the cooling water as a temperature rise.

In this system, the heating effect of the external steam has been multiplied. The amount of product water formed is proportional to the amount of external steam + number of effects. An index used to specify the steam economy is the gained output ratio (GOR), which is the ratio of the amount of distillate to the amount of heating steam used in the first evaporator. It varies from one plant to another, but as a first approximation it can be taken as the number of effects multiplied by 0.8.

From the point of view of steam economy, it is desirable to use as many effects as possible. But the capital cost increases with each added new unit. Often a compromise is needed. In general, if fuel is cheap, a fewer number of effects is used whereas in places where fuel is expensive a large number of effects is preferred (as much as 40 have been used).

2.5.6.2 Multi-stage flash distillation

The principle is simply that of evaporating seawater and obtaining pure water by condensation. This method differs from the multiple-effect distillation in that no evaporation of the seawater occurs during the heating process but rather the seawater is heated to its maximum temperature. The hot water is then admitted to the first flash chamber by an orifice, which reduces its pressure slightly. The pre-heated seawater evaporates immediately (flashes) under reduced pressure and is condensed on tubes cooled by incoming sea water thereby increasing the temperature of the seawater flowing towards the heat input section. The condensed water is collected in trays.

Original plants of this type had separate heat exchangers for each stage but modern units combine several stages in one casing. The excess saline water passes through an orifice into the second stage at a slightly lower temperature and pressure carrying out the second stage of cooling and evaporation. This process is repeated in each compartment at successively lower pressures and temperatures as the water flows through the remaining stages of the plant. The circulation rate of flow of saline water in a multi-stage flash distillation type plant must be very high because of the nature of flashing process.

The amount of water circulated through the plant is about 10 times the amount of water evaporated compared with multi-effect distillation where this ratio is about 2. It has been found that the best heat economy is obtained when the saline water temperature at the beginning of the flashing process is between 175°C and 200°C, the exact value depending upon other parameters. There is no direct relationship between the number of stages and the steam economy unlike in the case of multiple-effect distillation.

2.5.6.3 Vapour compression distillation

This process uses mechanical energy rather than heat energy for the performance of the evaporation–condensation cycle. The cycle is regenerative so that no heat rejection section is needed. Once the process is started it does not require any more heat energy. It needs only mechanical energy to keep the compressor running. The process may be made more efficient by pre-heating the brine going in.

2.5.6.4 Solar distillation

Solar distillation in nature is the process by which natural rainfall is produced. In a smaller scale, this process can be duplicated by having the seawater in a closed confined space covered with transparent roof (glass or polythene). The effect is similar to that of a green house. The roof permits solar energy to enter but prevents any evaporated water from leaving the enclosure. To distil 1 gpd requires approximately 1 m^2 under average conditions. The method can also be used to treat wastewater.

2.5.6.5 Electrodialysis

This process reduces salinity by transferring ions from feed compartments through membranes under the influence of an electrical potential difference. Saline feed water contains dissolved salts composed of positively and negatively charged ions. They will move towards oppositely charged electrodes immersed in the solution. The positively charged ions (Na$^+$) will move towards the cathode (negative electrical) while the negatively charged ions (Cl$^-$) will move towards the anode.

Special membranes, which are respectively permeable to cations and anions, are placed alternately, parallel to each other and separated by a small gap. The migration of ions takes place leaving alternate compartments with high and low ionic concentrations. Compartments with low concentrations contain the product water. This process reduces the salinity by about 40%. Passage through a second stack in series will reduce it by a further 40% and so on. It is especially good for treating brackish waters with low salinity.

The electricity required will depend upon the stack geometry, the salinity of water, the resistance of the membranes and the current density. The costs are less than for distillation although the efficiencies are low. Typical energy is in the range 5–20 amperes per cm^2 for 4,000–5,000 ppm saline water. Problems include scale formation, polarization and membrane fouling.

2.5.6.6 Ion exchange

This method uses the selective properties of granular exchange materials, which are organic resins containing H^+ and OH^-. These when placed in saline water exchange the unwanted anions and cations with H^+ and OH^-. Therefore, the water leaving the system is free from original dissolved ions and instead will have accumulated H^+ and OH^- equivalent to its original ionic concentration. Regeneration is necessary.

2.5.6.7 Solvent extraction

This method makes use of solvents, which absorb water preferentially to salts. The solvents include amines.

2.5.6.8 Freeze separation

The principle of purification is similar to that of distillation. The salts are removed by freezing. When seawater is frozen, the dissolved solids remain in liquid brine. Only the pure water freezes. It is also called 'cold distillation' and is attractive from an energy standpoint of view because latent heat of freezing is only about one-seventh that of vaporization. However, there is no great variation of freezing point with pressure variation and therefore, there is no equivalent to the multi-effect process in distillation.

Freezing can be accomplished by using water as a refrigerant or using a liquid hydrocarbon (butane), which will not mix with water. The latter is preferred because of design considerations and capital costs (a very high vacuum, and a large vapour compressor is required for the former). The method consists of vapourizing butane while in direct contact with saline water. The heat transfer causes cooling of the pre-cooled saline water, which forms an ice-brine slurry. The ice crystals are separated and washed to remove brine and melted to give product water. The melting of the ice converts the butane gas back into a liquid. Advantages include low energy cost and no corrosion.

2.5.6.9 Reverse osmosis

This method uses the property of an osmotic membrane, which permits water molecules to pass more readily than dissolved ions. When such a membrane is placed between solutions of different concentrations the water molecules will pass from the fresh to the saline until equilibrium is attained. This process is known as osmosis. The force that drives the water through the semi-permeable membrane is due to osmotic pressure. If a pressure in excess of this is applied to the saline solution, then freshwater will pass from the saline to the water, which is 'reverse osmosis'. The membrane used at present is cellulose acetate. It is weak and must be supported to withstand the high pressures. The flow rates of product water with present membranes are about 500 litres per day per m^2 of membrane.

2.6 WATER WITHDRAWALS

Water withdrawals may be considered as an indicator of the prosperity of a nation. As can be seen in Table 2.5 the industrial use is greater than the domestic use in Asia (mainly China and Japan), North America and Europe whereas the domestic use is greater than the industrial use in Africa, Caribbean and Oceania. Of all the regions, Asia has the highest withdrawals as a percentage of the renewable resources. The two major users of water are agriculture and industry. In terms of quantities, the requirements for domestic and municipal supplies (approximately 9% of total freshwater resources of earth) are far less than those for growing or producing food. Table 2.5 (WWDR3, 2009) shows the water withdrawals by region and sector. Forecasts for the future are likely to be on the increase partly due to unabated increase in population and partly due to increase in per capita consumption in developing countries.

Table 2.5 Water resources and withdrawals in 2000

Region	Renewable water resources (km^3/year)	Withdrawals for agriculture (km^3/year)	Withdrawals for industry (km^3/year)	Withdrawals for domestic use (km^3/year)	Withdrawals as a percentage of renewable resources
Africa	3,936	186	9	22	5.5
Asia	11,594	1,936	270	172	20.5
South America	13,477	178	26	47	1.9
Caribbean	93	9	1	3	14
North America	6,253	203	252	70	8.4
Oceania	1,703	18	3	5	1.5
Europe	6,603	132	223	63	6.3
World	43,659	2,663	784	382	8.8

Source: WWDR3, 2009.

2.7 VIRTUAL WATER AND WATER TRADE

The concept of virtual water was introduced in 1993 by the 2008 Stockholm Water Prize Laureate Professor John Allen. Virtual water (sometimes known as embedded water) refers to the water needed to produce goods and services imported or exported. For example, when a country imports goods (including food) produced in another country, the imported goods carry a hidden water cost. To produce 1 kg of wheat requires about 1.34 m^3 of water. Therefore, when a country (or region) imports wheat, it is also implicitly importing a proportionate quantity of water. This applies to any type of goods and services that require water for their production and transportation. It also means that the importing country is saving the amount of water that has been used for the production and transportation of the goods and services it imports. Virtual water trade is a relatively new concept but in recent years it has received attention from scientists, businesses and politicians.

Table 2.6 gives the global average water footprints for some typical foods. However, the actual water footprints vary from country to country depending upon the climate and the production process. For instance, growing wheat in France needs one-tenth of the water required to grow the same crop in Morocco (Mekonnen and Hoekstra, 2010). The water used during the growing season of a crop is based on crop evapo-transpiration, which is usually obtained from computer models. All computer models have some assumptions and evapo-transpiration values vary widely from place to place. Water required also depends upon whether the crop is rain-fed or irrigated. Hotter and drier countries have larger footprints compared with countries with temperate climate. For example, the water footprint of wheat grown in Slovakia is reported to be 465 litres per kg whereas for the same crop grown in Somalia it is 18,070 litres per kg (Chapagain and Hoekstra, 2004).

Table 2.6 The water footprint of some food items

Food/drink product	Global average water footprint (liters/kg)
Apple or pear	700
Banana	860
Beef	15,500
Beer from Barley	75/250 ml
Cheese	5,000
Chicken	3,900
Chocolate	24,000
Milk	250/250 ml
Rice	3,400

Source: https://waterfootprint.org/media/downloads/Hoekstra-2008-WaterfootprintFood.pdf

Virtual water has major impacts on global trade. Water-rich countries can export virtual water in the form of goods and services to water poor countries thereby increasing the water trade. Food security can be enhanced by virtual water, which can be considered as another water 'source'. For the period 1995–1999, the countries with the largest net virtual water export were United States, Canada, Thailand, Argentina and India. The countries with the largest net virtual water import in the same period were Sri Lanka, Japan, Netherlands, Republic of Korea and China (Hoekstra and Hung, 2002).

It is also a fact that developing countries produce basic food items with not much value added whereas developed countries produce manufactured goods which have a high water-cost. In developing countries, the water cost of producing food items is normally not added to the cost of the primary products because the water needed comes mainly from rainfall and sometimes from irrigation, which is not charged. In developed countries, the water cost is added to the cost of production. As a result, there appears to be a disparity in the virtual water trade between developed and developing countries.

There are also other shortcomings in the way virtual water is calculated. For example, it assumes that all sources of water, whether from rainfall or from irrigation have the same value. Also, whether the water that is saved as a result of importing virtual water is used economically in other less water intensive activities is questionable.

2.8 BOTTLED WATER

Bottled water is a lucrative business in the world today. Globally, it is estimated that about 55 million bottles are sold every hour. It can also be considered as a type of virtual water. Water from various sources packed into PET (polyethylene terephthalate) bottles has become a convenient source of drinking water despite their cost in comparison with the cost of tap water. Per unit volume, bottled water can be several thousand times the cost of tap water depending on the domestic water tariff structure. Still, people have got used to bottled water for convenience, perhaps quality and as a substitute for sugary drinks.

Different countries have different specifications for bottled water. The US Food and Drug Administration (FDA) considers bottled water as a packaged food product and by law the FDA's quality requirements must be at least as stringent as the Environmental Protection Agency standards for tap water. Bottled water cannot contain sweeteners or chemical additives (other than flavours, extracts or essences) and must be calorie-free and sugar-free. If flavours, extracts and essences, derived from spice or fruit, are added to the water, these additions must comprise less than 1% by weight of the final product.

The origin of most bottled water is groundwater and/or spring water. Some companies use purified tap water. One positive factor in bottled water is that there is hardly any metal (lead, copper etc.) contamination because the water is not conveyed through pipelines. Bottled water also does not contain chlorine, which is added to tap water during the treatment process. Excessive chlorine can lead to health complications. Despite some positive factors, there are also strong arguments to convince consumers that bottled water is no better than tap water from a well-managed domestic water provider.

An important factor to discourage consumers from using bottled water is the problem of disposal of used PET bottles. Plastic waste is not biodegradable and they tend to accumulate causing serious environmental problems. The Great Pacific Garbage Patch could become the final resting place for used PET bottles too.

2.9 WATER STRESS, WATER SCARCITY AND WATER RISK

Three indicators used to define the availability of water, or lack of it, are water stress, water scarcity and water risk. Water stress refers to the inability to meet human and ecological demand for fresh water. It is related to water availability, water quality, and accessibility of water including affordability. Due to insufficiency of infrastructure and unaffordability, water stress can occur even when sufficient resources are physically available.

Water scarcity refers to the lack of freshwater resources. It is human driven and depends on the volume of water consumption relative to the volume of water resources available in a given area. Therefore, an arid region with very little available water with no human consumption would not be considered as a water scarce region.

Water risk refers to the possibility of experiencing a water-related challenge such as water scarcity, water stress, flooding, infrastructure decay, drought etc. The magnitude of the risk is a function of the probability of occurrence of a specific challenge and its severity. The severity depends on the intensity of the challenge as well as the vulnerability and coping capacity of those affected.

All these definitions are subjective and may have different interpretations in different regions, cultures and lifestyles. For example, thresholds for clean drinking water and requirements for environmental freshwater ecosystems may have different values in different regions, cultures and lifestyles. A widely used quantitative definition of water stress (or water stress index) is that proposed by Falkenmark (1989) and Falkenmark et al. (1989), which state that a country experiences water stress if the amount of renewable water resources is below 1,700 m³ per person per year, and water scarcity if it falls below 1,000 m³ per person per year and absolute water scarcity if it falls below 500 m³ per person per year. Other similar indicators include defining water scarcity in terms of each country's water demand compared to the amount of water available, and an indicator that takes into account the water infrastructure such as desalination and recycling facilities available in the country and limiting water demand by consumptive use rather than total withdrawals, developed at the International Water Management Institute (IWMI) (Seckler et al., 1998).

According to a recent World Water Development Report, over 2 billion people live in countries experiencing high water stress, and about 4 billion people, representing nearly two-thirds of the world population, experience severe water scarcity during at least one month of the year (WWDR, 2019; Mekonnen and Hoekstra, 2016). The report also projects an increase in stress levels because of the increasing demand for water in the future. It is also highlighted that although the global average water stress is only 11%, 31 countries experience water stresses between 25% (which is defined as the minimum threshold of water stress) and 70%, and 22 countries are above 70% and are therefore under serious water stress (UN, 2018a).

2.10 WORLD WATER ASSESSMENT PROGRAMME (WWAP) AND UN-WATER

The World Water Assessment Programme (WWAP), founded in 2000, is the flagship programme of UN-Water, an initiative consisting of 28 members drawn from UN organizations. Organizations outside the UN system are partners in UN-Water. It was initially located in UNESCO headquarters in Paris, but later relocated to Perugia, Italy. WWAP monitors freshwater issues in order to provide recommendations, develop case studies, enhance assessment capacity at a national level and inform the decision-making process.

Its primary product, the *World Water Development Report* (WWDR), is a periodic comprehensive review providing information on the state of the world's freshwater resources. The basis for the WWDR sprang from the Rio Earth Summit of 1992 and the UN Millennium Declaration of 2000. The first four issues of WWDR have been published to coincide with the four World Water Forums held respectively in Kyoto, Japan (2003), Mexico City, Mexico (2006), Istanbul, Turkey (2009) and Marseille, France (2012). The first World Water Forum was held in Marrakesh, Morocco, in 1997, followed by the next eight held respectively in the Hague, Netherlands (2000), Kyoto, Japan (2003), Mexico City, Mexico (2006), Istanbul, Turkey (2009), Marseille, France (2012), Deigu, Korea (2015), Brasilia, Brazil (2018) and Dakar, Senegal (2022). The titles of the World Water Development Reports so far have been as follows:

- Water for people, Water for Life (WWDR1, 2003)
- Water: A Shared Responsibility (WWDR2, 2006)
- Water in a Changing World (WWDR3, 2009)
- Managing Water under Uncertainty and Risk (WWDR4, 2012)
- Water and Energy (WWDR, 2014)
- Water for a Sustainable World (WWDR, 2015)
- Water and Jobs (WWDR, 2016)
- Wastewater: An Untapped Resource (WWDR, 2017)
- Nature-Based Solutions for Water (WWDR, 2018)
- Leaving No One Behind (WWDR, 2019)
- Water and Climate Change (WWDR, 2020)
- Valuing Water (WWDR, 2021)
- Groundwater – Invisible Made Visible (WWDR, 2022)

Since 2014, the theme of the World Water Development Report and that of World Water Day have been harmonized in order to provide a deeper focus and in-depth analysis of a specific water-related issue every year.

These documents can be accessed at www.unesco.org/water/wwap/

Each WWDR gives some key messages and the ones from the last four reports (WWDR2019, WWDR2020, WWDR2021, and WWDR2022) are summarized below:

2.10.1 Highlights of the last four WWDRs

WWDR 2019 gives the following key messages:

- Access to safe, affordable and reliable drinking water and sanitation services are basic human rights.
- The wealthy generally receive high levels of service at very low price, while the poor often pay a much higher price for services of similar or lesser quality.
- Equitable access to water for agricultural production, even if only for supplemental watering of crops, can make the difference between farming as a mere means of survival and farming as a reliable source of livelihoods.
- Refugees and internally displaced people often face barriers in accessing water supply and sanitation services.

Overcoming exclusion and inequality

- International human rights law obliges states to work towards achieving universal access to water and sanitation for all, without discrimination, while prioritizing those most in need.

- Investing in water supply and sanitation in general, and for the vulnerable and disadvantaged in particular, makes good economic sense.
- Accountability, integrity, transparency, legitimacy, public participation, justice and efficiency are all essential features of 'good governance'.
- Responses that are tailored to specific target groups help ensure that affordable water supply and sanitation services are available to all.

WWDR (2020) highlights the sustainability of water resources under climate change. Extreme events such as floods and droughts are becoming more frequent in many regions of the world affecting food security, human health, energy production, economic development etc. The report also identifies water as a 'climate connector' that allows for greater collaboration and coordination across the majority of targets for sustainable development (2030 Agenda and its SDGs), climate change (Paris Agreement) and disaster reduction (Sendai Framework). It also refers to adaptation and mitigation as complementary strategies for managing and reducing the risks of climate change through water and suggests to embrace the two approaches through water as a triple-win proposal.

WWDR (2021) describes different approaches of valuing water considering five interrelated perspectives:

- Valuing water sources, in situ water resources and ecosystems
- Valuing water infrastructure for water storage, use, reuse or supply augmentation
- Valuing water services, mainly drinking water, sanitation and related human health aspects
- Valuing water as an input to production and socio-economic activity, such as food and agriculture, energy and industry, business and employment
- Valuing other socio-cultural aspects of water including recreational, cultural and spiritual attributes.

An important point highlighted in this report is that in the case of water there is no clear relationship between the price and the value of water. The price usually reflects attempts to recover cost and not the value delivered. Value can be measured in different ways and can be different among different stakeholders as well as within individual stakeholders. For economic transactions however, the traditional valuation follows the way most other products are valued. This approach needs re-evaluation in the light of knowledge, research and capacity development.

WWDR (2022) describes the importance of groundwater as a viable resource and emphasizes the importance of groundwater governance. Issues highlighted include the following:

- The importance of groundwater as a viable and affordable resource for the rural population in the world
- Groundwater as a critical resource for irrigated agriculture, livestock farming and other agricultural activities, including food processing
- The need to increase the agricultural productivity to meet future food supply through sustainable intensification of groundwater abstractions
- The need to coordinate, harmonize and share data as the first step in cooperation between neighbouring countries
- Developing and maintaining a dedicated groundwater knowledge base
- Protection of groundwater quality by effective regulation and strict enforcement

2.11 WATER FOOTPRINTS

Water footprint is defined as the total volume of water used in the production of goods and services consumed by an individual or community or produced by a business. In the context of a country, it is the volume of water used in the production of all goods and services consumed by all inhabitants of the country. The United States has a water footprint of 2,480 m^3 per capita per year; China about 700 m^3 per capita per year while the global average is 1,240 m^3 per capita per year (WWDR3, 2009). A country's internal water footprint is the volume of water used from domestic sources whereas the external footprint is the volume of water used in other countries to produce the goods it imports (virtual water).

Water footprints are classified as green, blue and grey. Green water footprint refers to water from precipitation that is stored in the root zone of the soil and evaporated, transpired or absorbed by plants. It is particularly relevant for agricultural, horticultural and forestry products. Blue water footprint refers to water from surface or groundwater and is either evaporated, absorbed into a product or taken from one body of water and returned to another, or returned at a different time. Irrigated agriculture, industry and domestic water use can each have a blue water footprint. Grey water footprint refers to the amount of fresh water required to assimilate pollutants to meet specific water quality standards. It considers point-source pollution as well as through runoff or leaching from the soil, impervious surfaces, or other diffused sources.

It is also possible for the water footprint of a country to fall outside the borders of that country. For example, about 10%, 20% and 77% of the water footprints in China, the United States and Japan respectively fall outside their national boundaries. It is also reported that the largest external water footprint of US consumption lies in the Yangtze River Basin, China (https://waterfootprint.org/en/water-footprint/what-is-water-footprint/).

A personal water footprint is the amount of water needed by a person for his/her food, clothing, shelter and lifestyle. It varies from country to country and within the country too. Globally, on average, a person consumes about 5,000 litres per day of water for his/her food and lifestyle. Obviously, a person living in a rich country consumes more water than a person living in a poor country. This amount of water may not necessarily be from sources within the country but may have hidden contributions from other countries too.

Water footprint statistics including national water footprints, international water footprints and product water footprints can be found in WaterStat, one of the world's most comprehensive water footprint data bases (https://waterfootprint.org/en/resources/waterstat/), the computations of which are based on the Global Water Footprint Assessment Standard.

2.12 WATER USE

Globally, surface water by far is the major source (about 71%) followed by groundwater (about 18%) for all types of uses (drinking, agriculture, energy production and industry). Water re-use is very limited and desalination is almost insignificant (about 0.34%) when all users are considered. For drinking water, the major source is surface water (about 48%) followed by groundwater (about 45%). For agriculture the major source is again surface water (about 71%) followed by groundwater (about 17%) whereas for energy and industry, the respective ratios are about 87% and 12%. The statistics quoted here are taken from the World Water Development Report 3 (2009).

2.13 TRENDS IN STREAMFLOW

From an analysis of 195 stream gauging stations worldwide, it has been shown (WWDR3, 2009) that 25% of stations in Africa having increasing trend while there is no increasing trend in Asia. Majority of the gauging stations (except Africa) show no trend whereas significant decreasing trends can be seen in Africa (50%) and Asia (38%). This observation appears to be contrary to the claims made by climate change proponents.

REFERENCES AND WEBSITES FOR WATER INFORMATION

Chapagain, A. K., and Hoekstra, A. Y. (2004). *Water Footprints of Nations* (Value of Water Research Report Series; No. 16). Delft: UNESCO-IHE Institute for Water Education.

Chikamori, H., Liu, H. and Daniell, T. (Editors) (2012). *Catalogue of Rivers for Southeast Asia and the Pacific – Volume VI.* UNESCO-IHP Regional Steering Committee for Southeast Asia and the Pacific, UNESCO-IHP Publication, March 2012, 99 pp.

Falkenmark, M. (1989). The massive water scarcity threatening Africa – why isn't it being addressed, *Ambio*, 18(2): 112–118.

Falkenmark, M., Lundquist, J. and Widstrand, C. (1989). Macro-scale Water scarcity requires micro-scale approaches: aspects of vulnerability in semi-arid development, *Nat. Resourc. Forum*, 13(4): 258–267.

Fernanda Castro (2017). Hidden precipitation, 31 January 2017. https://aprendera madeira.net/article/precipitacao-oculta

Food and Agricultural Organization (2017). *The Future of Food and Agriculture: Trends and Challenges,* Rome: FAO, www.fao.org/3/i6583e/i165 83e.pdf

Freeze, R. A. and Cherry, J. A. (1979). *Groundwater,* Prentice Hall, 604 pp.

Gleick, P. H. (1996). Water Resources. In: Schneider, S.H., Ed., *Encyclopedia of Climate and Weather,* Oxford University Press, New York, 817–823.

Harr, J. (1995). *A Civil Action,* Random House, 500 pp.

Hoekstra, A. Y. and Hung, P. Q. (2002). Virtual water trade – A quantification of virtual water flows between nations in relation to international crop trade, Research report 11, September 2002, IHE, Delft.

Ibbitt, R., Takara, K., Desa, M. N. M. and Pawitan, H. (Editors) (2002). *Catalogue of Rivers for Southeast Asia and the Pacific – Volume IV,* UNESCO-IHP Regional Steering Committee for Southeast Asia and the Pacific, UNESCO-IHP Publication, March 2002, 338 pp.

International Water Management Institute (IWMI), (2006). *Insights from the Comprehensive Assessment of Water Management in Agriculture,* Colombo, Sri Lanka.

Jayawardena, A. W. (2021). *Fluid Mechanics, Hydraulics, Hydrology and Water Resources for Civil Engineers,* Taylor and Francis Group, CRC Press, 894 pp.

Jayawardena, A. W., Takeuchi, K. and Machbub, B. (Editors) (1997). *Catalogue of Rivers for Southeast Asia and the Pacific – Volume II,* UNESCO-IHP Regional Steering Committee for Southeast Asia and the Pacific, UNESCO-IHP Publication, December 1997, 285 pp.

Lvovich, M. I. (1979). *World Water Resources and Their Future.* English translation edited by Raymond L. Nace. Washington, DC: American Geophysical Union.

Maria Fernanda Cárdenas, Conrado Tobón and Wouter Buytaert (2017). Contribution of Occult Precipitation to the Water Balance of Páramo Ecosystems in the Colombian Andes, *Hydrological Processes*, 31(24): 4440–4449. https://doi.org/10.1002/hyp.11374

Mekonnen, M. M. and Hoekstra, A. Y. (2016). Four billion people facing severe water scarcity, *Science Advances*, 2(2), doi.org/10.1126/sciadv.1500323.

Mekonnen, M. M. and Hoekstra, A. Y. (2010). A global and high-resolution assessment of the green, blue and grey water footprint of wheat. *Hydrol. Earth Syst. Sci.* 14(7): 1259–1276. CrossRefGoogle Scholar

Pawitan, H., Jayawardena, A. W., Takeuchi, K. and Lee, S. (Editors) (2000). *Catalogue of Rivers for Southeast Asia and the Pacific – Volume III,* UNESCO-IHP Regional Steering Committee for Southeast Asia and the Pacific, UNESCO-IHP Publication, May 2000, 268pp.

Sato, C., Haga, M., and Nishino, J. (2006). Land Subsidence and Groundwater Management in Tokyo, Special Feature on Groundwater Management and

Policy, *International Review for Environmental Strategies* Vol. 6, No. 2, pp. 403 – 424, Institute for Global Environmental Strategies.

Seckler, D. et al. (1998). *World Water Demand and Supply, 1990 to 2025: Scenarios and Issues*, International Water Management Institute (IWMI) Research Report 19, IWMI, Colombo, Sri Lanka.

Singh, V. P. (Editor) (2016). *Handbook of Applied Hydrology*, 2nd ed. New York: McGraw Hill Education, 1440 pp.

Tachikawa, Y., James, R., Abdulla, K. and Desa, M. N. M. (Editors) (2004). *Catalogue of Rivers for Southeast Asia and the Pacific – Volume V*, UNESCO-IHP Regional Steering Committee for Southeast Asia and the Pacific, UNESCO-IHP Publication, May 2004, 285 pp.

Takeuchi, K., Jayawardena, A. W. and Takahasi, Y. (Editors) (1995). *Catalogue of Rivers for Southeast Asia and the Pacific – Volume I*, UNESCO-IHP Regional Steering Committee for Southeast Asia and the Pacific, UNESCO-IHP Publication, October 1995, 289 pp.

Tison, L. J., ed. (1969). Land Subsidence – Proceedings of the Tokyo Symposium, September 1969, vols 1–2. IAHS Publs 88–89, pp. 661 http://iahs.info/Publi cations-News.do

Todd, D. K. (1980). *Groundwater Hydrology*, 2nd ed., Wiley, New York, 552 pp.

United Nations (2018a). *Sustainable Development Goal 6: Synthesis Report 2018 on Water and Sanitation*. New York: United Nations.

World Water Development Report 1 (2003). *Water for People, Water for Life*, World Water Assessment Programme, UNESCO.

World Water Development Report 2 (2006). *A Shared Responsibility*, World Water Assessment Programme, UNESCO.

World Water Development Report 3 (2009). *Water in a Changing World*, World Water Assessment Programme, UNESCO.

World Water Development Report 4 (2012). *Managing Water Under Uncertainty and Risk*, World Water Assessment Programme, UNESCO.

World Water Development Report (2014). *Water and Energy*, World Water Assessment Programme, UNESCO.

World Water Development Report (2015). *Water for a Sustainable World*, World Water Assessment Programme, UNESCO.

World Water Development Report (2016). *Water and Jobs*, World Water Assessment Programme, UNESCO.

World Water Development Report (2017). *Wastewater: An Untapped Resource*, World Water Assessment Programme, UNESCO.

World Water Development Report (2018). *Nature Based Solutions for Water*, World Water Assessment Programme, UNESCO.

World Water Development Report (2019). *Leaving No One Behind*, World Water Assessment Programme, UNESCO.

World Water Development Report (2020). *Water and Climate Change*, World Water Assessment Programme, UNESCO.

World Water Development Report (2021). *Valuing Water*, World Water Assessment Programme, UNESCO.

World Water Development Report (2022). *Groundwater – Invisible Made Visible*, World Water Assessment Programme, UNESCO.

www.unesco.org/water/wwap/
http://seer.cancer.gov/Publications/CSR7393/
www.agr.ca/pfra/water/groundw.htm
www.agric.gov.ab.ca/water/wells/index.html
www.agric.gov.ab.ca/water/wells/module1.html
www.angelfire.com/nh/cpkumar/hydrology.html
www.chi.on.ca/swmmqa.html
www.chula.ac.th/international/index_en.html
www.epa.gov/safewater/ars/arsenic.html
www.fluoridealert.org/f-arsenic.htm
www.lboro.ac.uk/departments/cv/wedc/education/dl.htm
www.nrdc.org/water/drinking/arsenic/aolinx.asp
www.pacinst.org/naw.html
https://waterfootprint.org/en/water-footprint/what-is-water-footprint/
https://waterfootprint.org/en/resources/waterstat/
www.rhodes.ac.za/institutes/iwr/
www.undp.org.vn/dmu/
www.worldwater.org/

Chapter 3

Chemistry of water

3.1 INTRODUCTION

Water is a precious resource essential for all forms of life. It is a chemical compound made of two of the universe's most abundant elements, hydrogen and oxygen. The chemical formula for water, H_2O, is better known than that of any other chemical compound but from the point of predicting its chemical properties, it is one of the least known substances on earth. Water is colourless, odourless, tasteless and transparent. In large volumes, liquid water appears to be blue because of the weak absorption of light at the red end of the visible spectrum. Though inert, it is a powerful universal solvent as well as a means of transporting many substances. In the human body, water flushes body waste (mainly urine), helps digestion to convert food to components needed for survival, lubricates joints, keeps mucosal membranes moist, regulates body temperature by sweating and respiration, forms saliva that helps digestion, allows body cells to grow, reproduce and survive and helps deliver oxygen all over the body.

Water has several unique properties. It has a density greater than that of its solid form. All substances which can exist in different states (solid, liquid and gas) have the following general relationship with respect to the density, ρ

$$\rho_{solid} > \rho_{liquid} > \rho_{gas}$$

In general, ρ_{soild} is of the same order of magnitude as ρ_{liquid} and ρ_{liquid} is about 1,000 times of ρ_{gas}. Water is the only exception. Water has the highest density at 4°C. At temperatures less than 4°C, water expands due to the formation of a hexagonal lattice structure. At temperatures greater than 4°C, expansion is due to the thermal motion of the molecules. Compressibility of water is very low and therefore considered as incompressible in fluid mechanics.

The hydrogen and hydroxyl ions in water determine the structure and biological properties of proteins and other cell materials. The 'wetness' of

DOI: 10.1201/9781003329206-3

water is due to the hydrogen bond in the water molecule. Water has the highest surface tension of all liquids except mercury.

3.2 CLASSIFICATION OF WATER

On a broad basis, water can be classified as follows:

- *Blue water* – Liquid water moving above and below ground surface. As it moves through the land phase of the earth, it can be re-used until it reaches the sea.
- *Green water* – Soil water (moisture) replenished by rainfall (or irrigation) and used up by plants and returned to the atmosphere by evapotranspiration. Green water becomes unproductive if evaporated from open water and bare soil.
- *White water* – That part of green water which is non-productive
- *Grey water* – Wastewater of poor quality, which can be re-used for some purposes.
- *Black water* – Heavily polluted water (usually with microbes) that are harmful for human use. They can be treated to acceptable quality at a high cost.

Water is the common name used for the H_2O molecule. Other names for water in chemistry but rarely used include

- Dihydrogen monoxide or DHMO
- Hydrogen hydroxide (HH or HOH)
- Hydrogen monoxide
- Dihydrogen oxide
- Hydric acid
- Hydrohydroxic acid
- Hydrol
- Hydrogen oxide
- Hydron hydroxide (polarized form of water, H^+ OH).

3.3 CHEMICAL COMPOSITION OF A MOLECULE

A molecule is a combination of several atoms. An atom is made of a nucleus containing protons with a positive charge and neutrons with no charge surrounded by electrons with a negative charge. Electrons are exchanged or shared to form molecules which can be from the same element or different elements. Compound is a type of molecule that can be formed from the same element or from different elements. For example, ozone (O_3) is a molecule but not a compound whereas CO_2 and H_2O are both molecules and compounds. In a molecule, two or more atoms stay together because of the mutual attraction between the positively charged protons from one atom and the negatively charged electrons from the other atom. This causes the covalent or ionic bonding that holds atoms or ions together.

The geometrical structure of a molecule can be linear such as in carbon dioxide (CO_2) and acetylene (C_2H_2) with a bond angle of 180° and therefore the atoms in a straight line, trigonal plane such as in ethylene (C_2H_4) and sulphur dioxide (SO_2) with a bond angle of 120°, and tetrahedral such as in water (H_2O) and methane (CH_4), and, pyramidal such as in ammonia (NH_3) with a bond angle of 109.5°. Of these, the water molecule and the sulphur dioxide molecule have bent geometrical structures.

3.4 BONDING OF SUB-ATOMIC PARTICLES TO FORM MOLECULES

There are four types of bonding that contribute to the formation of molecules. They are ionic bonding, covalent bonding, hydrogen bonding and van der Waal bonding. Ionic bonds are formed through the exchange of valence electrons between atoms, typically a metal and a non-metal. Valence electrons occupy the outermost shell (furthest from the nucleus) of the atom. The transfer of valence electrons allows the ions to obey the octet rule and become more stable. The octet rule states that an atom is most stable when there are eight electrons in its valence shell. Atoms with less than eight electrons tend to satisfy the duet rule, having two electrons in their valence shell. By satisfying the duet rule or the octet rule, ions become more stable. An atom that loses one or more valence electrons becomes a positively charged ion (cation) while an atom that gains electrons becomes a negatively charged ion (anion). For example, when a sodium (Na) atom loses one electron, it has one extra proton in the nucleus making it a positively charged ion Na^+. Similarly, when a chlorine atom (Cl) gains an electron, it has an extra electron, making it a negatively charged ion (Cl^-). The oppositely charged ions attract each other to form ionic compounds, which in this example is NaCl. The presence of two oppositely charged ions results in a strong attractive force between them referred to as an ionic bond or electrovalent bond. Ionic bonds are formed when the differences in electronegativity between atoms are large whereas covalent bonds are formed when the differences in electronegativity between atoms are small. Electronegativity is the tendency of an atom in a molecule to attract the shared pair of electrons towards itself. Electronegativity holds electrons tightly. In the periodic table, fluorine is the most electronegative element and caesium is the least electronegative element. In an ionic bond, electrons are transferred from one atom to another.

In covalent bonding, electrons are shared between atoms. They may be single covalent in which case one pair of electrons is shared by two atoms, double covalent in which case two pairs of electrons are shared between two atoms or triple covalent in which case three pairs of electrons are shared between two atoms. Depending on the polarity, they can be non-polar covalent bonding or polar covalent bonding. When two electrons of the same atom are shared, the difference in electronegativity is zero and there is no polarity found in the molecule. Examples include the molecules H_2, N_2, O_2 etc. Non-polar covalent bonds share electron pairs equally between atoms and they occur only when the atoms are identical to each other. On the other hand, a polar covalent bond is formed when different electronegative non-metallic atoms share their bonding electron pair between themselves. If one end of the molecule has a positive charge and the other end has a negative charge, the molecule is polar. If the charge is evenly distributed around the central atom, the molecule is non-polar. Because of different electronegativities, the atoms share a bonding electron pair unequally and the bonding electron pair tends to stay closer to the more electronegative atom. Bonds between atoms with a difference in electronegativity less than 0.4 are considered to be non-polar covalent. The smaller the difference between electronegativity values, the more likely atoms will form a covalent bond. A large difference between electronegativity values is seen with ionic bonds. Covalent bonding does not necessarily require that the two atoms be of the same element, only that they should be of comparable electronegativity. Polar bonds are intermediate between pure covalent bonds and ionic bonds. They form when the electronegativity difference between the anion and the cation is between 0.4 and 1.7. Most compounds having covalent bonds have relatively low melting and boiling points. Compounds formed by covalent bonding do not conduct electricity due to lack of free electrons.

A hydrogen bond is a type of polar covalent bond formed when a pair of electrons is unequally distributed between the hydrogen atom and another atom. It is an electrostatic attractive force between a partially positively charged hydrogen atom of one molecule and a partially negatively charged atom of the same molecule or other molecules. It is also considered as the attraction between oppositely charged dipoles. Of the three types of bonds, ionic bond is the strongest and hydrogen bond is the weakest with covalent bond in between. However, hydrogen bond is stronger than van der Waal bond, which is the attraction of intermolecular forces between molecules. There are two kinds of van der Waals bonds, the weak 'London dispersion forces' and the stronger 'dipole–dipole forces'. They are distance-dependent interactions between atoms or molecules and bonds formed by the attraction of two electrically polarized molecules.

3.5 GEOMETRICAL STRUCTURE OF A WATER MOLECULE

Water is a polar covalent molecule because of unequal sharing of electrons between hydrogen and oxygen atoms. Other polar molecules include ethanol, ammonia, sulphur dioxide and hydrogen sulphide. Water molecule contains two hydrogen atoms and one oxygen atom. One electron each from the hydrogen atom share an electron from the oxygen atom, forming a covalent bond. Covalent bonds occur when atoms share electrons instead of donating them. As oxygen is more electronegative than hydrogen, oxygen pulls the electrons closer towards its nucleus, giving oxygen a partial negative charge and hydrogen a partial positive charge. The electronegativity value of hydrogen is 2.1, while that of oxygen is 3.5. The atomic structure of the water molecule is bent as shown in Figure 3.1.

The water molecule is electrically neutral, but the positive and negative charges are not distributed uniformly. The hydrogen nucleus is bound to the central oxygen atom by a pair of electrons shared between them. In chemistry, this shared pair of electrons is called a covalent chemical bond. Such covalent bonds occur when atoms share electrons to satisfy the octet rule (eight electrons in the outer shell). The eight electrons in the oxygen atom are not distributed uniformly. In the water molecule, there is slightly more negative charge at the oxygen end, and a compensating positive charge at the hydrogen end. The resulting polarity is largely responsible for water's unique properties.

The chemical bonds between hydrogen and oxygen atoms in a water molecule are polar covalent. Water readily forms hydrogen bonds with

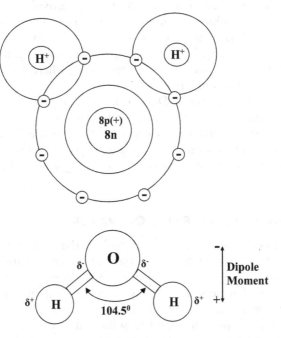

Figure 3.1 Structure of the water molecule (p denotes protons and n denotes neutrons).

other water molecules. One water molecule may form a maximum of four hydrogen bonds with other molecules. Hydrogen bond is the most important intermolecular force in the water molecule. It allows the hydrogen of one molecule to interact with the oxygen of another.

In the water molecule, only two of the six electrons in the outer shell of oxygen atom are used as a covalent bond. The remaining four electrons are organized into two non-bonding pairs. The negatively charged four electron pairs around the oxygen atom tend to arrange themselves as far from each other to minimize the repulsive force. The two non-bonding pairs remain closer to oxygen atom, exerting a stronger repulsive force against the two covalent bonding pairs, resulting in the two hydrogen atoms getting closer together. This results in a bent structure of the water molecule with a H–O–H bond angle of 104.5°. The hydrogen atoms are close to the two corners of a tetrahedron centred on the oxygen. At the other two corners are a lone pair of valence electrons that do not participate in the bonding. In a perfect tetrahedron, the atoms would form a 109.5° angle, but due to the repulsion between the lone pair been greater than the repulsion between the hydrogen atoms the bond angle is reduced to 104.5°.

A dipole occurs when positive and negative charge in sub-atomic particles (protons and electrons) or cations and anions are separated from each other by a small distance. Dipoles occur due to the differences in electronegativity between two chemically bonded atoms. In most molecules, the centres of positive and negative charges coincide. Carbon dioxide and methane fall under the category, which are non-polar molecules. They have zero dipole moments. A dipole moment which is a vector is the product of the distance d (m) between charges multiplied by the partial charge δ^+ and δ^- (Coulomb). The unit of the dipole moment is the Debye (D), where 1 Debye is equivalent to 3.34×10^{-30} Coulomb.m.

Water molecule is a dipole. Due to the bent structure of the molecule, the dipole moment is non-zero and has a magnitude of $1.84D$. The oxygen side of the molecule carries a net negative charge while the hydrogen side carries a net positive charge. The charges are partial meaning they do not add up to unity to become a proton or an electron. Dipoles can occur in magnetism too.

3.6 PHYSICAL PROPERTIES OF WATER

The basic physical properties of water include the density, the boiling and freezing points, the latent heats of vaporization/condensation and that of freezing/melting. Vaporization and freezing are energy absorbing whereas condensation and melting are energy releasing. The numerical values of some of these properties are given in Table 3.1. Normally, the phase change takes place from vapour to liquid to solid and vice versa. Sublimation is the direct transition from solid to vapour without going through the

Table 3.1 Some physical properties of water

Property	Magnitude	Unit
Density (liquid water at 4°C)	1000	kg per m³
Density (ice at 4°C)	917	kg per m³
Density (water vapour at 100°C)	0.7267	g per litre
Molar mass	18.01528	g per mole
Freezing/Melting point	0	°C
Boiling point	100	°C
Refractive index	1.333	
Viscosity (at 20°C)	0.001	Pa.s
Latent heat of freezing/melting	80	cal per g
Latent heat of vaporization/condensation	540	cal per g
Specific heat capacity	4.2	joules per gram °C
Dielectric constant (at 20°C)	80	
pH(at 25°C)	7	
Thermal conductivity	0.598	W per m°K
Bulk modulus of elasticity (at 20°C and at atmospheric pressure)	2.18×10^9	Pa
Surface tension	0.0731	N per m
Vapour pressure (saturated at 100°C)	100	kPa

intermediate state of liquid whereas deposition is the direct transition from vapour to solid.

The speed of light in water is three-quarters of that in air. This speed difference causes light to refract (change direction) as it passes through the air–water interface. The speed of sound in water is 4.3 times that in air at 20°C. In dry air it is 344 m/s, in freshwater it is 1,483 m/s and in seawater it is 1,522 m/s.

Water has a very high specific heat capacity and a high heat of vaporization (Table 3.1). It is an excellent solvent because of its polarity and high dielectric constant. It has a high capillary action because of the strong adhesive and cohesive forces. Because of the hydrogen bonding between water molecules, water also has a high surface tension. Pure water is an insulator but the presence of salts (especially NaCl) can make water conducting. Water is a rare substance that has its density in solid form less than that in liquid form. That is why lakes and rivers freeze from the top down, with ice floating on water. Most substances that exist in three phases have their densities in solid form greater than those in liquid form and the densities in liquid form greater than those in gaseous form. Water has the second highest specific heat capacity of all known substances. Ammonia has the highest specific heat.

3.7 CHEMICAL PROPERTIES OF WATER MOLECULE

One cubic centimetre of liquid water contains about 3.4×10^{12} molecules having diameters of about 3×10^{-8} cm (3 Angstroms). The hydrogen and oxygen atom each have three isotopes – ^1H (protium), ^2H (deuterium), ^3H (tritium) and ^{16}O (oxygen), ^{17}O and ^{18}O. Protium has a single proton in its nucleus and it does not decay; deuterium has a proton and a neutron in its nucleus thus having a mass of two; tritium is radioactive and through beta decay transforms into helium. It has two neutrons and one proton in its nucleus and has a mass of three. ^{16}O has eight protons and eight neutrons and is abundant in nature (about 99.7%). ^{17}O has eight protons and nine neutrons and is less abundant whereas ^{18}O is an environmental isotope with eight protons and ten neutrons in the nucleus. The hydrogen atom has a positively charged proton and a negatively charged electron. The oxygen atom has a nucleus having eight positively charged protons and eight neutrons surrounded by eight electrons. The arrangement of electrons in the water molecule is electrically asymmetric.

The chemical bond between hydrogen and oxygen atoms in water molecules is polar covalent. A polar bond refers to an ionic bond where the electrons are not equally shared and the electronegativity values are slightly different. The oxygen side of the water molecule has a net negative charge while the two hydrogen atoms have net positive charges. Other molecules with polar bonds include hydrogen fluoride, sulphur dioxide and ammonia. Water is amphoteric, meaning that it can act as both acid and base (alkaline). It is a hydrophilic substance that has a special affinity for water. Substances that repel water are known as hydrophobic whereas those attracted to water but do not dissolve in water as known as hygroscopic.

When a metal is dissolved in water an alkaline solution is formed. For example,

$$Na_2O\,(Sodium\ oxide) + H_2O \rightarrow 2NaOH\,(Sodium\ hydroxide) \qquad (3.1)$$

When non-metals are dissolved in water, acids are formed. For example,

$$SO_3\,(Sulphur\ trioxide) + H_2O \rightarrow H_2SO_4\,(Sulphuric\ acid) \qquad (3.2)$$

Acids neutralize salts forming water. For example,

$$2NaOH\,(Sodium\ hydroxide) + H_2SO_4\,(Sulphuric\ acid)$$
$$\rightarrow Na_2SO_4\,(Sodium\ sulphate) + 2H_2O \qquad (3.3)$$

The water molecule ionizes as

$$H_2O \rightleftharpoons H^+ + OH^- \qquad (3.4)$$

An excess concentration of H^+ ions gives acidic property whereas an excess concentration of hydroxyl ions (OH^-) gives an alkalinity. When they are equal, the condition is known as neutral. Hydrogen ion concentration pH is defined as

$$pH = \log_{10}\frac{1}{[H^+]} = -\log_{10}[H^+] \tag{3.5}$$

In a perfect neutral solution at 25°C,

$$[H^+] = [OH^-] = 1 \times 10^{-7} \text{ moles} \tag{3.6}$$

The pH of a neutral solution is therefore 7. The square brackets indicate ionic concentration.

The hydrogen and hydroxyl ions in water determine the structure and biological properties of proteins and other cell materials. Though inert, it is a powerful universal solvent as well as a means of transporting many substances.

3.8 UNIQUE PROPERTIES OF WATER

Hydrogen bonds affect the boiling and melting points as the greater hydrogen bond results in increased intermolecular attraction. In general, compounds having more hydrogen bonds have higher boiling and melting points. Water is also an ideal solvent due to its polar bonds. Water molecules also attract each other due to hydrogen bonding (dipole–dipole interaction).

Water has a very high surface tension that allows objects with a higher density to float on the surface rather than to sink. Surface tension is the tendency of molecules of a liquid to be attracted more towards one another at the surface of the liquid than to the air above it. This is due to the strong hydrogen bond in the water molecule.

Water's polar bonds allow it to dissolve many substances, making it an ideal solvent. Water which is a polar molecule is bent, with the negatively charged oxygen on one side and the pair of positively charged hydrogen on the other side of the molecule. H_2O molecules attract each other through the special type of dipole–dipole interaction known as hydrogen bonding. The small negative charge near the oxygen atom attracts nearby hydrogen atoms from water or positively charged regions of other molecules. Similarly, the small positive charge near the hydrogen side of each water molecule attracts other oxygen atoms and negatively charged regions of other molecules.

Capillary action is the movement of water (or any liquid) within the pores of a porous medium due to adhesion, cohesion and surface tension. Adhesive forces are caused by the hydrogen bond between unlike atoms, cohesive forces by the hydrogen bonding between like atoms and surface tension by hydrogen bonding for water to be attracted to itself and away from other materials. In a glass tube (SiO_2), the hydrogen bond causes the water to stick to the glass surface due to the hydrogen bond between the hydrogen atom and the oxygen atom in glass. In the same way, water rises against gravity in xylem tubes whose cell walls are made of cellulose, which is a polymer ($(C_6H_{10}O_5)_n$ where n represents the degree of polymerization). Cellulose and lignin are the main compounds that give the structural shape of plants.

Capillary action, adhesion, cohesion and surface tension, which are all caused by hydrogen bonding, have a very important role in plant life. Water and nutrient circulation in plants is analogous to blood circulation in mammals. The former is driven by pressure and chemical potential gradients whereas the latter is driven by the heart which acts as a pump. Most of the water absorbed and transported through plants is driven by negative pressure generated by the transpiration from the leaves. This system is sustained because of the cohesive nature of water through forces generated by hydrogen bonding. When water is contained in tiny capillaries, the hydrogen bond allows water columns to sustain high tensions capable of transporting water to tree canopies some 100 m above the soil base.

In plants, the root system absorbs the water with dissolved nutrients aided by the osmotic pressure. The plant roots have a higher concentration than the soil and therefore the water flows into the roots. The roots then convey the water to the stem aided by capillary action, which has its limits. The building of plant matter starts in the leaves in the canopy through photosynthesis, which requires water. The canopy in some plants can be quite high and may sometimes exceed 100 m above the base of the plant. The process of transporting water and nutrients to the canopy and transporting processed 'food' from the canopy to other parts of the plants is via the xylem and phloem.

Xylem is one of the important tissues of vascular system of plants. Vascular plants are land plants with lignified tissues for conducting water and minerals throughout the plant. Lignin is a complex polymer that forms the structural material to support tissues. It is the second most abundant plant biopolymer after cellulose and accounts for about 30% of plant biomass. Vascular plants have special tissues such as xylem and phloem. Xylem is a type of permanent tissue that does not decay easily and found in wood and bark. Most plants belong to the vascular type but mosses and algae are non-vascular. The main activity of this tissue is to transport minerals and water from roots and carry them to other parts of the plants. Xylem tissues are present in roots, stems and leaves. Structurally, xylem tissues are tubular-shaped with no cross walls. Phloem tissues are tubular and elongated, having walls with thin sieve tubes. They carry the nutrients and 'food' from the leaves to the growing parts of the plant. The water transport in the xylem is unidirectional upwards whereas that in the phloem it is bidirectional. The former is due to hydrogen bonding. The words xylem and phloem both have Greek origins with meanings wood and bark, respectively.

Chapter 4

Drinking water – Global perspective

4.1 INTRODUCTION

Water is life as there is no life without water. It can be thought as the blood of the earth. The unfortunate inconvenient truth is that safely managed drinking water is not available for about 2.1 billion people (29% of the world population) on earth and that the people who are deprived of having access to safely managed drinking water live in low-income countries. The World Health Organization (WHO) has concluded that water-borne disease is the leading killer in the world (Berman, 2009; www.voanews.com/arch ive/who-waterborne-disease-worlds-leading-killer). In this context, safely managed water means water available on premises when needed and free from contamination.

Hand washing as a precaution against the transmission of the corona virus is promoted and encouraged by all health authorities across the world, but the big question is where is the water for 29% of the world population? Of the roughly 818 million children worldwide who lack basic hand washing facilities at school, more than one-third are in sub-Saharan Africa (https://news.un.org/en/story/2020/08/1070072). About 40% of the global population lives without basic hand washing facilities at home (Kisaakye et al., 2021). The inconvenient and unfortunate truth is that the recurrent annual deaths resulting from unsafe drinking water is approximately the same as the death rate due to current COVID-19, which hopefully is not a recurrent pandemic. Unfortunately, this problem is not receiving the same attention that climate change is receiving despite the fact that there have been no deaths as a direct result of climate change. This issue is highlighted in a recent publication by the author (Jayawardena, 2021).

4.2 SOME STATISTICS ON ACCESS TO DRINKING WATER

According to the key facts from WHO, the following statistics can be summarized:

- In 2017, 71% of the global population (5.3 billion people) had access to a safely managed drinking water service.
- 90% of the global population (6.8 billion people) used at least a basic service. Basic service in this context means an improved drinking water source within a round-trip of 30 minutes to collect water.
- 785 million people lack even a basic drinking water service. This includes 144 million people who are dependent on surface water.
- Globally, at least 2 billion people use a drinking water source contaminated with faeces.
- Contaminated drinking water is estimated to cause 485,000 diarrheal deaths each year.
- By 2025, half of the world's population will be living in water-stressed areas.
- In least developed countries, 22% of health care facilities have no water service, 21% no sanitation service and 22% no waste management service.

This information is available in the following WHO website: (www.who.int/news-room/fact-sheets/detail/drinking-water, June 14, 2019) Furthermore, (Hannah Ritchie and Max Roser, 2019),

- Unsafe water is responsible for 1.2 million preventable deaths each year.
- 6% of deaths in low-income countries are the results of unsafe water sources.
- 666 million (9% of the world) do not have access to an improved water source.
- 2.1 billion (29% of the world) do not have access to safe drinking water.

This information is available in the following website: (https://ourworldindata.org/water-access#unsafe-water-is-a-leading-risk-factor-for-death)

4.3 WATER SCARCITY

Water scarcity can arise from physical lack of water, poor quality or due to lack of capacity for developing and maintaining a reliable supply. Arid and desert areas suffer from physical lack of water and such areas also experience pollution problems. In such situations, home-grown techniques such as rainwater harvesting and groundwater exploitation would be better appropriate technologies than conventional water supply technology. Since such areas are sparsely populated, achieving individual household self-sufficiency is more favoured and should be encouraged than traditional water distribution systems where the conveyance cost can be excessively high. It is also important to introduce low-cost water filters, which can be used in individual households. A dollar invested in improving access to safe drinking water is estimated to increase the GDP by $3–14, depending on the type of investment. It is projected that by 2025, 1.8 billion people will be living in absolute water scarcity and two-thirds of world population could be living under water-stressed conditions. In this context, absolute scarcity means the annual per capita water availability is less than 500 m³, water scarcity means when it drops below 1,000 m³ and water stress means when it drops below 1,700 m³ (www.un.org/waterforlifedecade/scarcity.shtml). It is also projected that by 2050, 4.8–5.7 billion people will be living in potential water-scarce areas at least one month in a year (Burek et al., 2016).

4.4 INTERNATIONAL INITIATIVES

The first International Decade for Clean Drinking Water, 1981–1990 was set up at the 1977 'United Nations Water Conference' at Mar del Plata with the aim of making access to clean drinking water available across the world.

In the 1980s, of the 1.8 billion people living in the rural areas of developing countries, only one in five had access to clean water. Over half a billion of the children under 15 years old did not have clean water. In developing countries, one hospital patient in four suffered from an illness caused by polluted water. Although the decade focused on safe water and sanitation for all by 1990, only 1.2 billion people were fortunate to have access to safe drinking water and about 770 million to safe sanitation by the end of the decade. In 2017, at least 1.2 billion people worldwide were estimated to drink water contaminated with faeces and over 2 billion people did not have access to basic sanitation facilities such as toilets or latrines (www.un.org/en/global-issues/water).

As a follow-up, the United Nations General Assembly adopted a draft resolution proclaiming 2005 to 2015 as the 'International Decade for Action – Water for Life' with the aim of halving by 2015 the proportion of people who are unable to reach or afford safe drinking water and who do not have access to basic sanitation with emphasis on the participation of women in water-related development efforts. From 1990 to 2015, 2.1 billion people gained access to a latrine, flush toilet or other improved sanitation facility. While the team of the United Nations Office to Support the International Decade for Action 'Water for Life' 2005–2015 have received accolades for their efforts to collect and disseminate water-related information, over 663 million people still drew water from an unimproved source in

2015. An unimproved drinking water source in this context means a source that does not protect against contamination.

One of the key Sustainable Development Goals (SDGs) is SDG6 on safe drinking water and sanitation. It involves health, dignity, environmental sustainability and the survival of the planet and requires efforts at regional, national and international levels. Universal, affordable and sustainable access to 'water, sanitation and hygiene' (WASH) is the focus of the first two targets of SGD6. On July 28, 2010, the UN General Assembly adopted a resolution recognizing 'the right to safe and clean drinking water and sanitation as a human right that is essential for the full enjoyment and all human rights'. Regions and nations therefore have an obligation to work towards achieving universal access to safe drinking water and sanitation for all human beings without any discrimination while prioritizing those most in need.

The SDG6 also defines accessibility to safely managed drinking water as water available at the premises when needed that is not contaminated by faecal and chemical matter. In terms of access, there is no legal standard for physical accessibility, but a maximum of 30 min round-trip travel time to a location of managed water supply is considered as a norm. For basic sanitation, the facilities should not be shared and available at the premises. On the affordability issue, while it is not a human right to expect free access to safe and clean water, disconnection of water services due to lack of means is considered as a violation of human rights. Sometimes water rights may impinge on human rights. Water right is a legal issue and can be taken away from an individual, but human rights cannot be taken away from an individual.

The sustainable development goals (SDGs) were set in 2015 by the United Nations General Assembly and are intended to be achieved by the year 2030. In 2017, 2.2 billion people did not have access to safe drinking water, 4.2 billion people did not have access to safe sanitation, and 3 billion people worldwide did not have hand washing facilities at home (https://sdgs.un.org/goals/goal6). This has been replaced by the SDG6 Global Acceleration Framework in partnership with over 30 UN entities and 40 international organizations as a contribution to deliver the goals by 2030.

4.5 DISPARITY IN PREVENTABLE DEATHS DUE TO UNSAFE WATER AND SANITATION

Table 4.1 illustrates the disparity in preventable deaths due to unsafe water and sanitation across some selected low-income and high-income countries. In low-income countries, the deaths due to unsafe water as a percentage of the total deaths range from about 5.75% to 14.45% whereas in high-income countries the corresponding range is from about 0% to 0.04%. The worldwide average is about 2.2%. Based on these figures, there is a

Table 4.1 Death rates due to unsafe water and sanitation in some selected countries in 2017

Country	No. of deaths due to unsafe water (per 100,000)	Percentage of deaths due to unsafe water sources	Percentage of population with access to improved water source	No. of deaths due to unsafe sanitation	Percentage of deaths due to unsafe sanitation	Percentage of population with access to improved sanitation
Chad	130.5	14.45	50.8	16,517	11.25	12.1
South Sudan	118.42	9.86	58.7	7,514	7.73	6.7
Madagascar	96.05	10.57	51.5	15,674	8.47	12.0
Niger	93.74	10.73	58.2	14,624	8.45	10.9
Angola	83.06	7.94	49.0	9,616	5.21	51.6
Somalia	75.1	6.06	31.7	6,792	4.66	23.5
India	67.46	5.75	94.1	328,720	3.32	39.6
China	0.32	0.04	95.5	2,405	0.02	76.5
Japan	0.11	0.03	100	104	0.01	100
United States	0.05	0.01	99.2	577	0.02	100
Luxembourg	0.04	0.01	100	< 1	0.01	97.6
UK	0.03	0.01	100	6	0	99.2
Italy	0.02	0.01	100	26	0	99.5
Netherlands	0.02	0.01	100	5	0	97.7
Ireland	0.01	0.0	97.9	2	0.01	90.5
World average	16.97	2.2	90.95	774,241	1.38	67.53

Source: Extracted from https://ourworldindata.org/water-access#unsafe-water-is-a-leading-risk-factor-for-death

1,000-fold difference in death rates between developing (low-income) countries and developed (high-income) countries. Similar trends can be seen in the death rates due to unsafe or lack of proper sanitation. In developing countries, it is in the range of 3.32–11.25% whereas in developed countries the corresponding range is 0–0.02% with a world average of about 1.38%, that is, greater than a 1,000-fold difference. The percentage of population with access to improved sanitation ranges from a high of 100% to a low of about 6.7% (South Sudan) in 2017 (https://ourworldindata.org/sanitation#unsafe-sanitation-is-a-leading-risk-factor-for-death).

Improved sanitation in this context means having access to a piped sewer system, septic tanks, ventilated improved pit (VIP) latrines, pit latrines with slab and composting toilets, limited service means improved facility shared with other households, unimproved facility means facilities which do not separate excreta from human contact, and no service means open defecation. In terms of numbers, deaths due to unsafe drinking water worldwide are about 1.2 million per year and due to unsafe sanitation worldwide are about 775,000 per year, which are approximately 2.2 % and 1.4% respectively of global deaths. In low-income countries, they are about 6% and

5% respectively (https://ourworldindata.org/water-access#unsafe-water-is-a-leading-risk-factor-for-death; https://ourworldindata.org/sanitation)

The health of a nation depends upon the level of cleanliness of the domestic water supply. It is a problem that is often ignored or sidelined by the developed countries as it is only a problem of the poor and the developing countries. Coupled with the accompanying sanitation problem, the situation can be disastrous if allowed to continue. Unlike a major flood or an earthquake, which affect a small region with high population density, the effect of the lack of safe drinking water is spread over vast areas with relatively low population densities. From the media point of view, such widespread and prolonged suffering receives much less attention compared with that received for high-impact type of disasters such as earthquakes and major floods.

Although the numbers of people with access to improved water sources have increased in recent years, over 666 million people still did not have access to improved water sources by 2015. South and East Asia and sub-Saharan regions are among the worst affected. In this context, having access to an improved water source does not necessarily guarantee that the water is free from contamination. It is estimated that 29% of the world population does not have access to safe drinking water (https://ourworldindata.org/water-access).

Access to improved water sources is highly correlated with income levels in any region (Table 4.2). Although the per capita GDP is an indicator of the income level in a country, it may sometimes give a distorted image particularly when the income is concentrated in a small fraction of the overall population with a high degree of inequality in income distribution.

Table 4.3 gives some statistics of water availability in some selected low-income countries with the lowest percentages of population served. (Extracted from Drinking water source – The World Factbook (cia.gov)).

Table 4.2 Share of the population with access to improved drinking water versus annual GDP per capita, 2015

Region	Share of population with access to improved water	Annual per capita GDP (US$)
Low-income countries	65.64%	1,844
Low middle-income countries	89.19%	6,054
Middle-income countries	92.15%	10,300
Upper middle-income countries	95.49%	15,126
High-income countries	99.54%	41,901
World average	90.95%	14,778

Source: Extracted from https://ourworldindata.org/grapher/improved-drinking-water-vs-gdp-per-capita?tab=table

Table 4.3 Some statistics of water availability in some selected low-income countries

Country	Improved services			Unimproved services		
	Urban (%)	Rural (%)	Total (%)	Urban (%)	Rural (%)	Total (%)
Chad	86.7	46.6	55.7	13.3	53.4	44.3
Eritrea	73.2	53.3	57.8	26.8	46.7	42.2
Kiribati	–	–	71.6	–	–	28.4
World average	96.5	84.7	91.1	3.5	15.3	8.9

4.6 VALUING WATER

Water, which has no substitute, is a basic human need as there is no life without water and access to water and sanitation is a basic human right. In addition to drinking water, food production also depends on water which can be thought of as the blood of the earth. The value of water is unquantifiable in monetary terms. It is also important to realize that the value of water is much more than its price.

The cost of providing safe drinking water includes financial, social, environmental and health components. Benefits include economic, social, environmental, food security and health, some of which cannot be easily converted into monetary terms. For example, lives saved by providing clean water cannot be assigned a monetary value in a traditional cost–benefit analysis. It is also important to consider the possible costs that may be incurred if projects to provide safe drinking water are not carried out. As illustrated in Table 4.1, the cost of not providing safe drinking water would be preventable deaths. In this context, instead of the traditional cost–benefit analysis, an approach in which cost-effectiveness, which considers the lives saved and people served, would be more appropriate.

As highlighted in World Water Development Report 2021, which was released on the World Water Day (March 22), valuing water involves valuing the water sources and ecosystems, valuing water infrastructure for water storage, use, reuse or supply augmentation, valuing water services, which include drinking water, sanitation and related human health aspects, valuing water as an input to production and socio-economic activities such as food production, energy and industry, business and employment and other socio-cultural values of water, including recreational, cultural and spiritual attributes. Good governance plays a crucial role in valuing water infrastructure. Traditional approaches of valuing water are based on cost–benefit analysis in which there is a tendency to over-estimate benefits and under-estimate costs. In particular, the value of water assigned to food production is low compared to other uses and also, the values attached to staple food production are much lower than the corresponding values attached to high-value crops such as flowers.

Even the rich countries which hitherto have not experienced the lack of safe drinking water would have realized the importance of safe drinking water during the recent water shortages in Texas where over 1.4 million Texans faced water disruptions due to winter storm 2021, and in Flint in Michigan where the water has been contaminated with lead and possibly Legionella bacteria during 2014–2019 (Masten et.al., 2016). It is also reported that rural Alaskan homes without piped water use only 5.7 litres per capita per day compared to the WHO standard of 20 litres per capita per day and far below the average consumption of 110 litres per capita per day in some parts of Canada (Rural Alaskans struggle to access and afford water – McGill University). More recently, it has been reported that water utilities and regulators in the United States have identified 56 new contaminants in drinking water over the past two years, a list that includes dangerous substances linked to a range of health problems such as cancer, reproductive disruption, liver disease and much more (www.theg uardian.com/environment/2021/nov/03/us-tap-water-contaminants-discov ery-radioactive-material-pesticides?utm_medium=email&utm_source=rasa _io&PostID=40947334&MessageRunDetailID=6918170138).

4.7 WATER CRISES AND GDP

It has been reported that securing water for all by 2030 will cost just over 1% of global GDP (Strong and Kuzma, 2020) and that the economic benefits far outweigh the costs. The share of population with access to improved water ranges from about 65% in low-income countries to about 99% in high-income countries with a world average of about 90% (Table 4.2). It is also reported that failing to implement better water management policies would result in regional GDP losses by 2–10 % by 2050 (Strong and Kuzma, 2020). More than 10% of the global population who live mostly in low-income countries will need more than 8% of their annual GDP to deliver sustainable water management. This can be achieved only with assistance from financial institutions and governments of high-income countries. Since water crises cascade beyond national boundaries, the governments around the world should pay attention to and invest in water security regardless of national or political boundaries.

4.8 WHAT CAN AND SHOULD BE DONE?

There are no insurmountable technical problems in implementing solutions to the world's water crises. The problem is finding money to implement the solutions. This requires political will as well as seeing it not as a burden but as an opportunity. Economic losses due to lack of safe drinking water and sanitation in Africa is estimated to be about 5% of GDP (3rd UN World Water Development Report, 2009). It is an inconvenient and sad truth that many of the people who most need water filters are those who are least able to afford them. It is difficult to expect the high-income countries to solve a problem of low-income countries but from a humanitarian point of view, the former can offer a helping hand to the latter. WHO also has highlighted the importance of solidarity in the context of fighting the corona virus. High-tech and expensive water filters which such people cannot afford will not solve the problem. Therefore, attempts should be made to introduce low-tech inexpensive filters, which can be constructed with minimum instructions and with locally available materials. Nature-based solutions need to be promoted. Some possible options are given below.

4.8.1 Rainwater harvesting

Rainwater harvesting has been practiced in both developed and developing countries to augment their supplies of freshwater for domestic use. However, the quality of harvested rainwater is not always clean as one would expect because of the possible contamination enroute to the storage tank and within the storage tank. Contaminants may include bird droppings, insects, wind-blown dirt etc. Rainwater is slightly acidic and therefore can dissolve metallic substances from metallic roofs and storage container surfaces. Typical heavy metal contaminants include lead and zinc, which are harmful to human health. Microbial contamination can be significantly reduced if harvesting is done after the initial flush of rainwater. Storage containers should be covered to prevent mosquito breeding and to prevent sunlight reaching the water that can promote algal growth. Harvested rainwater can be used as a non-potable source for domestic use and also as a potable source after some treatment. Ideally, the harvested rainwater can be used as the raw water in mini slow sand filters.

4.8.2 Slow sand filtration

Slow sand filters operate with the combined action of sedimentation, adsorption, straining and biological and microbial filtration. They not only remove suspended matter in water but also remove harmful bacteria present in water once the filter has ripened, which is achieved when a gelatinous layer of bacteria, algae, fungi and higher organisms known as schmutzdecke grows over the filter. Construction of mini slow sand filters can be carried out without much skill and would be ideal in remote sparsely populated areas. With some initial guidance on the construction and maintenance, slow sand filters can be installed in individual houses.

4.8.3 Xylem filters

A more recent low-cost filtration technique practiced in India makes use of plant xylem. Xylem is the porous tissue that conducts fluid in plants.

Based on a study carried out by MIT scientists (Lee et.al., 2013; Boutilier et.al, 2014), it has been demonstrated that filters made of pine sapwood removed *E. coli* bacteria from polluted water. *E. coli* have typical diameters of the order of 1 μm. The process begins with cutting cross-sectional discs from tree branches after removing the bark, soaking them in hot water, drying and dipping in room temperature water. The discs are then soaked in ethanol to keep the xylem membranes from sucking and sticking together when drying. After drying, the discs are placed in cylindrical filtration devices. The filtration process is then similar to that of a slow sand filter. Tests carried out in India have shown that the xylem filters remove over 99% of *E. coli* and rotavirus from contaminated water. Since raw materials are widely available, fabrication process is simple once the initial instructions are given to individual users. Xylem filters are able to effectively reject particles with diameters larger than 100 nm. The principle of the xylem filter is based on the mechanism that transports water from the root system to the plant cells.

4.8.4 Solar evaporation

Water purification by evaporation using solar energy is a possible solution to an increasing demand for clean water, but the downside is that the production rate is small. Still, in remote areas it is an option for producing potable water for individual houses.

4.8.5 Bioinspired hydrogel

A new technique that does not require evaporation has been recently tested (Xu et al., 2021; Corless, 2021) using solar absorber gel (SAG) that is inspired by the pufferfish's (sometimes called blowfish) ability to inflate and deflate itself with water in response to dangers from predators. In the context of water purification, the process is equivalent to absorbing contaminated water and spitting out clean water. The purifier consists of a macroporous heat-responsive hydrogel coated with polydopamine (PDA) and sodium alginate, which helps to repel salts. A hydrogel is a three-dimensional network of hydrophilic polymers that can expand and hold large amounts of water while maintaining their physical structure due to cross-linking of individual polymer chains. When immersed in contaminated water, the SAG absorbs large quantities of water while repelling contaminants such as salts, bacteria and other pollutants. When exposed to sunlight, the gel gets heated releasing clean water. The SAG expels the clean water by a phase change at a lower critical solution temperature without having to go through the evaporation process. Thus, its energy efficiency is much higher than that in the evaporation process.

4.8.6 Diverting military spending to provide safe drinking water

The military spending as a percentage of the GDP ranges from 1.2% in low-income countries, 1.9% in middle-income countries and 2.4% in high-income counties to a high value of 20.9% in Eritrea (https://data.worldb ank.org/indicator/MS.MIL.XPND.GD.ZS?view=chart). Military spending is also promoted by arms traders in high-income countries. Rather than selling hardware to kill people, it would be a noble practice if at least part of the resources in arms trade can be diverted to selling equipment and methods to save lives.

4.8.7 Research and development, and education

Addressing this problem requires no high-tech solutions. Nature-based infrastructure, which preserves ecosystems and where the traditional cost–benefit analysis is replaced with cost effectiveness, needs to be promoted as it is difficult if not impossible to quantify in monetary terms the losses or savings of human lives. Research and development can enhance the provision of fair and equitable share of water to all human beings as well as increasing the productivity of water. In addition to research and development, other means of educating the general public on the importance and sustainability of water considering cultural and religious practices as well as the value of water, which may differ from region to region and even from person to person need be adopted. It is also ironical to note the suggestion that wealthier lenders should provide financial support to the most vulnerable countries on condition that the money is used to increase climate resilience when 29% of the world population does not have access to safe drinking water. The result of this sad inconvenient truth is about 1.2 million preventable deaths per year worldwide. The lack of sufficient attention to this problem is further highlighted in the American Society of Civil Engineers (ASCE) infrastructure report card for drinking water that received a grade of C in 2021 while that for wastewater it has been D+!

REFERENCES

Berman, J. (October 29, 2009). WHO: Waterborne Disease Is World's Leading Killer (www.voanews.com/archive/who-waterborne-disease-worlds-leading-killer)

Boutilier, M. S. H., Lee, J., Chambers, V., Venkatesh, V. and Karnik, R. (2014). Water filtration using plant xylem. *PLoS One* 9(2): e89934. https://doi.org/10.1371/journal.pone.0089934

Burek, P., Satoh, Y., Fischer, G., Kahil, M. T., Scherzer, A., Tramberend, S., Nava, L. F., Wada, Y. et al. (2016). *Water Futures and Solution – Fast Track Initiative (Final Report)*. IIASA Working Paper. IIASA, Luxembourg, Austria: WP-16-006.

Corless, V. (2021). Inspired by the pufferfish, this hydrogel purifies water using nothing but sunlight, *Advanced Science News, Environment, Food and Water*, March 31, 2021. www.advancedsciencenews.com/inspired-by-the-pufferfish-this-hydrogel-purifies-water-using-nothing-but-sunlight/

Jayawardena, A. W. (2021). An inconvenient truth about access to safe drinking water, *International Journal of Environment and Climate Change*, 10 (11): 158–168.

Kisaakye, P., Ndagurwa, P. and Mushomi, J. (2021). An assessment of availability of handwashing facilities in households from four East African countries, *Journal of Water, Sanitation and Hygiene for Development*, 11 (1): 75–90. https://doi.org/10.2166/washdev.2020.129.

Lee, J., Boutilier, M. S. H., Chambers, V., Venkatesh, V. and Karnik, R. (2013). Water filtration using plant xylem, arXiv:1310.4814[physics.flu-dyn], pp11.

Masten, S. J., Davies, S. H. and Mcelmurry, S. P. (2016). Flint water crisis: what happened and why? *Journal of the American Water Works Association*, https://doi.org/10.5942/jawwa.2016.108.0195

Ritchie, H. and Roser, M. (2019). Clean water (https://ourworldindata.org/water-access#unsafe-water-is-a-leading-risk-factor-for-death)

Strong, C. and Kuzma, S. (2020). It Could Only Cost 1% of GDP to Solve Global Water Crises, World Resources Institute, Washington DC, January 21, 2020.

World Water Development Report (2021). *Valuing Water*, World Water Assessment Programme, UNESCO, p. 187.

World Water Development Report 3 (2009). *Water in a Changing World*, World Water Assessment Programme, UNESCO, p. 318.

Xu, X., Ozden, S., Bizmark, N. Arnold, C. B. Datta, S. S. and Priestley, R. D. (2021). A Bioinspired Elastic Hydrogel for Solar-Driven Water Purification, Advanced Materials, Wiley Online library.

Chapter 5

Domestic water – Supply and demand

5.1 INTRODUCTION

Water is essential for all forms of life. It has contributed to the elimination of hunger through irrigation and diseases through public water supplies. The basic demand for this precious resource can be classified under six areas, namely, domestic, agricultural, industrial, recreational, transportation and energy. In ancient times, people used to live in places where there was water such as in or near riverbanks. In the modern days, people want water to be brought into their places of living. This requires the necessary infrastructure to find sources of water, harness and store if necessary, treat and bring potable water to the places of living. It is also important to ensure that the source of water has adequate quantity to meet the domestic requirements as well as the quality that meets the health requirements since poor quality water can carry pathogenic pollutants. This chapter describes the different sources of water, types of domestic demand for water, collection and storage and associated constraints and problems.

5.2 SOURCES OF DOMESTIC WATER

There are several sources of water for domestic use that need to satisfy the following important requirements:

- The source should be near the demand area.
- Water should be available in sufficiently large quantities.
- Water should be clean.
- The elevation of water level with respect to the service area should be sufficiently high to make use of gravity flow wherever possible.

They include rainwater, surface water, groundwater, imported water, recycled water and seawater. How these sources can be made to provide domestic water is briefly described below.

5.2.1 Rainwater

Rainwater is widely used in rural areas in developing countries as a source of water and is being encouraged by agencies such as the WHO and UN Centre for Appropriate Technology because of the low level of technology involved. It has a long history dating back to Roman villas. In the present days, rainwater is used as a potable source of water in countries such as Australia, Bermuda, Israel, Thailand, Kenya, Tanzania and the United States (California and Hawaii). Because of its importance to the International Drinking Water Decade, an International Conference on Rainwater Cistern Systems was held in Hawaii in June 1982. Basically, there are two types of rainwater catchment systems (RWCS) as described below.

5.2.1.1 Prepared catchments

Usually, the catchments should be upland ones to enable a clean supply. Rainwater is stored in reservoirs constructed by damming a valley. Water loss by seepage is prevented by covering with a plastic sheet or by applying a sealant at the bed. Potential yield is estimated from the runoff characteristics. Usually, the runoff is of the order of about 40% of the rainfall.

The quality of rainwater from upland catchment is good, soft and free from man-made pollutants. However, sterilization is preferred if kept for a long time in view of microbial growth.

5.2.1.2 Catchwaters

The function of a catchwater is to provide yield from another catchment. It can be at an elevation higher than the reservoir in which case the transfer of water is by gravity, or at a lower elevation in which case the water needs to be pumped into the reservoir. Catchwaters are usually ineffective in increasing drought yield but assist in refilling rates.

5.2.1.3 Rooftop catchments

Rainwater is collected by rooftops and stored in cisterns. The first flush of such collected water is discarded as it could be contaminated with dust and other particles contained in the atmosphere and on the collecting surfaces as well as bird droppings. Usually, rooftop catchments are suitable in dry regions and in atoll[1] islands where there are no substantial groundwater resources. For instance, in Bermuda, which is a coral atoll, it is required by

---

1 Ring-shaped coral reef enclosing lagoon.

law to have facilities to collect water using rooftop catchments for use by those living in the building.

Rooftop catchment systems require little or no technology. The main problem for the designer is to design a system to provide potable water to the user at the most economical cost. In most instances, RWCS will not provide the total water requirements. In such situations, the optimum combination with another source must be sought. The variables involved in the design are the rainfall, the rooftop area (roof area projected on a horizontal plane), the storage capacity and the demand. Of these, the rooftop area is pre-determined from other considerations. Rainfall is a random process and therefore the only other variable that has to be determined by the designer is the capacity of the storage tank. It is estimated on the basis of the per capita consumption and the percentage of the per capita consumption that is expected to be provided by the RWCS.

In Bermuda, the design code specifies a mandated ratio of the tank capacity to roof area (Rowe, 2010). For example, a roof area of 200 m^2 should have a tank capacity of about 68 m^3. These specifications are unrealistic in urban environments with high population density.

Storage tanks are constructed either underground or overground. Underground tanks keep the water cool but require energy for lifting. Overground tanks require the necessary structural support and need to be covered.

The quality of RWCS water is good, soft and acidic if collected on a clean surface and kept covered and sealed. However, it lacks the minerals the body needs from a nutritional point of view.

5.2.2 Surface water

5.2.2.1 Streams and lakes by continuous draft

(i) Streams

This method is suitable for perennial rivers in which the minimum flows are greater than the maximum demand for the season. Abstraction may be either by direct tapping from the river itself or indirect tapping from a nearby water table. In the latter case, filtration is automatically achieved. Intake structures are required for both types of tapping, but the cost of these will be small in comparison with head-works required for impounding reservoirs. The usual problems associated with this type of sources are that

- the quality is not satisfactory because all sewage and other effluents are usually discharged into the river,
- the flow is not sufficient for all the time and
- the flow usually carries a large amount of silt.

(ii) Lakes

Lakes are natural reservoirs and the water in lakes is generally clearer than in rivers because of the settling of silt and other suspended matter. However, they are not found everywhere. Abstraction from lakes is done by pumping or by gravity flow if the elevation is high. Some details of natural lakes are given in Chapter 2.

5.2.2.2 Rivers and streams dammed to form impounding reservoirs

Impounding reservoirs are used to store water during high flow periods and release during low-flow periods. They are constructed in upland areas by damming a watercourse. There are many advantages of having an impounding reservoir as a source of potable water supply. For example,

- being usually at a high elevation, distribution to treatment works and to consumers (sometimes) can be by gravity,
- runoff, which is stored in the reservoir will improve its clarity, turbidity etc. when allowed to remain in the reservoir for a period of time,
- indirectly serves as a means of regulating flow downstream and
- can be used for other activities, for example, power generation, flood control, aquaculture, recreation etc.

5.2.3 Groundwater

5.2.3.1 Natural springs

When the water table surfaces due to geological formations, a natural spring appears. Artesian springs usually have a large tributary watershed at some distance from the spring and the water appears usually under pressure. They go through the earth's natural filtration system. Spring water also contain healthy minerals such as silica, magnesium and calcium.

Natural springs can be found in many parts of the world. Some are hot water springs that attract tourists. Florida in the United States has over 700 natural springs, of which Wakulla, Manate and Silver Springs are better known and Wakulla Spring, which flows into Wakulla River is claimed to be the largest and deepest spring in the world. Thousand Springs along the Snake River in Idaho is also a large natural spring in the United States. Natural spring water is used for bottled water in France, Norway, Italy, UK, the United States and Germany.

5.2.3.2 Wells

Wells derive their water from aquifers. Shallow wells usually derive water from unconfined aquifers (water table) whereas deep wells derive water from confined aquifers. Shallow wells are easily vulnerable to pollution and are often the main source of water supply in many rural areas. They usually dry up during the dry season. Deep wells, on the other hand, are more reliable and can provide large quantities of water as they penetrate into deep confined aquifers. They are suitable for public supplies or for large private users. The quality of water from deep wells is generally good but hard. Wells may also be classified according to the method of construction. Shallow wells are constructed by digging, boring, driving or jetting whereas deep wells are drilled.

(i) Dug wells

This is the oldest form of wells and probably is the oldest source of potable water in the world. Depths of dug wells may be up to about 20 m, with diameters ranging from 1 to 10 m. Dug wells are used mainly for individual supplies. Excavation is usually by hand.

A properly constructed dug well in a water-bearing area may yield about 2,500–7,500 m^3 per day, although most dug wells yield much smaller amounts.

(ii) Bored wells

Bored wells are constructed by boring with augers, which may be hand-operated or power-driven. Diameters range from 15 to 20 cm, and depths up to about 15 m. The yield is usually small. So is the cost of construction.

(iii) Driven wells

A driven well consists of a series of connected pipes driven by impact. Diameters are in the range of 3–10 cm while the depths may range from 15 to 30 m. In a driven well, water enters the well through a drive point at the lower end of the well. If the well is more than 10 m deep, the water table must be at a depth of approximately 3–5 m from the surface in order to provide sufficient drawdown without exceeding the suction limit.

Driven wells yield very small quantities (in the range of about 100–250 m^3 per day). They are therefore suitable for individual domestic supplies, as observation wells, temporary water supplies and for dewatering during construction. A well-point system is a practical example of a series of driven wells connected by a suction header to a single pump. Driven wells are confined to soft formations containing no rocks or gravel.

(iv) Jetted wells

Jetted wells are constructed by the cutting action of a jet of water. A casing, which is lowered as the jet washes away the soil, carries the water and the material loosened by the jet up to the surface. Usually, the diameters range from 3 to 10 cm and depths up to about 15 m. They have a very small yield. Because of the portability of the equipment, this type is suitable for observation wells, exploratory wells and well point systems.

(v) Horizontal wells

(a) *Infiltration galleries*: An infiltration gallery may be thought of as a horizontal well, which intercepts and collects groundwater by gravity flow. Historical examples of infiltration galleries include qanats, found in Iran, Afghanistan and Morocco that are about 3,000 years old. The longest qanat is 29 km long and found near Zarand in Iran where over 22,000 qanats provide about 75% of the water used in the country. Discharges from qanats vary but seldom exceed 100 m^3 per hour. Similar horizontal wells can be found in the grape valley, Turpan in the Xinjiang Province of China. Infiltration basins are areas where flood waters or polluted water are detained in basins for infiltration into the water table to take place. Water may then be pumped via a well.

(b) *Horizontal pipes*: Horizontal perforated pipes can be used in sloping grounds to catch aquifer water. These may also be used for draining slopes saturated with water as a measure to prevent landslides.

(c) *Collector wells*: Collector wells are suitable for areas near rivers. A collector well lowers the water table and thereby induces the infiltration of surface water. Therefore, much larger quantities of water can be extracted from collector wells than would be available from the aquifer.

A collector well consists of a concrete caisson about 5 m in diameter sunk into the aquifer by excavating the inside. After reaching the required depth the bottom is sealed by a concrete plug. Perforated pipes of 15–20 cm diameter are jacked hydraulically into water-bearing area through holes in the caisson in radial directions. These may extend to 50–100 m radially.

Large flow area in a collector well results in low flow velocities, which minimize erosion and sand transport. Surface water is purified by the natural filtration process. Although the initial cost of a collector well is high, the maintenance costs are usually lower than those for a vertical well.

5.2.3.3 Well construction and development

Shallow wells are more vulnerable to pollution and therefore need to be well protected. Most deep wells are constructed by drilling. Percussion method (Cable tool method) and Rotary method are widely used depending upon the formation. The construction procedure depends upon the site conditions.

(i) Percussion method (cable tool method)

By this method, wells of 8–60 cm in diameter and depths of up to 600 m can be drilled in consolidated rocks. It is not suitable for loose sand and gravel formations.

(ii) Rotary method

Deep wells of up to 45 cm in diameter can be drilled through unconsolidated formations by the rotary method. The method consists of rotating a hollow drill bit through which a mixture of clay and water (called drilling mud) is forced. Material broken down by the drill bit is carried upwards to the surface by the rising mud. The mud also forms a clay lining on the walls and therefore no lining is normally required.

After a well has been drilled, it must be developed for maximum yield. This involves placement of casing and screens, and gravel packing. The well casing seals out surface water and any undesirable water entering the well. It is driven to the full depth of the well. Cementing the annular space surrounding the casing further protects the well from pollution.

Screens are placed usually in unconsolidated formations to prevent the entry of fine sand into the well. They should at the same time offer the least resistance to the flow of water. Gravel packing is done in the annular space around the well. The gravel is poured through pipes, which extend to the bottom and which can be withdrawn after the gravel is placed.

5.2.4 Imported water

Imported water refers to water that is transported from areas of abundance to areas of shortages, for example, from Guangdong Province of China to Hong Kong.

5.2.5 Recycled water

Recycled wastewater can be used in agriculture and industry with minimum treatment although with the present level of technology, wastewater can be made biologically and chemically acceptable for human consumption too. However, there is still a psychological stigma towards the use of recycled water as a source of potable water. Indirect potable reuse (IPR), whereby treated wastewater is blended with ground or surface sources which receives further treatment and eventually ends up as drinking water, has become increasingly common. In such cases, the wastewater after tertiary treatment is discharged to storage reservoirs (surface or groundwater) where it is kept for periods of six months or more to assure public fears about 'toilet to tap' concerns.

In recent years, wastewater is considered an 'untapped resource' rather than as a waste product. This paradigm shift is highlighted in WWDR2017. Absolute necessity due to lack of alternative sources, health and environmental considerations, and cost are some of the factors driving the need for water reuse whereas public acceptability is a distracting factor. Countries like Australia, Israel, Namibia and Singapore, as well as some states in the United States including California, Virginia and New Mexico are already drinking treated wastewater.

Singapore, which has a high population density and lacks land as catchments to collect rainwater, has undertaken a scheme to reclaim water, which their Public Utilities Board names as NEWater. The system involves microfiltration, which removes microscopic particles including some bacteria, reverse osmosis, a process by which undesirable contaminants are removed followed by ultraviolet disinfection. Chemicals are added to restore the pH balance. The product is a high-grade recycled water. It is used mainly in industry but is also added to reservoirs to blend with raw water. NEWater is more energy-efficient and cost-efficient than desalination. At present, there are five NEWater plants supplying up to 40% of Singapore's current water needs. By 2060, NEWater is expected to meet up to 55% of Singapore's future water demand. NEWater surpasses WHO drinking water standards.

Namibia, the most arid country in southern Africa, which produces 35% of the water needs for the capital city Windhoek by reclaiming wastewater, has been drinking recycled water since 1969. With increasing population, the demand for water is likely to exceed the supplies at least in certain regions and cities and during certain times. Coupled with advances in wastewater treatment technology, reclaimed water will become a major resource in the future.

5.2.6 Seawater – desalination

This is the most abundant source of water on earth (approximately 97%). Although very large in quantity, the quality of seawater is not good enough for domestic or agricultural use (except for flushing and cooling purposes). However, by an expensive process of purification, it is possible to convert seawater into freshwater. This topic is discussed in Chapter 2 under desalination.

5.3 DOMESTIC DEMAND

This arises out of human's biological requirements for sustaining life and the amenities required for living a lifestyle of the present civilization.

5.3.1 Human's biological requirements

An adult's body consists of about 60% of its weight in water; approximately 26.5 litres inside cells, 12 litres flowing between cells and 3.5 litres in the blood (total of 42 litres). In order to maintain metabolism, an adult needs approximately 2.5 litres of water per day. This is made up as follows:

1.3 litres in drinks
1.2 litres in solid food (0.85 litres of free water + 0.35 litres of metabolic water)

Food + oxygen = CO_2 + energy + water

This intake can vary from place to place depending upon the environmental conditions. For instance, in hot climates, a greater amount of intake is needed to prevent dehydration.

The approximate outflow of 2.5 litres of water from a human body in a temperate climate is as follows:

0.4 litres expired from lungs
0.1 litres in faecal matter
1.5 litres in urine
0.5 litres in respiration (or perspiration)

In hot climates respiration may be as high as 11 litres, that is, 22 times.

Since a reduction in the body's water of anything over 10% is fatal, the same amount of water must be taken in to maintain proper metabolism.

The figure of 2.5 litres is slightly misleading because it reflects only what a human directly consumes. It is important to note that the food chain which provides the energy also requires large amounts of water.

In composition, most food substances contain large amounts of water. For example, lettuce, cucumber and spinach contain about 95% water (in fact, cucumber contains more water than sea water – the weight of dry matter in a given weight of cucumber is less than the same in sea water); bamboo shoots, mushrooms and tomatoes contain about 90% water; potatoes contain about 80% water; bread about 33%; dried beans about 10%.

If an adult needs about 1 kg of food by dry weight per day, this may mean 1 kg of bread per day for a simple vegetarian diet. Bread is made from wheat, which has a transpiration ratio of 500. (Transpiration ratio is defined as the ratio of the weight of water circulating through the plant in a growing season to the weight of dry matter produced during the same period.) Assuming that 2 kg of wheat is required to make 1 kg of bread, 1,000 kg of water is needed for 1 kg of bread, that is, 1 m³.

Animal proteins lengthen the food chain and increase the water requirements. For example, the water requirements for a diet of 0.5 kg of beef + 1 kg of vegetable matter may be worked out as follows:

It takes about two years to raise a beef cow and it may yield about 350 kg of beef in two years, which is approximately 0.5 kg/day over the two-year period. Therefore, to have a continuous supply, it is necessary to have one beef cow per person continuously.

A beef cow's food requirements are approximately 10–15 kg of alfalfa per day and about 50 litres of water per day. The transpiration ratio of alfalfa is about 800. Therefore 10 kg will require 8,000 kg of water (8 m³). It can be seen that the water cost of 1 kg of meat is approximately 16 times that of 1 kg of vegetable matter. In summary, it can be said that 1–9 m³ of water is needed to sustain life for a naked human being.

5.3.2 Human's requirements demanded by the present-day living habits (Personal hygiene, waste disposal, washing clothes, kitchen, car, garden, swimming pool etc.)

In the ancient times, people lived as nomads roaming from place to place and lived near lakes and rivers. Their water requirements were only for their survival. But since the present-day civilization began, humans began to harness resources not only for survival but also for improving the quality of life. The standards of living began to change. Humans looked for more conveniences, resulting in a complete change of lifestyle. Instead of going to the place of natural occurrence of water, humans devised ways of bringing water to their places of living. The society also demanded a minimum of hygienic standards such as cleanliness of body, clothing, kitchen utensils, place of living, among other things.

All these contributed to a heavy demand of domestic water. The present-day domestic water consumption rates are highly varied. They change from place to place, season to season and even day to day. They also depend upon the availability and cost. For instance, in a place where water for domestic purposes is provided on a communal basis, the consumption will depend upon the distance of the consumer to the standpipe. If it is closer, the consumption will be higher. Cost is also another factor. In many countries, water is not metered. Therefore, there is no incentive to conserve or to save. On the other hand, if water is metered and if there are different tariffs for different rates of consumption, many instances of unnecessary use and wastage could be avoided.

Typical domestic consumption rates vary from 10 to 20 litres per capita per day (lpcd) in under-developed countries to about 100–250 lpcd in developed countries. Kuwait has the highest water consumption rate, over 800 lpcd, followed by the United States, which has a rate of about 575 lpcd.

Japan has a rate of about 250 lpcd whereas Nigeria has a rate as low as 30 lpcd. The highest consumption frequency is for the consumption rate range of 105–115 lpcd. Of the water consumed at home, the distribution in a developed area is as follows (Memon and Butler, 2001):

Toilet	31%
Washing machine	20%
Bath	15%
Kitchen	15%
Wash basin	9%
Shower	5%
Car and garden	4%
Dishwasher	1%
Total	100%

The highest household consumption of water is for toilet flushing. In Kuwait, the consumption rates are very high because of two factors. The climate is very dry and therefore more water is needed for even normal living. Secondly, the water in Kuwait comes from seawater through desalination. Since the country is oil rich, the cost of water is insignificant for their economy. In the United States, particularly in California, the consumption rates are high because of higher standards of living. For instance, automatic washing machines, garbage grinders, lawns and swimming pools consume large amounts of water. The older toilets in the United States used 30 litres per flush, which has been reduced to about 13.2 litres per flush during the period 1980–1992. Currently, manufacturers are mandated to reduce toilet flushing capacity to less than 6 litres per flush. Certain brands of toilets have flushing capacities as low as 4.8 litres per flush. The water consumption in household appliances has also gone down over the years as a result of more efficient machines.

In addition to these demands at the consumer's end, allowance must be made for leakages from pipes, plumbing and taps. As a general guideline for estimating demands, the following figures may be assumed:

20 lpcd	Absolute minimum
150 lpcd	Average for a city
250 lpcd	Tropical and semi-tropical countries
(It can be seen that the biological need is only a fraction of the actual demand.)	

It is also a fact that as the consumption increases, the general health of the community also improves. It is also true that wastage of water also increases as the per capita consumption increases.

5.4 COLLECTION AND STORAGE OF WATER

In order to cater to fluctuations in supply, water from any source needs to be harnessed and stored. The method of harnessing and storage will to a great extent depend upon the type of source used. The three main viable sources of water for a public water supply are surface water, groundwater and seawater. Imported water is also an important source, but since it is in a readymade form, it has already gone through the stages of harnessing and storage.

Surface water is stored in impounding reservoirs if the supply is not uniform and insufficient in dry weather. Storage during the wet season is necessary for release during the dry season. On the other hand, if the source is a perennial river carrying large quantities of water at all times, then there is no need for storage facility.

Groundwater is stored in aquifers, which are natural underground reservoirs. Seawater also does not need any storage facility since most desalination plants are located near the sea. Therefore, storage reservoirs are confined to supplies where the source is a stream or a river.

5.4.1 Reservoirs

The basic function of a reservoir is to stabilize the flow of water either by regulating a varying supply or by satisfying a varying demand.

5.4.1.1 Types of reservoirs

- Direct supply (impounding) reservoirs
- Regulating reservoirs – released when demand downstream is high
- Pumped storage reservoirs – stored by pumping from a river
- Distribution reservoirs (service reservoirs)

The first three types are for long-term storage whereas the last is for short-term storage. There are also other types of reservoirs in water resources engineering, for instance, detention reservoirs and flood plain reservoirs.

5.4.1.2 Reservoir site characteristics

Some preferred site characteristics are:

Geology   - Sub-surface must be impermeable to prevent water seeping.
            - Undesirable soluble salts must be absent.
            - Silting should be minimal.

Topography   - Narrow valley will shorten the length of dam.
             - Rapid widening at the upstream of the dam.
             - Steep side slopes throughout the basin to reduce surface
               area per unit volume.
Land value   - Submerging area should not be expensive.
             - Should be free from residential and other developments.
Environment - Vegetation free
             - Should be free from organic matter which will cause
               objectionable odour, colour etc.
             - Pollution-free catchment
             - Good quality water

In practice, no site will have all these requirements, but it may be possible to improve the existing conditions by artificial means. Trees and shrubs that would be submerged should be cleared before inundating a reservoir because they would cause organic pollution by decaying and could interfere in the inlet and outlet works. They would also be objectionable from an aesthetic point of view.

5.4.1.3 Reservoir physical characteristics

The most important physical characteristic of a reservoir is its storage capacity. It is obtained by integrating the area–elevation curve, which can be obtained from topographical maps. Storage capacity in practice is obtained as the average area within two contours multiplied by the contour interval. They are then summed to obtain the cumulative storage capacity. The area–elevation curve and the storage–elevation curve can be combined into one diagram.

5.4.1.4 Reservoir yield

Yield is the amount of water that can be drawn off from a reservoir in a given interval of time. It is dependent upon the inflow and therefore will vary from year to year. A hydrological study of the catchment is therefore necessary for estimating reservoir yield. The maximum yield that can be guaranteed during a critical dry period is called the safe (firm) yield. Any available water in excess of the safe yield is called the secondary yield. With full development of storage, the safe yield will approach the average yield.

   In order to obtain the reservoir yield, the capacity of the reservoir and the inflow must be known. The reservoir capacity is obtained by comparing the cumulative inflows and demands over a given period of time. It is equal to the maximum deficiency, which is the difference between the cumulative demand and the cumulative inflow:

Storage = maximum deficiency = max Σ (demand – inflow)

This may be obtained either graphically or arithmetically. The graphical procedure is that proposed by Rippl (1883). It is referred to as the mass diagram. It assumes that the reservoir is full at the beginning of the dry period. The reservoir is gradually emptied after that until the end of the dry period when the reservoir will start to refill again. The method consists of plotting the cumulative inflow and cumulative demand (which is normally a straight line) and drawing lines parallel to the demand line through the high points of the inflow curve. The maximum departures between these lines and the inflow curve will give the storage required for that period. By drawing lines at different high points, it is possible to obtain the storage required for the entire period of record. It is important to note that a line drawn through a high point must intersect the inflow curve at some point when extrapolated. Otherwise, the reservoir will never fill again.

The mass diagram procedure can also be used to determine the yield when the storage is known. In this case, the slope of the line drawn through a high point with the maximum departure not exceeding the capacity of the reservoir will give the yield.

5.4.1.5 Losses from reservoir storage

Reservoir yield may be modified by evaporation, bank storage, seepage out of the catchment area and silting. With increased area of water surface, the evaporation rates will be increased. There is a gain in the direct precipitation over the surface, but the net effect in an arid area is a loss due to evaporation. Seepage losses will depend upon the geology of the underlying soil. In porous soils, this may be significant.

Loss of reservoir storage and useful life due to silting can be significant and must be taken into consideration in the planning stage.

5.4.1.6 Reservoir sedimentation

Reservoir sedimentation has three stages: the first stage involves sediment formation by soil erosion, man-made modifications in the environment such as construction activities and rock fragmentation; the second stage involves sediment transport, which is usually by rainfall and runoff; and the final stage involves sediment deposition in reservoirs, stream beds and over-banks.

The sediment carried by streams can be classified as wash load and bed load depending upon the grain size. Wash load consists of grain sizes finer than the bed material (the limiting value being arbitrarily chosen as the grain size of which 10% of the bed mixture is finer) and is usually washed away without leaving any trace in the stream. It depends only on the sediment supply and therefore cannot be estimated except by actual measurements.

Bed load refers to the sediment particles moving in the bed layer, which is defined as a flow layer of two-grain diameters thick immediately above the bed. Particles moving outside the bed layer are called the suspended load and the transport mechanism is by suspension.

The quantities of sediments carried into a reservoir depend upon the hydrograph shape – more runoff implies accumulation of more sediments. However, it is difficult to establish the exact relationship between runoff and sediment inflow. Therefore, empirical relationships based upon sediment sampling surveys are often made use of. A curve giving the sediment load as a function of the catchment area (Fleming, 1965) suggests a relationship of the form

$$y = Ax^n \qquad\qquad (5.1)$$

where y is the sediment load, x is the catchment area and A and n are empirical constants.

There are also relationships of the form

$$y = kQ^n \qquad\qquad (5.2)$$

where Q is the stream flow and k and n are empirical constants.

In order to obtain the total sediment load, something of the order of 10–20% of the suspended load estimated empirically from relationships of the form referred to above is added as bed load, which cannot normally be measured.

It should be noted that a sediment rating curve is much less reliable than the corresponding stream flow rating curve. Usually, the first storm after a long period of drought carries a large amount of sediments. Also, the sediments carried by a large storm are proportionately more than the sediments carried by smaller storms. This means that the sediments carried in a single flood may be greater than the total sediments carried for a year under normal flow conditions.

5.4.1.7 Sediment yield

Sediment yield refers to that part of the gross erosion that is transported. Gross erosion is the total amount of material eroded.

(i) Methods of estimation

(a) *Sediment sampling*: Sediment sampling method requires the sediment rating curve, which is obtained by sampling, and the flow duration curve, which is obtained from flow records. The sediment rating curve gives the relationship between the sediment discharge and the flow discharge. The flow duration curve gives the relationship between the magnitude of flow with the percentage of the time it is equalled or exceeded. The procedure is as follows:

- Separate the flow duration curve into a certain number of class intervals.
- Compute the range of each probability interval and the mid-point of each interval.
- Determine the flow discharge from the flow duration curves.
- Using the sediment rating curve, determine the sediment discharge for each flow discharge.
- Compute the expected sediment discharge as (sediment discharge) × (probability interval).
- Compute the total sediment in a year as (number of days in the year) × (sum of the expected sediment discharges).
- Divide by the sediment (catchment) area.

(b) *Universal soil loss equation*: This equation is based on statistical analysis of data from 47 locations in 24 states in the United States and is of the form

$$A = RKLSCP \tag{5.3}$$

where

A is the computed soil loss in Ton/unit area/year
R the rainfall factor
K the soil erodibility factor
L the slope-length factor
S the slope steepness factor
C the crop-management factor
P the erosion-control practice factor

The rainfall factor R accounts for differences in rainfall intensity–duration–frequency for different locations; soil erodibility factor accounts for the susceptibility of a given soil to erosion; slope-length factor accounts for the increase in quantity of runoff that occurs as the distance from the top of the slope increases; slope steepness factor accounts for the increase in velocity of runoff as slope steepness increases; crop management factor accounts for the crop rotation used. Typical values of these factors can be found in tables, maps and graphs (e.g. pp. 269–274, Yang, 1996).

(c) *Use of sediment yield equations*:

$$\text{Sediment yield} = \frac{\text{Sediment accumulated}}{\text{Trap efficiency}} \tag{5.4}$$

where 'trap efficiency' is the fraction of the sediment inflow, which gets deposited in the reservoir. In a different form, it is of the form

$$\text{Trap efficiency} = \frac{\text{Sediment retained}}{\text{Sediment inflow}} \qquad (5.5)$$

and is a function of the reservoir capacity – inflow ratio. For large reservoirs, it is nearly 100% whereas for small reservoirs, it is almost insignificant. The reason been that large reservoirs retain the water for long periods of time and therefore there is sufficient time for the sediments (particles in suspension in particular) to settle down. In small reservoirs, water is not retained for long periods of time. Figure 5.1 shows the relationship between the trap efficiency and the capacity inflow ratio as obtained by Brune (1953).

(d) *Use of sediment delivery ratios (SDR)*: All sediment material eroded is not delivered to the downstream point. Sediment delivery ratio (SDR) is defined as

$$\text{SDR} = \frac{\text{Sediment transported to a particular point}}{\text{Gross erosion in the catchment upstream of the point}} \qquad (5.6)$$

and it is a function of the

* sediment source (catchment versus channel),
* magnitude and proximity of source (distant versus nearby),

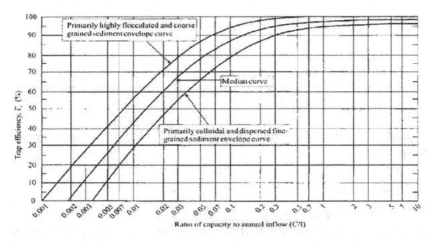

Figure 5.1 Trap efficiency versus capacity-inflow ratio as per Brune (1953).

- characteristics of the transport system (channel network, slope etc.),
- frequency, duration and intensity of the erosion producing storms,
- texture of soil material,
- potential for sediment deposition (presence of depressions etc.) and
- catchment characteristics (area, slope and shape).

(ii) Where sediments get deposited

Large particles and most of the bed load are trapped in the delta while fine sediments tend to accumulate at the face of the dam.

(iii) Density of deposited sediments

Most sediment transport estimates give the sediment load as a weight; for example, tons per day and tons per year. To estimate the loss of storage due to sediment deposition, the sediment loads by weight must be converted to sediment loads by volume. The conversion requires the density of sediments, which depend upon

- the sediment size,
- the depth of deposit,
- the degree of submergence and
- the length of time the sediments have been deposited (consolidation).

An empirical equation proposed by Lane and Koelzer (1943) has found wide use in estimating density variation in time. It is of the form

$$w_t = w_1 + k \log(t) \tag{5.7}$$

where

w_t is the density after t years
w_1 the density after 1 year
k a constant for each class of sediments and operating conditions to represent consolidation.

Typical values of the constants are given in the Table 5.1. All the values are in kg/m^3.

(iv) Sediment control

All reservoirs will someday be filled with sediments. Preventive measures can, however, extend the useful life of a reservoir. There are instances where

Table 5.1 Parameter values for Eq. 5.7

Reservoir operation	SAND 0.05 mm		SILT 0.005–0.05 mm		CLAY < 0.005 mm	
	w_i	k	w_i	k	w_i	k
Sediments always submerged or nearly submerged	1488	0	1040	91.2	480	256
Normally a moderate drawdown	1488	0	1184	43.2	736	170
Normally a considerable drawdown	1488	0	1264	16.0	960	96
Reservoir normally empty	1488	0	1312	0	1250	0

Source: Lane and Koelzer (1943).

reservoirs have been filled with sediments in very short periods of time (a reservoir on Solomon River near Osborne, Kansas, was filled in one year). Sediment deposition may be retarded by

- selecting a site where the normal sediment inflow is low,
- increasing the size of the reservoir so that the life span is justifiable,
- adopting upstream land use and soil conservation measures such as terracing, strip cropping, contour ploughing, check dams in gullies etc.,
- having vegetation and
- providing facilities for silt ejection periodically.

However, it should be noted that when the sediment equilibrium is disturbed by upstream control, there would be adverse effects in the downstream reaches until a new sediment equilibrium is attained.

(a) *Reservoir safety reviews:* In many countries, reservoir safety review is a requirement and must be regularly carried out by a competent engineer. Over the years, the design characteristics may change, and the performance of the reservoir and the spillway in particular should be reviewed.

5.4.1.8 Compensation water

This is the water that must be discharged below a reservoir to compensate for the riparian rights of downstream users. Each country has different laws to protect the rights of the downstream users. The current practice is to provide water sufficient to prevent any adverse environmental impact downstream. This has to be assessed by considering several factors such as downstream agricultural and industrial needs, downstream domestic needs, aquatic life and fish migration (if tidal conditions exist).

A guideline is that compensation water need not be greater than the flow, which is normally exceeded for 90% of the time; that means very nearly dry weather flow.

5.4.2 Water intakes

The factors that have to be considered in the selection of the location of an intake are the following:

- The location should be free from fast currents which may damage the intake.
- The ground should be stable.
- Approach to the intake should be free from obstacles.
- The intake should be below the surface for receiving cooler and clearer water. It should also be well above the bottom of the reservoir (or river).
- The intake should be located upstream of any sources of pollution.
- The intake should not interfere with aquatic life.

5.4.2.1 Types of water intakes

Direct intakes: These are suitable for deep waters where the embankments are stable.

Canal intakes: These require masonry chambers with openings built into the canal bank. Due to reduction in the flow area, increased velocities are expected around the structure.

Reservoir intakes: These are often built into the dam of the impounding reservoir.

River intakes: The head in a river intake is usually raised by building a weir across the river. The advantages are that it can maintain a constant head and that the weir may be used as a device for measuring the flow. The disadvantages are that it is very often expensive to build a weir across a wide river, that it would affect navigation and that it would cause upstream silting.

Floating intakes: This type is a compromise for wide rivers in which the water levels and the flow changes from one bank to the other. Advantages are that a constant suction head can be maintained and that there is flexibility to move around into deeper areas. The disadvantage is that the structure is vulnerable to damage by floods.

Elements of an intake include

- Bell mouth strainer or cylindrical strainer
- Strainer structure with arrangements for its protection

- Raw water gravity pipes or channel
- Gate or sluice valve
- Foot valve
- Suction pipe for the low lift pump

5.4.2.2 Design of intakes

The following factors need to be taken into consideration in the design of intake structures:

- Entrance losses should be kept to a minimum.
- The open end should be provided with a screen and gates or valves to regulate flow.
- Reliability of operation and water quality.

If the fluctuation of the river stage is not high, the intake structure may be located on-shore, thereby simplifying both the design, construction and maintenance.

Most intakes are located at low levels compared to the distribution works and therefore resort is often made to pumping. Pumping stations must be located on-shore and be well protected against possible floods. They should be easily accessible. The suction head should not be greater than 5–6 m including friction when the river (or reservoir) is at its lowest stage. To meet this requirement, it is sometimes necessary to install the pumps in a well.

REFERENCES

Brune, G. M. (1953). Trap efficiency of reservoirs, *Transactions, American Geophysical Union*, 34: 407, http://dx.doi.org/10.1029/TR034i003p00407

Fleming, G. (1965). Design curves for suspended load estimation, *Proc. ICE*, 43: 1–9.

Lane, E. and Koelzer, V. (1943). Density of sediments deposited in reservoirs. 9, St Paul US Engineer District Sub-Office, Hydraulic Laboratory, University of Iowa.

Memon, F. and Butler, D. (2001). Water consumption trends and domestic demand forecasts, Watersave Network, Second meeting, December 4, 2001 (Web presentation).

Rippl, W. (1883). The capacity of storage – reservoirs for water supply (including plate), Minutes of the Proceedings of the Institution of Civil Engineers, vol. 71, issue 1883, pp. 270–278.

Rowe, M. P. (2010). Bermuda's water supply, Part 1 – Rainwater harvesting in Bermuda, May 2020, 36 pp.

World Water Development Report (2017). *Wastewater – An Untapped Resource*, World Water Assessment Programme, UNESCO.

Yang, C. T. (1996) *Sediment Transport – Theory and Practice*, McGraw Hill.

Chapter 6

Drinking water – From source to tap

6.1 INTRODUCTION

Before the drinking water appears at the tap, there are many processes it has to go through. These include identifying a source of water, collection and storage of the water, treatment of the water, storing the treated water to cater to variations in demand and distributing it to the users. Chapter 5 described the collection and storage aspects and this chapter aims at describing the remaining key processes. A schematic diagram of the key processes in a typical water supply system is shown in Figure 6.1. The treatment process, which is the most important one from a health point of view, consists of several sub-processes such as screening, coagulation, flocculation, sedimentation, filtration, disinfection, fluoridation, hardness, taste, odour and colour removal and aeration. The treatment processes need to be carried out to ensure the quality of the treated water is safe for human consumption as a source of drinking water. Guidelines for the required standards are established by national water authorities as well as by the World Health Organization (WHO). In order to cater to varying demands, the treated water needs to be stored in service reservoirs usually located at higher elevations or in overhead tanks to enable distribution by gravity. It is also important to ensure that there is sufficient pressure when the water reaches the consumers' taps. In highly urbanized cities where the population is concentrated in high-rise buildings, the water delivered from service reservoirs may have to be pumped up to rooftop tanks. There are four types of distribution systems: (i) dead-end or tree distribution, (ii) grid iron, (iii) circular or ring and (iv) radial. Finally, a collection and disposal system for the wastewater from the consumers' ends needs to be provided.

DOI: 10.1201/9781003329206-6

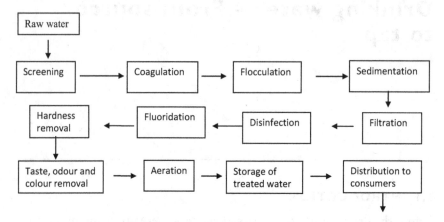

Figure 6.1 Typical flow chart of a water supply scheme.

6.2 TREATMENT PROCESSES

6.2.1 Screening

Screening is the first step in water treatment plants. All raw waters contain solid impurities such as leaves, wood chips, twigs and aquatic plants, which if allowed to go through the next stages of treatment can cause clogging and sometimes damage to equipment. It is a mechanical process of trapping solid material by allowing the raw water to pass through screens, which may be of the bar type, perforated plates or drum type with a mesh of fine metallic fabric. For surface raw waters, the screen openings typically range from 20 to 40 mm. They should be cleaned periodically. With raw water from groundwater sources, there is no need for screening.

6.2.2 Coagulation

All waters contain suspended particles. The smallest particles known as colloids such as silt and clay remain as separate entities (stable state) in the water in a dispersed state. Most colloids present in water are negatively charged, stabilized and therefore repel each other by the action of electrostatic forces. The forces acting are the electrostatic forces, which are repulsive, and the van der Waals forces (intermolecular forces on neutral particles), which are attractive. Repulsive forces are large. Electrostatic repulsive forces can be reduced using coagulants, which are positively charged. Colloids cannot be settled by gravity. Coagulation is the destabilization of colloids by the addition of chemicals that neutralize the negative charge. The chemicals are known as coagulants of high valence value.

The main coagulant used is aluminium sulphate (commonly referred to as alum) in both solid ($Al_2(SO4)_3.nH_2O$ (n = 14–21)) and liquid ($Al_2(SO4)_3$) forms, which react with the natural alkalinity of calcium carbonate present in water to produce aluminium hydroxide, which coagulates to form flocs according to the chemical reaction described by

$$Al_2(SO_4)_3 + 3Ca(HCO_3)_2 = 2Al(OH)_3 + 3CaSO_4 + 6\ CO_2 \qquad (6.1)$$

If the water does not have sufficient alkalinity, additional alkali such as hydrated lime, sodium hydroxide or sodium carbonate is added. For coagulation, the pH control is important and the optimum range is 6.5–7.5.

Other coagulants in use include polyaluminium chloride (PAC, $Al_k(OH)_l.Cl_m$), which is a liquid, polyaluminium silicate sulphate ($Al_k(OH)_l(SO_4)_m(SiO_2)_n$), a liquid, sodium aluminate ($NaAlO_2$), a liquid, polyaluminium chlorosulphate ($Al_k(SO4)_l.Cl_m.(OH)_n$) a liquid, ferrous sulphate ($FeSO_4.7H_2O$), a solid, ferric sulphate ($Fe_2(SO_4)_3$), a liquid, and ferric chloride ($FeCl_3$), a liquid. Properties of these are given in Table 7.3 of Twort et al (1994). Coagulation is essentially a chemical process.

Coagulation is the process of thickening of particles in a colloid. Examples include thickening of milk to form yogurt, blood platelets thickening to seal a wound, pectin (a fibre found in fruits such as pears, apples, guavas, plums, gooseberries and oranges and other citrus fruits) thickening to form a jam, and gravy-thickening as it cools. In haematology, coagulation refers to clotting of blood.

Coagulation destroys the process by which tiny particles repel each other and promotes their consolidation to bigger sizes by clumping together. Bigger particles are easier to separate from the liquid. These larger 'clumps' of particles are called micro-flocs, which are still invisible to the naked eye. Rapid mixing ensures adequate dispersion to promote particle collisions. The water surrounding them would be clear if the particles have been neutralized. If not, more coagulant should be added. However, too much coagulants can oversaturate with a positive charge, which will lead to repulsive forces.

The use of coagulants for treating water goes way back to around 2000 BC when the Egyptians used almonds, smeared around vessels, to treat river water. Coagulant aids include lime ($CaCO_3$) and caustic soda ($NaOH$).

6.2.3 Flocculation

Flocculation is the agglomeration of destabilized (reduced charge) particles into larger size particles known as flocs by a process of gentle mixing. Collision of smaller flocs causes them to produce larger flocs. It is a physical process preceded by a chemical process of adding chemicals to neutralize the negative charge. The flocs can then be removed by sedimentation or floatation. The clumping of particles can be enhanced by the addition of flocculants, which are mostly long-chain polymers with low charge. They enhance van der Waal forces and hydrogen bonding between particles. Flocculation depends on physical factors such as turbulence, chemical factors such as ionic concentration and biological factors such as bacterial population. Coagulation neutralizes the charges on the particles whereas flocculation enables them to bind together, making them bigger, so that they can be more easily separated from the liquid.

6.2.4 Sedimentation

Sedimentation is the process of removal of small particles naturally by gravity or artificially by mechanical assistance. Depending on the quality of raw water, sedimentation can be done without coagulation and flocculation, but in general it is done after coagulation and flocculation. When it is done prior to coagulation, it minimizes the need for coagulation and flocculation. The effectiveness of sedimentation depends on the size and weight of sediments. Particles with specific gravities less than or equal to that of water will remain in suspension. Sedimentation is done in different types of sedimentation tanks such as horizontal flow tanks, multi-layer tanks, radial flow tanks and settling tanks.

6.2.4.1 Horizontal flow tanks

Horizontal flow tanks, which are rectangular in shape and allow water to flow horizontally, are the simplest. During the passage of water from the influent section to the effluent section, heavy particles settle down to the bottom of the tank. Baffle walls are sometimes fitted to prevent short-circuiting and to reduce influent flow velocities, thereby increasing the effective length of travel of particles. The sludge collected at the bottom should be removed periodically in order for the process to continue. Maintenance costs are low in the case of rectangular sedimentation tanks. They are also suitable for large-capacity plants.

6.2.4.2 Multi-layer tanks

A multi-layer tank is a variation of the horizontal flow tank and the process is the same except that multiple decks are built in the tank. Water is passed from one layer to the next until the sediment is properly separated.

6.2.4.3 Radial flow tanks

Radial flow tanks can be of radial flow type or spiral flow type. Tanks are circular and the influent is sent through a central pipe. Sediments collected at the bottom are raked using mechanical sludge scrapers (raking arms), which move around the circular base continuously at slow velocities and removed through sludge pipes at the bottom. In spiral flow type, which has a rotating raking arm, water enters the tank through two or three vertical slits and move along the circumference of the tank. Water moves at very low velocities to allow suspended particles to settle down. Circular tanks are costlier to construct and maintain compared with rectangular tanks but have higher clarification efficiency.

6.2.4.4 Hopper bottom tanks

The bottom of hopper bottom tanks has a hopper shape and the flow of water is in the vertical direction upwards and downwards. Water enters from top into a deflector box located at the top and reverses its direction around the deflector box after flowing downward inside the deflector box. Heavy particles follow the water at the time of reversing its direction and settle at the bottom.

6.2.4.5 Settling tanks

Settling tanks have inclined bases at the bottom. The flow of water can be multi-directional. Effectiveness is dependent on the size of the tank, the depth of the water and the placement of the inclined plates at the bottom. The flow of the water can move in multiple directions depending on the sedimentation needs.

6.2.4.6 Ballasted sedimentation

Ballasted sedimentation, which relies on the application of high-molecular-weight polymers, is preferred when additional flocculation is needed to help with coagulation. Such polyelectrolytes increase particle density, which promotes separation. Fine sand and bentonite can also be used as ballasting agents. The process involves adding coagulants to form large flocs, flocculation, adding ballast and polymer and ballasted flocculation.

6.2.5 Filtration

Filtration is the process by which particles contained in water are trapped by passing through a porous medium. The porous medium commonly used is sand. There are different types of filtration units in existence, the difference depending upon the mode of action. The three common types are slow sand filters, rapid gravity filters and pressure filters. There are also packaged treatment plants that combine filtration with other treatment processes usually for small supplies.

6.2.5.1 Slow sand filtration

In the history of water treatment, slow sand filters were the first biological purifiers of surface waters contaminated by pathogenic bacteria. They were first introduced in the United Kingdom circa 1850 with no significant advancement over time. They still remain effective and maintain certain advantages over rapid gravity filters despite the replacement of slow sand filters with rapid gravity filters in some situations. Slow sand filtration is the simplest low-cost method for surface water treatment that can provide a safe and acceptable drinking water comparable with water produced by complex chemical coagulation and rapid gravity filtration treatment plants.

Slow sand filtration is practised worldwide and has a long history and still considered as one of the best and least cost-effective methods of communal or public water treatment. For example, the largest slow sand filter plant in the United States is in Denver, Colorado, with six filter beds occupying 4.2 hectares producing up to 170,000 m³/day since 1901. In the UK, there are over 200 slow sand filtration units, some of which produce over 30,000 m³/day with the largest at Ashford Common Works rated at 410,000 m³/day. Much of London receives Thames River water treated by sedimentation, micro straining, slow sand filtration and disinfection by chlorination. Similar facilities are used in other countries too as it is the most cost-effective means of providing safe drinking water to communities.

Slow sand filters are so called because the rate of filtration through the filter media is slow. The rate of filtration is about one-twentieth of that for rapid gravity filters. The main process of filtration involves the passage of water through a thick sand bed. Thus, the effectiveness of the filter depends very much on the characteristics of the sand. An effective filtration is able to remove some or almost all of the suspended or colloidal matters, bacteria, turbidity, colouring matter, taste and odour.

The historical importance of slow sand filtration can be illustrated in the cholera outbreak in Hamburg in 1892 where some immigrants who were carriers of cholera landed. This outbreak caused some 16,000 cholera infections and some 8,600 fatalities. But in Altona, a town a short distance downstream of Hamburg where slow sand filters have been in operation was spared. The source of water for both these cities has been River Elbe. Despite the vivid illustration of the effectiveness in purifying water bacteriologically, slow sand filters have been gradually replaced with rapid gravity filters with the use of coagulation, flocculation, sedimentation and above all disinfection. The need for large areas for slow sand filters is also another factor.

Advantages of slow sand filters

- No coagulation needed
- Simple equipment
- Materials of construction readily available

- Supervision is easy
- Effluent is less corrosive and of uniform quality than chemically treated water
- Effective bacterial removal
- Cost and ease of construction and operation
- Conservation of water
- Disposal of sludge which contain little chemicals compared to rapid gravity filters

Disadvantages of slow sand filters

- Initial cost is high because of large area and construction material required
- Not suitable for turbid waters
- Less flexibility in operation
- Less effective in removing colour
- Poor results with water of high algal content
- Operating conditions must be favourable, that is, sunlight, not too low temperature
- Slow rates of filtration

Slow sand filtration should not be preceded by chlorination because chlorine will negate the bacterial activity.

(i) Mechanics of filtration

When a new filter is put into operation, the upper layer of sand grains gets coated with reddish brown sticky deposits of partly decomposed organic matter together with iron, manganese, aluminium and silica. This coating tends to absorb organic matter existing in the colloidal state. After about two to three weeks of operation, a film of algae, bacteria and protozoa is formed at the topmost layer of sand. Together with finely divided suspended matter, plankton and other organic matter in the raw water, this thin layer (Schmutzdecke) acts as an extremely fine-meshed strainer. In the autotrophic zone, the growing plant matter breaks down organic matter, decomposes the plankton and uses up available nitrogen, phosphates and carbon dioxide, providing oxygen in their place. The filtrate thus becomes oxidized. In the heterotrophic zone, the bacteria multiply to complete the breakdown of organic matter, leaving only simple inorganic matter and unobjectionable salts. The bacteria not only breakdown organic matter but also destroy themselves to maintain a balance of life native to the filter.

After water enters the filter box and during several hours of storage above the sand surface, flocculation and sedimentation of large particles take place. During the passage of the water through the filter medium, straining, filtration and adsorption take place simultaneously. In the presence of sunlight during daytime, the storage effect of water above the filter medium will have some bacterial effect as well as settling of microorganisms with solids. The main activity of slow sand filters takes place in the top layer of sand known as the schmutzdecke where bacteria, algae, protozoa and rotifers are entrapped. Organic compounds of raw water in this layer are broken down by organisms in this layer, resulting in physical, chemical and biological improvements of the effluent water quality. Bacterial removal is in the range from 98 to 99.5%, while *E. coli* are reduced by a factor of about 1,000, and virus removal is even better. Slow sand filters are also more efficient in removing helminths and protozoa than any other process. Any remaining *E. coli* and other undesirable pollutants can be removed by chlorination following the slow sand filtration. The schmutzdecke is the most important component of a slow sand filter as it traps, breaks down and digests all microorganisms passing through in addition to straining minute suspended matter in the water.

The main forces that keep the particles attached to the sand grains are electrostatic forces, van der Waal forces (a distance-dependent force between atoms and molecules) and adherence, which in combination is referred to as adsorption. Clean quartz sand has a negative charge and attracts positively charged colloidal matter such as crystals of carbonates and floccules of iron, aluminium hydroxide as well as cations (positively charged) of iron, manganese, aluminium and other metals. Organic colloidal particles, anions of nitrates, phosphates and bacteria usually have a negative charge and are therefore repelled. This is why the slow sand filters are not effective until the filter has ripened. During the ripening process, positively charged particles accumulate on some of the sand grains causing a reversal of electrical charge. Adsorption on such grains is able to remove negatively charged impurities such as colloidal matter of animal or vegetable origin and anions of nitrates and phosphates until oversaturation leads to reversal of electrostatic charge. This process of reversal of charges, once started, continues throughout the life of the filter bed. Although clean sand of rock origin carries a negative charge, natural sand has some positive charge from groundwater flowing through them. Such sands need less time for ripening. Mass attraction becomes more effective in holding the particles to surfaces once contact has been made since the distances of separation are small.

During the ripening period, particles of organic origin get deposited on the surfaces of sand grains, which become breeding grounds for bacteria and other microorganisms. This becomes the schmutzdecke on the surface of the filter medium. It consists of algae and other living matter including plankton, diatoms, protozoa, rotifers and bacteria. The bacteria multiply selectively using the organic matter as food and oxidize part of the food to provide the energy for their metabolism and for cell material growth. On the surface of the filter bed, the thin layer of slimy schmutzdecke acts as the biological filter while the medium below acts as a mechanical filter.

Bacterial population is limited by the amount of organic material present in the raw water and their growth is accompanied by dye-off, which in turn generates organic matter available at lower depths. The degradable organic matter is finally converted into water, carbon dioxide and harmless sulphates, nitrates and phosphates. The bacterial activity is pronounced at the upper layers of the filter medium and gradually decreases in the downward direction. For efficient operation of the filtering process, the flow rate should be kept constant and continuous. Sufficient oxygen (above 3 mg/l) should be available and the temperature should not be allowed to fall too much. If the oxygen content becomes negative, anaerobic decomposition can take place, producing undesirable products that affect odour and taste such as hydrogen sulphide and ammonia.

The processes that occur in the schmutzdecke are quite complex and not well understood. The first stage is one of mechanical straining of most suspended matter in a thin dense layer in which the pores may be very small (less than about a micron). The thickness of this layer increases (to about 25 mm) with time from the beginning of the filtration process until the flow rates become unacceptably small. The most important benefit of this layer is its ability to trap bacteria and viruses, which gradually decrease with the depth of filter medium. For the aerobic actions to take place, a certain amount of dissolved oxygen in the raw water is essential. Ripening may take days or weeks depending heavily on the ambient temperature. Until the filter is ripened, the effluent water is unsafe for drinking.

The two important factors that contribute to the filtering mechanism are that the particles must move and intercept the sand grains and they must get attached, which is assisted by the random Brownian motion that brings particles close to the sand grains. It is followed by sedimentation assisted by gravity, which brings particles downwards to the top surface of grains. The attachment of the particles to the grain surfaces is assisted by electrostatic and molecular forces, which are sensitive to the electrostatic charges on the sand grains, which in turn depend on the pH of raw water. Low pH is preferred for virus and *E. coli* removal.

An important property of slow sand filters is adsorption resulting from electrical forces, chemical bonding and mass attraction, which take place over every surface of sand particles in contact with water. The surface area of sand particles in a slow sand filter is extremely large – of the order of about 15,000 m^2/m^3 of the filter medium. During the passage of water through the filter medium, every particle of impurity as well as bacteria and viruses get attached to the surfaces of sand particles by mass attraction or electrical forces. The result is the formation of a slimy layer consisting of living organisms as well as other fine particles of impurities, which is similar to the schmutzdecke but without the larger particles, and algae, which would have got trapped in the upper layers. This living layer continues to some 40 cm of depth with the greatest activity near the surface where food is plentiful. The mass of microorganisms, bacteria and protozoa feed on the adsorbed impurities as well as on each other. The food is mainly organic particles carried by the water. The sticky coating holds them together until they are broken down, consumed and converted to cell material, which in turn is converted to inorganic matter such as water, CO_2, nitrates and phosphates and carried down with water. The available food decreases with depth. Algae build up cell material from simple minerals such as water, CO_2, nitrates and phosphates according to

$$nCO_2 + nH_2O + Energy \rightarrow (CH_2O)_n + nO_2 \qquad (6.2)$$

Energy is derived from the oxidation of organic matter. Reverse reaction occurs when algae die and their cell material is consumed by bacteria.

$$(CH_2O)_n + nO_2 \rightarrow nCO_2 + nH_2O + Energy \qquad (6.3)$$

The relative magnitudes of Eqs. 6.2 and 6.3 govern the growth and decay of algal population. In spring and summer in temperate climates, the reaction in Eq. 6.2 dominates implying increasing oxygen and decreasing CO_2. Increase in oxygen is an advantage, but a decrease in CO_2 may cause bicarbonates to dissociate into carbonates and CO_2 according to

$$Ca(HCO_3)_2 \rightarrow CaCO_3 \downarrow + CO_2 + H_2O \qquad (6.4)$$

resulting in the precipitation of $CaCO_3$ that can clog the filter.

(ii) *Limitations of slow sand filter units*

Slow sand filters are not suitable for turbid waters (turbidity > 30 NTU), which can clog the 'schmutzdecke' if the raw water contains more than a certain amount of suspended matter. Turbidity can also limit the amount of light passing through, which is required for bacterial activity. Turbid water may be pre-treated by passing them through plain sedimentation tanks with no coagulating chemicals added. Bacterial removal is about 99%. They are suitable for raw water with relatively low colour (10–15 Hazen units). WHO guidelines give a maximum of 15 TCU (true colour units – same as Hazen units).

(iii) *Design considerations*

The basic construction of a slow sand filter consists of a concrete basin, about 3–4 m deep. It should preferably be covered to prevent dirt falling onto the filter bed and to minimize algal growth, with sufficient head space for cleaning purposes. There should be valves to control the influent and effluent flows as

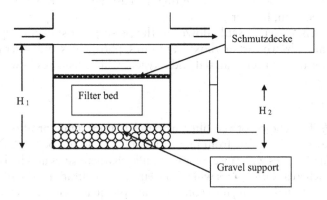

Figure 6.2 A typical slow sand filter unit.

well as water level indicators to measure head losses. The filter should also have an under-drain system to collect the effluent water (Figure 6.2).

(iv) Structure of a slow sand filtration unit

A typical slow sand filter consists of a sand bed, a gravel layer and an under-drain system (Figure 6.2). The filter works by a combination of straining and microbiological filtration of which the latter is more important. In the filter medium, three zones can be identified. The top thin layer of sand known as the schmutzdecke where the microbial filtration takes place, an autotrophic zone a few mm below the schmutzdecke, and a heterotrophic zone, which extends some 300 mm into the filter bed.

The basic components of a slow sand filter consist of a raw water reservoir with its water level above that of the upper surface (typically 1–1.5 m) of the filter medium, a filter bed made of fine sand, an under-drainage system to collect the filtered water and to provide structural support for the filter and a system of control valves to control the flow of water so that the head above the filter does not fall below a predetermined value. The first three components are contained within a rectangular box with a depth ranging from 2 to 4 m. The under-drainage system consists of graded gravel that supports the filter medium with connections to divert the filtered water to a receiving reservoir. Above the under-drainage system is the sand layer (typically about 0.6–1.2 m) with the effective size[1] ranging from 0.15 to 0.35 mm and a uniformity coefficient[2] of about 2–5, above which the raw water lies to a

1 Effective size = Particle size 10% finer by weight = d_{10}
2 Uniformity coefficient =

$$\frac{Particle\ size\ 60\%\ finer\ by\ weight}{Particle\ size\ 10\%\ finer\ by\ weight} = \frac{d_{60}}{d_{10}}$$

typical depth of about 1–1.5 m. During the passage of the raw water through the filter medium, heavier particles settle down and lighter particles coalesce, making them easier to settle down. In the presence of sunlight, algae begin to grow and absorb carbon dioxide, nitrates, phosphates and other nutrients from water to form cell material and oxygen, which dissolves in water.

(v) Filter media

(a) *Sand:* The choice of the filter material is important for effective filtration. They should be free from impurities and should not be too large or too small. Smaller sizes will cause poor hydraulic characteristics and higher risk of clogging whereas larger sizes will not be effective in filtration. The size of sand should be selected, taking into consideration the quality of the raw water. For example, coarse sand for relatively good water and fine sand for water with suspended solids and heavy bacterial pollution. Lower rates of filtration and finer sand should be used when bacterial contamination of raw water is high. Filter bed should not be less than 0.7 m, preferably 1.2–1.4 m. Sometimes, a layer of activated carbon of about 0.1 m thick is incorporated into the filter bed to adsorb any remaining traces of taste and odour-producing substances. Since the activated carbon would have become saturated with impurities after some time, it has to be removed and replaced during re-sanding.

The sand should be hard, durable, free from impurities and free of soluble substances. Typical effective size ranges from 0.15 to 0.35 mm with a uniformity coefficient between 2.5 and 3.5. In selecting the sand sizes, due consideration should be given to the raw water quality and the filtration rate. Typical thickness of the filter layer should be in the range of 0.6 to 0.9 m. It is better to start with a deeper layer that will cater for a number of scrapings before refilling with clean sand again. There should be no room for short-circuiting of raw water. Filter sand should be washed before placing on the filter bed.

(b) *Gravel:* Gravel should be graded to prevent the penetration of sand and yet provide free flow of water. Typical depths of graded gravel range from 15 cm layer passing a 80 mm screen but held on a 25 mm screen, followed by 5 cm layer passing a 25 mm screen but held on a 10 mm screen, followed by 5 cm layer passing a 10 mm screen but held on a 5 mm screen. Gravel should be placed above the under-drain and below the sand but not within 0.6 m along the side walls of the filter.

The function of the gravel layer is to support the sand layer and to prevent the penetration of sand and yet allow free flow of water towards the under-drain. Typical gravel sizes range from about 3.5 mm at the top to about 50 mm at the bottom placed in layers.

(vi) Operation

The rate of filtration is controlled by the influent and effluent control valves. The depth of water should not be less than 1.2 m. At the beginning of operation, the filtration rate will be high, but the effluent will not be clean enough until the schmutzdecke is ripened which can take up to three weeks. As the schmutzdecke builds, the filtration rate will gradually decrease. When the water level drops to the level of the top of the sand layer, the filter needs cleaning.

The filters should be covered to minimize the growth of algae, but this greatly increases the cost. If a cover is provided, then it should have a clearance to enable scraping and replacing sand.

The filtration (or flow of water through sand) that has a low flow velocity is considered as laminar and obeys Darcy's law with the flow velocity proportional to the head loss $H_1 - H_2$ as in Figure 6.2.

Gauges should be installed to measure the loss of head. Care should be taken not to operate a slow sand filter under a negative head, which will cause air to get in. Each filter should have a main drain and a number of lateral drains to collect the filtered water. Tile pipes are usually used.

The bed of the filter should be completely drowned, that is, no air should be allowed. It is not a trickling filter used in sewage works.

When the bed is new, a head of raw water of only 5–7.5 cm is sufficient to maintain the design flow. But as the schmutzdecke gets developed, the friction increases and therefore more head is required to pass the design flow. When the head increase has reached a predetermined value (about 1 m), the bed must be cleaned.

(vii) Rate of filtration

The effectiveness of filtration is generally proportional to the depth and fineness of the filter sand and inversely proportional to the rate of filtration. As a guideline, when the depth of sand is about 0.9 m, the rate of filtration should not exceed about 0.15 m/h. Typical rates of filtration are about 2–5 m/d (440–1,100 gpd/m^2). Otherwise, some bacteria may penetrate through the filter. The best criterion to decide on a permissible filtration rate is to check the bacterial quality of the effluent. In suspicious situations, disinfection of the effluent by chlorination should be done. The filtration area can be expressed as the hourly rate of water to be treated divided by the filtration rate, which can be expressed on the basis of Darcy's law.

When the loss of head reaches the permissible maximum, the filtration should be stopped and about 0.5–2.5 cm of top sand of the filters should be scraped. The scraped sand may be washed and reused. A filter should be scraped once every two to three months.

6.2.5.2 Rapid gravity filtration

Rapid gravity filters derive their name because of the high rates of filtration under the action of gravity. The rates of filtration are about 20 times those of slow sand filters. They work on principles which are quite different from those of slow sand filters. The raw water is pre-treated with coagulation, flocculation and sedimentation before passing through rapid gravity filters. Purification process takes place within the filter bed implying a thicker filter bed is preferable. The chemical coagulants used in the pre-treatment stage deposit as a coating on the sand grains, which is positively charged and the negatively charged matter in suspension gets attracted and deposited on the sand grains. Natural electrical charges of sand grains are supplemented by the electro-kinetic charges by the high rate flow of water. The resulting adsorption is the most effective part of the filtration process. Pre-treatment is a pre-condition for effective filtration. The size of the sand particles in the filter medium is also important. If it is too small, the head loss is too high and if it is too large, the filtration is poor. The advantages of rapid gravity filters include small area, flexibility, high rate of filtration and no swapping or removal of sand. The disadvantages include the need for pre-treatment and backwashing, cost of equipment as well as the low bacterial removal. Rapid gravity filters produce physically clean water but may not remove bacteria and viruses. Disinfection is essential.

Rapid gravity filters have the following features:

- Purification of raw water takes place within the filter bed.
- There is no schmutzdecke formed. If the bed is thin, the purity of effluent is reduced.
- It is believed that the purification takes place throughout the bed depth, both physical and bacteriological.
- The chemical coagulants used in preparing the water prior to filtration deposits as a coating on the sand grains. This coating is positively charged and the negatively charged matter in suspension gets attracted and deposited on the sand grains.
- This theory is supported by the fact that a new clean filter does not function very efficiently. The unit must ripen before it is effective, that is, the sand grains must become coated with electrically charged deposits before organic matter can be removed from water.
- The size of the sand particles is also important. There is a maximum and a minimum size. If it is too small, the interstices are too small and the head loss is high. If it is too large, the filtration is poor.

The advantages of rapid gravity filters include the small area needed, flexibility of adding more units as and when needed and high rate of filtration and no swapping or removal of sand. The disadvantages include the need for pre-treatment, need for expensive equipment for backwashing and that bacterial removal is not very effective. Post-treatment by disinfection is necessary.

The filter medium normally consists of sand with effective grain size of 0.4–1.5 mm and uniformity coefficient of 1.3–1.8. (For slow sand filtration they are of the order of 0.15–0.35 mm and 2.5–3.5, respectively). Anthracite, which is made from coal, which is lighter than sand and more angular than sand, giving a greater surface area per unit volume, can also be used as the filter medium. Because it is lighter (approximately half the weight of sand), the amount of backwashing water (and pressure) needed is very much reduced. They can be cleaned much easier than sand when backwashing. Anthracite is useful for portable filters because the density is about half that of sand. Crushed coconut shell and pumice (light porous lava stone) can also be used.

Below the filter medium is a gravel support layer, which has dual functions. It supports the sand bed as well as prevents the fine sand from entering the under-drain. A typical gravel support layer consists of several layers of gravel each about 50 mm thick with the largest (about 20 mm) at the bottom and the smallest (about 2 mm) at the top with a total depth of about 300 mm. New designs use media retention plates that are porous moulded plates made from high-density polyethylene fitted to the under-drain block as substitutes for gravel layers. The rates of filtration are in the range of 5–15 $m^3/m^2/h$ (120–360 $m^3/m^2/day$). Modes of action of rapid gravity filters consist of straining, sedimentation, adsorption and bacterial and biochemical, which are the same as in slow sand filters, but their relative importance is different because of the coarseness of sand. For example, straining is not so effective because of the large pore size, sedimentation is not so effective because of the higher filtration rates, adsorption is effective because impurities and sand grains have opposite charges. Bacterial and biochemical filtration is not so effective because the residence time is very short. Rapid gravity filters produce physically clean water but not necessarily microbiologically clean water. Disinfection is therefore necessary.

(i) Design details

The design parameters are grain size of filter, thickness of filter, depth of supernatant water and rate of filtration, which are interrelated and which must be selected based on experience on existing plants. When no experience is available, they are based on the results of a pilot plant. Cost is another parameter. Typical size of a unit ranges from about 5 × 7 m to 10 × 15 m. Size is limited by the amount of backwashing water needed to clean the filter bed.

(ii) Backwashing

Backwashing is the process of sending pressurized air and water upwards through the filter. The filter bed should be first drained until the water level is a few centimetres above the top of the bed. Air is sent up to break up the crust (scum) or dirt from sand grains. The water becomes dirty because of the agitation and the separation of dirt from sand grains. Next, water is sent upwards at high velocity that is sufficient to expand the bed and cause agitation so that the sand grains are washed. They should be uniformly spread. Otherwise, they might break the sand bed or overturn part of it. Proper inspection is needed during backwashing.

The velocity of backwashing water must be high enough to produce an expansion of the bed to loosen accumulated dirt. Upward flow rates are of the order of 100 gpm/m², which makes the bed expand up to about 30%. If backwashing is not done properly, the filter bed can shrink. Temperature has considerable effect upon backwashing. At 21°C, the rate of flow for the same expansion is about half the flow rate at 0°C.

Backwash water, which is dirty, should be discharged to waste. Wash water should be filtered and sterilized. Raw water is not used. Wash water should not be greater than 1.5–2 % of filtered water.

Filters can run without backwashing for between 8 hours and 3 days depending upon the quality of raw water. Backwashing is done only one unit at a time usually about one filter per day. This is desirable because the equipment necessary is not duplicated.

Equipment needed for rapid gravity filters include compressed air storage cylinders, pumps and a wash water tank.

Typical backwash rates for 20% expansion are as follows:

$d = 0.5$ mm 20 m³/m²/h at 20°C
$d = 1.0$ mm 56 m³/m²/h at 20°C
$d = 1.2$ mm 73 m³/m²/h at 20°C.

For higher backwash water temperatures, the rates are higher.

(iii) Washwater tank arrangement

It is also possible to use backwash water from the filter units themselves provided sufficient head (say 1.5–2 m) is available. But the operating filter units then must supply enough water for backwashing. For this reason, at least six filter units are necessary when backwashing is by this method.

The filter bed should have a supporting gravel layer, which will not permit the grains from entering the under-drain system (holes in pipes).

For instance, for filter sand of 0.7–1.0 mm, effective size would require four gravel layers:

0.15 m	2–2.8 mm
0.10 m	5.6–8 mm
0.1 m	16–23 mm
0.2 m	38–54 mm
Total gravel pack – 0.55 m.	

The wash water is collected in troughs. The distance the wash water has to travel horizontally to troughs should be about 1.5–2.5 m. The troughs are set with their top 0.5–0.6 m above the unexpanded sand bed.

Their cross-sectional areas are computed on the basis of critical flow assumption.

When fine sand is used (< 0.8 mm), the scouring force of the backwashing water may not be sufficient to scour the grains. After a long run, sticky organic matter adheres on to the grains. These can be cleaned by providing compressed air before backwashing typically at about 30–50 m/h. They loosen the dirt and the subsequent water carries them away. A separate pipe system is then needed. For small treatment works, this is not recommended.

(iv) Plant layout

Units must be sufficient to provide the required filtrate when one filter bed is taken out of service for cleaning. The units must be arranged compactly so that the necessary influent and effluent lines are as short as possible. Allowance should also be made for future expansion.

6.2.5.3 Pressure filters

Pressure filters are identical to rapid gravity filters in principle, but their sand beds are placed in horizontal or vertical steel cylinders, which can withstand pressures up to about 10 atm. The filter bed in both cases is flat. Filtration takes place under pressure without the dissipation of pressure head. The only head loss is that due to frictional losses through the sand bed and inlet and outlet controllers. The filtered water at exit will have sufficient pressure and can be used for backwashing without the need for a separate backwash water tank. Pressure filters may be installed in a pumping line or a gravitational line without loss of pressure. Advantages include no great loss of pressure, compact and portable and can be factory assembled. Disadvantages include the difficulties in treatment of water under pressure such as coagulation, sedimentation and application of chemicals and the inability to observe the washing, cleaning and replacement of sand and loss of effectiveness of filtration due to excessive pressure. Other factors being the same, vertical pressure filters are preferred to horizontal ones. The quality of the product water from a pressure filter is not as good as from gravity filters. Pressure filters are used in swimming pools.

6.2.5.4 Microfiltration

Microfiltration employs a membrane that allows particles of a certain size to pass through or be trapped. Membranes can remove particles finer than media filtration. They have pore sizes in the range of 0.1–10 μm and trap certain bacteria and suspended solids while letting some viruses pass through. Although the viruses are smaller than the pore sizes, they can attach themselves to bacterial biofilm, allowing some viral contaminants to pass through. They have very high effectiveness in removing protozoa and a moderate effectiveness in removing bacteria but not effective in removing viruses and chemicals. Microfiltration can be used for refining petroleum, drinking water treatment, wastewater treatment, separation of oil/water emulsions, processing dairy products, for sterilizing beverages and pharmaceuticals as well as for pre-treating water prior to reverse osmosis.

6.2.5.5 Ultrafiltration

Membranes used for ultrafiltration have finer pore sizes ranging from 0.01 to 0.1 μm and can therefore trap finer particles. They can trap everything micro-filters trap as well as viruses, silica, proteins, plastics, endotoxins and smog and/or fumes. They are highly effective in removing protozoa and bacteria, moderately effective in removing viruses and not so effective in removing chemicals. Ultrafiltration requires a pressure higher than that required for microfiltration and can be used for treating wastewater, separating oil/water emulsions, diafiltration (removal or separation of components based on their molecular size) in pharmaceutical biotechnology, chemical process separation and diafiltration, concentrating proteins, clarifying fruit juices, removing pathogens from milk, for making cheese, as well as for pre-treating water prior to reverse osmosis.

6.2.5.6 Nanofiltration

Nanofilters have pore sizes ranging from 0.008 to 0.01 μm and are highly effective in removing protozoa, bacteria and viruses and moderately effective in removing chemicals. Microfiltration, ultrafiltration and nanofiltration involve physical separation that is highly dependent on the pore size of the membranes.

6.2.5.7 Reverse osmosis

Osmosis is the spontaneous movement of solvents in a solute solution from a lower concentration to a higher concentration until equilibrium. It is a process of diffusion, which is the movement of molecules from a region of lower concentration to a region of higher concentration assisted by the concentration gradient across a semipermeable membrane. In plant biology, the semipermeable membrane allows the solvent with a lower concentration to pass through to the other side that has a higher concentration until the osmotic pressures across the membrane becomes equal. It does so through a semipermeable membrane that allows some substances to pass through while not others. Osmosis also helps the kidneys to purify blood. It allows the passage or diffusion of water or other solvents while blocking the passage of dissolved solutes. Pure water can flow in either direction. Osmosis can be slowed down, stopped or reversed depending upon the pressure applied on the high concentration side of the membrane.

Reverse osmosis is the opposite process of making the solvent of a highly concentrated solution to pass through a membrane to that of a lower concentration under the action of applied pressure. It will leave a higher concentration of the solute on one side and the solvent on the other side. Reverse osmosis is one of the methods used to extract fresh water from sea water, brackish water and contaminated water. Fresh water will pass through the membrane while the solutes will remain. The applied pressure should be greater than the osmotic pressure. The semipermeable membrane allows the water to pass but prevents the passage of ions (Na^+, Ca^+, Cl^-) or larger molecules such as glucose and bacteria.

Reverse osmosis has a long history, but its application for desalination started in the 1950s. With the developments in the production of efficient membranes, reverse osmosis has become one of the main techniques of converting sea water to fresh water with relatively less energy. The technique is also nowadays used to convert wastewater to potable water in accordance with the 'toilet-to-tap' concept that carries a strong psychological stigma. Other applications of reverse osmosis include separation of sugary concentrate from water in sap, filtering out undesirable elements like some acids and control of alcohol content in the wine-making industry, among others. A by-product of reverse osmosis is the possibility of recycling the high pressure used in reverse osmosis. The remaining pressure and the high flow can be used to drive turbines and motors, thereby harvesting energy.

The membranes are usually cellulose acetate and polyamide composites. After use for some time, the membranes will compact or foul as a result of compression and deposition of substances from feed water such as calcium carbonate, fine colloids, oxides of metals and silica. They can be restored by periodic cleaning with acid.

6.2.6 Disinfection

6.2.6.1 Disinfection systems

Disinfection is a necessary treatment process to prevent pathogens (especially bacteria) from being carried to drinking water. It is particularly important for surface waters and for groundwater subject to faecal contamination. Although chemical disinfection of a faecally contaminated drinking water supply will reduce the overall risk of disease but may not necessarily render the supply safe. For example, chlorine disinfection of drinking water has limitations against protozoan pathogens – in particular Cryptosporidium – and some viruses. Disinfection efficacy may also be unsatisfactory against pathogens within flocs or particles. Chemical disinfection usually results in the formation of chemical by-products, which can pose a health risk. However, such risks are small in comparison with the risks associated with inadequate disinfection.

6.2.6.2 Distillation

Distillation is a phase change process of heating water to the boiling point and then collecting the water vapour, leaving many of the contaminants behind. It is highly effective in removing, protozoa, bacteria, viruses and effective in removing common chemical contaminants such as arsenic, barium, cadmium, chromium, lead, nitrate, sodium, sulphate and many organic chemicals.

6.2.6.3 Ultraviolet (UV) radiation (with pre-filtration)

Ultraviolet treatment, a non-chemical disinfection process, with pre-filtration uses ultraviolet light to disinfect water or to reduce the amount of bacteria present. It is suitable for disinfecting water free of suspended matter, turbidity and colour and is done usually after filtration. The process is highly effective in removing protozoa, bacteria and viruses but not effective in removing chemicals. The wavelength range for UV is from 15 to 400 nm (nm = 10^{-9} m) but the range for peak disinfection lies within 255–265 nm. UV radiation is emitted by a low-pressure mercury arc lamp, biocidal, and between wavelengths of 180 and 320 nm. UV disinfection is used in small portable water treatment facilities. The exact mechanism of bacterial killing by UV disinfection is not very clear but believed to be irreversible inactivation of the DNA molecules of microorganisms, thereby making them unable to replicate. The rate of disinfection depends upon the wavelength, and the product of the intensity of radiation and the time of exposure. UV radiation is provided by UV lamps.

6.2.6.4 Chlorination

Among the available chemical disinfection methods, chlorination is by far the most widely used method for domestic water supplies. Other less widely used disinfectants include ozone (discussed below) and potassium permanganate ($KMnO_4$), which can control colour, taste and odour that cannot be controlled by chlorine. Bromine and iodine are also disinfectants but rarely used in water treatment. Chlorination can be achieved by using liquefied chlorine gas, sodium hypochlorite solution or calcium hypochlorite granules and on-site chlorine generators. Liquefied chlorine gas is withdrawn from pressurized containers and dosed into water by a chlorinator, which controls and measures the gas flow rate. Sodium hypochlorite solution is dosed using a positive-displacement electric dosing pump or gravity feed system. Calcium hypochlorite is applied after dissolving in water and mixed with the main supply. They all form hypochlorous acid (HOCl) and hypochlorite ion (OCl^-).

A widely used technique in chlorination is breakpoint chlorination in which sufficient chlorine dose is rapidly added to oxidize all the ammonia nitrogen in water and to allow free residual chlorine to remain in the water to protect against re-infection. Superchlorination/dechlorination is a technique in which a large dose of chlorine is added to effect rapid disinfection followed by reduction of excess free chlorine residual. Removal of excess chlorine is important to prevent taste problems. Marginal chlorination in which simple dosing of chlorine is used to produce the desired level of residual chlorine and is applicable to water supplies of high quality.

Although chlorination is used mainly for microbial disinfection, chlorine also acts as an oxidant that can remove or assist in the removal of some chemicals. For example, decomposition of easily oxidized pesticides, oxidation of dissolved species to form insoluble products that can be removed by subsequent filtration and oxidation of dissolved species to more easily removable forms.

The effectiveness of chlorination primarily depends on the concentration of chlorine and the contact time. Disinfection rate can therefore be considered as proportional to the product of concentration and contact time. It is also related to the temperature. At lower temperature, bacteria kill tends to be slower. However, chlorine is more stable at lower temperatures and residual chlorine will remain in water for a longer time, thereby compensating for the slower rate of bacterial kill. Subject to other factors being the same, chlorine is more effective at higher temperatures. The chemistry of chlorine can be described by the following reactions:

$$Cl_2 + H_2O \rightarrow HOCl \text{(Hypochlorus acid)} + HCl \text{(hypochloric acid)} \quad (6.5)$$

These products are weak and dissociate as

$$HOCl \rightarrow H^+ + OCl^- \text{(Hypochlorite ion)} \qquad (6.6)$$

$$HCl \rightarrow H^+ + Cl^- \qquad (6.7)$$

The above residuals can be derived from calcium hypochlorite ($Ca(OCl)_2$) or sodium hypochlorite (NaOCl).

Almost all raw waters have impurities and the reactions of chlorine with such impurities interfere with the formation of free chlorine residual. Chlorine will first react with reducing agents such as organic matter, nitrites, iron, manganese and ammonia to form chloramines and chloroorganic compounds before forming residual chlorine. Adding more chlorine results in reactions with ammonia to form chloramines according to the following equations:

$$NH_3 + HOCl \rightarrow NH_2Cl + H_2O \qquad (6.8)$$

Further addition of chlorine will form dichloramines ($NHCl_2$) and trichloramines (NCl_3) according to

$$NH_2Cl + HOCl \rightarrow NHCl_2 + H_2O \qquad (6.9)$$

$$NHCl_2 + HOCl \rightarrow NCl_3 + H_2O \qquad (6.10)$$

The pH of water also affects the disinfecting action as it determines the ratio of HOCl to OCl. After the chloramine level reaches a minimum, the addition of more chlorine produces free residual chlorine and the point at which this occurs is known as the 'breakpoint'. After the breakpoint, an increase in chlorine dose will produce a proportionate increase in free residual chlorine. The combined chlorine residuals are not as effective as free residual chlorine. For poor quality water, addition of chlorine beyond the breakpoint, a process known as superchlorination, results in free residual chlorine levels higher than needed. In such situations, dechlorination is done by adding substances that react with residual chlorine. The compounds that can be added include thiosulphates, hydrogen peroxide and ammonia. Similar effects can be achieved by passing the water through a bed of activated carbon.

An important negative effect of chlorination is the formation of chloroorganic compounds known as trihalomethanes (THMs) when chlorine reacts with organic compounds originating from decayed vegetation contained in runoff from forested catchments and lakes with high algal growth. They include bromoform, bromodichloromethane, dibromochloromethane and chloroform. Such compounds are known to be carcinogenic and chloroform is the most common among such compounds. Exposure to chloroform comes from showering to elevated levels of chlorine in tap water. Options for avoiding this negative effect include prevention of THMs in the first place, removal of THMs from treated water using activated carbon and to use a disinfectant other than chlorine such as ozone. THMs are not found in raw water. The guideline value for chloroform is 0.3 mg/litre, bromoform 0.1 mg/litre, dibromochloromethane 0.1 mg/litre and bromodichloromethane 0.06 mg/litre.

6.2.6.5 Ozonation

Ozone gas, produced by passing dry air or oxygen through a high voltage electric field, is a powerful oxidant that can oxidize organic chemicals. The power requirement is of the order of 2–2.7 kWh/100 g of O_3 produced. Since ozone is toxic, direct discharge of unused ozone to the atmosphere should be avoided. Instead, what is not dissolved is passed through an ozone destructor and released to the atmosphere. The usual practice is to convert the spent ozone to oxygen by heating the gas to 350°C to decompose. The ozone-enriched air is dosed into water by means of porous diffusers at the base of a contactor tank typically about 5 m deep and allowed to be in contact for about 10–20 min. The doses required, which depend on the type of water, are typically in the range 2–5 mg/litre. For example, for oxidation of organic chemicals, a residual of about 0.5 mg/litre after a contact time of up to 20 min is typical. According to WHO guidelines, a residual concentration of 0.2–0.4 mg/litre maintained for 4 min is sufficient for viral disinfection. The performance depends on achieving the desired concentration after a given contact time. It is believed to be more effective than chlorine in killing viruses. However, its half-life is too short to be effective in ensuring residual disinfection.

Ozone reacts with natural organics to increase their biodegradability. To avoid undesirable bacterial growth in the distribution system, ozonation is normally used with subsequent treatment such as filtration followed by a chlorine residual to remove remaining biodegradable organics. Ozone oxidizes colour, some substances responsible for taste and odour, iron and manganese. Some by-products such as formaldehyde and organic peroxides of ozonation can pose a health hazard, which can be avoided by adding granular activated carbon beds. In recent times, ozone is used together with other oxidants such as hydrogen peroxide known as peroxones. Ozone is chemically unstable, toxic and cannot be stored. It is a strong oxidant with no taste or odour problems. The cost of ozonation is about three times that of chlorination.

Other disinfectants that can be used include calcium hypochlorite powder (commonly known as bleaching powder), which contains about 30–35% of releasable chlorine, calcium hypochlorite granules that contains about 65–70% weight/weight of chlorine and sodium hypochlorite solution that contains about 14–15% weight/weight of available chlorine. The former is used when liquefied chlorine is unavailable or too expensive.

6.2.7 Fluoridation

Fluoride is an ion of the element fluorine, which is present naturally in groundwater and is an important substance to protect teeth. Optimal levels of fluoride can reduce the incidences of tooth decay significantly. The best way of providing fluoride for all is to add it to the public water supply although it is now contained in some brands of toothpaste. The recommended optimal fluoride concentration in water supplies in tropical climates (temperature between 26.3 to 32.5°C) is about 0.7 mg/litre (Table 8.1, AWWA, 1995). In temperate climates, the optimal concentration can be bit higher because people drink less water in temperate climates than in tropics. Intake of excessive fluoride leads to a disease known as fluorosis, resulting in mottling of teeth, especially in children. In more severe cases, teeth will darken and pitting will occur, leading to tooth cavity. The chemicals used in adding fluoride to drinking water supplies are sodium fluoride (NaF, available in powder or crystal form), fluorosilicic acid (H_2SiF_6, available in liquid form) and sodium fluorosilicate (Na_2SiF_6, available in powder or fine crystal form).

In some groundwater sources, the level of fluorine may exceed the safe level (about 1.5 mg/litre). In such situations, defluoridation should be done by chemical precipitation or adsorption. Hydrated lime precipitates as calcium fluoride, which requires an optimum pH of 12. Aluminium sulphate coagulation reduces fluoride levels by 10–60%. Adsorption media uses granulated activated carbon (GAC), bone char, serpentine, activated bauxite and activated alumina.

6.2.8 Colour, taste and odour

Drinking water ideally should be colourless. However, due to the presence of soil, clay, organic matter and iron and other metals as well as industrial wastes, raw water can display some colour. The acceptability criterion is somewhat subjective but many can detect colours above 15 TCUs (True Colour Units) in a glass of water.

Cool water is generally more palatable than warm water. Taste and odours may be caused by the presence of inorganic material, microorganisms and chemicals. Taste and odours caused by disinfectants and disinfection by-products (DBPs) can be controlled by a careful operation of the disinfection process. High water temperature promotes the growth of microorganisms that may increase taste and odour problems. Aeration and granulated activation carbon (GAC) can control odours due to the presence of hydrogen sulphide whereas biological nitrification can control ammonia. Potassium permanganate ($KMnO_4$), first used in London in 1910, together with activated carbon can be used to remove hydrogen sulphide, iron and manganese.

6.2.9 Control of corrosion and scale formation

Corrosion generates an electric current that flows through the metal. Impurities in water cause one spot in the pipe to act as an anode and another spot to act as a cathode. At the anode, ferrous iron (Fe^{2+}) breaks away from the pipe and goes into solution in water, which ionizes by losing two electrons that travel to the cathode. Water molecules dissociate into H^+ and OH^- and the Fe^{2+} combines with two OH^- radicals to form ferrous hydroxide ($Fe(OH)_2$). Two H^+ ions near the cathode pick up two electrons from the iron atom to form H_2, as hydrogen gas. These reactions, which form ferrous hydroxide, leave an excess of H^+ near the anode and an excess of OH^- radicals at the cathode. This change in normal concentrations of H^+ and OH^- accelerates the rate of corrosion. With the dissolved oxygen in water, ferrous hydroxide is converted to ferric hydroxide, which precipitates as iron rust.

Corrosion affects water quality as it can leak toxic metals (mainly lead and copper) and form iron deposits that protect bacteria and microorganisms. It also shortens the life of plumbing systems that adds an economic cost.

Corrosion increases with increasing levels of dissolved oxygen, total dissolved solids and temperature while it decreases with increasing levels of pH and alkalinity. pH and alkalinity can be adjusted using lime or caustic soda. Flow velocity, which increases the production of dissolved oxygen and bacteria, which in turn produce carbon dioxide and hydrogen sulphide also increases the rate of corrosion. Certain types of bacteria which produce carbon dioxide and hydrogen sulphide also increase the rate of corrosion. Corrosion can be localized or uniform and also depends upon the type of metal. It is also important to note that these factors interact with each other.

Formation of scales on the interior of pipes can prevent corrosion as it separates the water from the pipe surfaces. Scale-forming compounds include magnesium carbonate ($MgCO_3$), calcium carbonate ($CaCO_3$), calcium sulphate ($CaSO_4$) and magnesium chloride ($MgCl_2$). As there is a limit to how much water can hold chemicals in solution, the point at which no more chemicals can be dissolved is called the 'saturation point'. With calcium carbonate, the saturation point depends on the pH of water and temperature.

6.2.10 Removal of hardness

Hardness is caused by dissolved calcium and expressed as the equivalent quantity of calcium carbonate. In daily life, it is detected by its inability to form lather with soap and the formation of scales in boiling systems such as kettles and boilers. Less than 50 mg/litre of $CaCO_3$ is considered as soft water and over 300 mg/litre considered as very hard water. Sometimes, hardness can also be caused by the presence of dissolved magnesium but to a lesser extent. By international standards, the maximum permissible level of hardness in household water is about 500 mg of calcium carbonate per litre. There is no established health-based criterion for the hardness level acceptable for drinking water. There is, however, some indication that very soft waters may have an adverse effect on mineral balance and incidences of cardiovascular diseases. One possible negative effect is that a hardness equivalent to 200 mg/litre may cause deposition in the distribution system. WHO gives more details about the guidelines for hardness in drinking water (WHO, 2003). Soft water is corrosive.

Hardness can be broadly classified as calcium hardness and magnesium hardness or as carbonate hardness and non-carbonate hardness. Calcium hardness is caused by calcium carbonate, calcium sulphate and calcium chloride. Total hardness in each case is the sum of calcium hardness and magnesium hardness or carbonate hardness and non-carbonate hardness.

In groundwater, calcium hardness is caused by calcium dissolved in water when water flows through limestone deposits and magnesium hardness is caused when water flows through dolomite (calcium magnesium carbonate, $(CaMg(CO_3)_2)$ and other magnesium salts. Non-carbonate hardness is caused by calcium sulphate, calcium chloride for calcium and magnesium sulphate and magnesium chloride for magnesium. When water is boiled, carbon dioxide is released. The calcium and magnesium bicarbonate salts precipitate to form respective carbonates. These can be seen as deposits in kettles and other kitchen utensils. Carbonate hardness is sometimes referred to as 'temporary hardness' because it can be removed by boiling whereas non-carbonate hardness, which cannot be removed, is referred to as 'permanent hardness'. It is desirable to have the magnesium hardness less than 40 mg/litre.

Hardness can be reduced by various softening processes. Lime $(Ca(OH)_2)$ softening or lime soda ash (Na_2CO_3) softening are the widely used approaches.

The basic chemical reactions in the lime softening process for carbonate hardness are as follows:

$$Ca(HCO_3)_2 + Ca(OH)_2 \rightarrow 2CaCO_3 \downarrow + 2H_2O \qquad (6.11)$$

$$Mg(HCO_3)_2 + Ca(OH)_2 \rightarrow CaCO_3 \downarrow + MgCO_3 + 2H_2O \qquad (6.12)$$

$$MgCO_3 + Ca(OH)_2 \rightarrow CaCO_3 \downarrow + Mg(OH)_2 \qquad (6.13)$$

For non-carbonate calcium hardness with soda ash (Na_2CO_3), they are

$$CaSO_4 + Na_2CO_3 \rightarrow CaCO_3 \downarrow + Na_2SO_4 \qquad (6.14)$$

$$CaCl_2 + Na_2CO_3 \rightarrow CaCO_3 \downarrow + 2NaCl \qquad (6.15)$$

For non-carbonate magnesium hardness with lime $(Ca(OH)_2)$ softening, they are

$$MgCl_2 + Ca(OH)_2 \rightarrow Mg(OH)_2 \downarrow + CaCl_2 \qquad (6.16)$$

$$MgSO_4 + Ca(OH)_2 \rightarrow Mg(OH)_2 \downarrow + CaSO_4 \qquad (6.17)$$

This process produces non-carbonate salts $(CaCl_2; CaSO_4)$ and hence non-carbonate hardness, which can be removed by adding soda ash (same reactions as above):

$$CaCl_2 + Na_2CO_3 \rightarrow CaCO_3 \downarrow + 2NaCl \qquad (6.18)$$

$$CaSO_4 + Na_2CO_3 \rightarrow CaCO_3 \downarrow + Na_2SO_4 \tag{6.19}$$

Excess $CaCO_3$ and $Mg(OH)_2$ can precipitate and plug the filter under-drain and stick to the walls of distribution pipes. $Mg(OH)_2$ can form scales inside boilers and water heaters. The process of controlling this deposition is known as necarbonation, which reduces scale formation. The reactions are as follows:

$$CaCO_3 + CO_2 + H_2O \rightarrow Ca(HCO_3)_2 \tag{6.20}$$

$$Mg(OH)_2 + CO_2 \rightarrow MgCO_3 \downarrow + H_2O \tag{6.21}$$

An alternative to lime treatment is ion exchange, which in general can be either cation (+) exchange or anion (–) exchange and can be used when impurities are soluble and ionized. In hardness removal in water, the ions that cause hardness are mainly calcium (Ca^+) and magnesium (Mg^+), which are exchanged with sodium (Na^+) ions that do not cause hardness. Cation exchange resins are used to remove positively charged contaminants while anion exchange resins are used to remove negatively charged contaminants. Resins exchange hardness-causing ions (Ca^+, Mg^+) are used with ions that do not cause hardness (Na^+). For carbonate hardness, the reactions are as follows:

$$Ca(HCO_3)_2 + Na_2X \rightarrow CaX + 2NaHCO_3 \tag{6.22}$$

$$Mg(HCO_3)_2 + Na_2X \rightarrow MgX + 2NaHCO_3 \tag{6.23}$$

and, for non-carbonate hardness the reactions are as follows:

$$CaSO_4 + Na_2X \rightarrow CaX + Na_2SO_4 \tag{6.24}$$

$$CaCl_2 + Na_2X \rightarrow CaX + 2NaCl \tag{6.25}$$

$$MgSO_4 + Na_2X \rightarrow MgX + Na_2SO_4 \tag{6.26}$$

$$MgCl_2 + Na_2X \rightarrow MgX + 2NaCl \tag{6.27}$$

Regeneration is according to

$$CaX + 2NaCl \rightarrow CaCl_2 + Na_2X \tag{6.28}$$

$$MgX + 2NaCl \rightarrow MgCl_2 + Na_2X \tag{6.29}$$

In these chemical reactions, X represents the ion exchange material. Ion exchange resins used for hardness removal include natural zeolites (sodium aluminium silicate) and synthetic polystyrene resins. Ion exchange can remove temporary hardness as well as permanent hardness by allowing the hard water to pass through a column of resins continuously. As the water passes through the column, sodium ions come off the resins and go into the water while calcium ions come out of the water and stick to the resin.

The advantages of ion exchange over lime soda ash treatment include the fact that it is cheaper than the lime soda ash method, particularly for non-carbonate hardness removal, that it removes all hardness, that the only chemical used is sodium chloride and that there is no possibility of equipment failure. Once the entire hardness has been removed, the treated water can be blended with water with some degree of hardness so that the final product water has the desired degree of hardness. Soft water is not ideal as it is corrosive.

6.2.11 Aeration

Water at 20°C and 1 atm pressure when exposed to air tend to reach an equilibrium dissolved oxygen concentration of about 9.1 mg/litre and about 0.5 mg/litre of carbon dioxide. Aeration is the process of adding oxygen to water, which can be achieved by using a simple cascade or diffusion of air into water, which is the most common. Alternatively, compressed air can be diffused through a system of submerged perforated pipes. Aeration, in addition to oxygenation of water, can also cause precipitation of iron and manganese. Air stripping (transferring of volatile components of a liquid into an air stream) can be used for the removal of volatile organics (e.g. solvents), some taste- and odour-causing compounds and radon. Air stripping needs to provide the necessary contact between air and water. Air strippers are usually packed into towers and operate with counter-flow of water and air. Water enters from the top and air is ventilated through the bottom. The counter-flow removes particles from the water into the air, a process known as volatization. Water reaches the bottom and air exiting may require emission control.

Other methods of water treatment include solar evaporation and the use of bioinspired hydrogel (see Sections 4.8.4 and 4.8.5).

6.3 SLUDGE DISPOSAL

All treatments result in some amount of sludge, mostly alum sludge and lime sludge, at the end of the treatment process, which should be disposed of in an environmentally friendly manner. The difficulty in separating the solids from water is due to the chemical bonding of aluminium hydroxide flocs with water. The easiest option is to discharge into rivers if there are no objectionable environmental consequences. Other options include discharge into sewage treatment provided it does not interfere with the sewage treatment processes, surface spreading on land and lagooning on shallow drying beds. In the latter option, dewatering happens by evaporation and infiltration. Typical depths of lagoons are about 2.5 m with the depth of sludge about 1 m. Another option is to press the sludge into thick cakes, which can be disposed on land. The water in the sludge can also be removed by centrifuging.

6.4 SERVICE RESERVOIRS

After the source water has been treated to the required level, it has to be temporarily stored in service reservoirs to meet fluctuating demands, to provide sufficient pressure, as well as to supply during failures of treatment works. They are for short-term storage and located at higher elevations to enable distribution to consumers by gravity. They may also be underground reservoirs, sometimes with several compartments, or elevated tanks. Choice depends on the economics. Elevated storage is generally more expensive. Construction can be with reinforced concrete, pre-stressed concrete or steel depending on the type of reservoir. The reservoirs should be covered to prevent contamination as well as to block UV radiation and sunlight that promotes algal growth. Ventilation of the air above the water surface should be provided to maintain fresh supply of air and for temperature control with fluctuating water levels in the reservoir. Structural design should take into account all types of loadings such as water and structure dead load and wind, water pressure, earth pressure, negative and positive air pressures, thermal loads and loads on the roof including dynamic effects in earthquake-prone areas. In general, service reservoirs are built with at least two compartments so that one can be drained for maintenance. The most cost-effective shape of a reservoir is circular in plan, but the area of land required is greater. However, circular tanks are less suitable for sub-division.

The overall capacity should be sufficient to cater for about 18–24 h of demand, which is usually high in the mornings and evenings. Allowance should also be made for emergency use such as for fire-fighting. It is also important to have a contingency storage to cover intermittent source operations, breakdowns at sources and loss of supplies after major bursts. On the negative effect of high storage is the possibility of deterioration of water quality due to decay of disinfectant residuals and growth of disinfectant by-products such as trihalomethanes (THMs). Before delivering water to consumers, water should be allowed to stay in the tank (particularly in concrete tanks) for at least 7 days to allow absorption into the concrete. Samples of water from each compartment of the reservoir must be tested frequently according to the specifications of the relevant water authority. Flat roofed concrete reservoirs are usually covered with earth and grass for appearance and heat insulation.

Reservoir piping normally comprise of inlets, outlets, drawdown, over-flow and drainage pipes with associated valves. Before putting into service, reservoirs should be tested for water tightness, cleaned and disinfected.

6.5 DISTRIBUTION SYSTEM

From the service reservoir, water is distributed to the consumers using four different distribution systems as described below. The basic requirements of a distribution system include maintaining good quality of water during transmission, providing sufficient pressure at the consumers' ends, providing sufficient pressure and quantity of water during fire-fighting, water tightness and laid at a reasonable distance away from sewage lines.

6.5.1 Dead-end system

Components of a dead-end system include service reservoir, main line, sub-mains and branches with dead ends similar to the branches of a tree. Service connections are given from the branches. The arrangement is as shown in Figure 6.3. The advantages include simplicity, economy and a smaller number of cutoff valves. Disadvantages include non-availability of water to an area when a pipe providing water to that area is under repair and water remaining stagnant, which can lead to sedimentation and bacterial growth. Water at dead ends should be discharged periodically to avoid this problem, which leads to wastage of treated water. The pressure needed for fire-fighting may be inadequate since the water is supplied by one branch only. If more connections are added to the branches, the pressure may become too low.

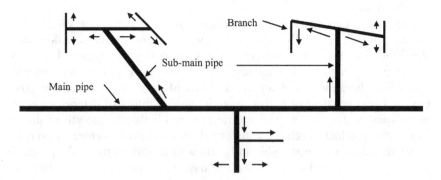

Figure 6.3 Dead-end distribution system.

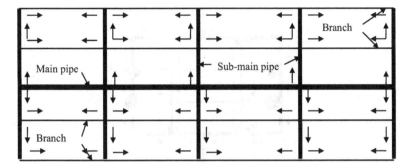

Figure 6.4 Grid iron distribution system.

6.5.2 Grid iron system

Grid iron system is suitable for well-planned cities with a rectangular layout and consists of main lines, branch lines and sub-mains all interconnected as shown in Figure 6.4. There are no dead ends in this system, allowing water to reach from more than one direction and therefore no stagnation of water. Advantages include no stagnation, provide service to areas where a pipe is under repair as water can flow from the other direction and high pressure availability during fire-fighting. Disadvantages include cost of pipe laying as more pipe lengths are needed, the need to have more control valves and the difficulties in hydraulic calculations.

6.5.3 Ring system

Ring system, sometimes called circular system, is suitable for cities with well-planned streets and roads. In this system, the main pipeline is laid around the service area and the branch lines are laid normal to the ring and connected with each other as shown in Figure 6.5.

Advantages include no stagnation of water, easy for repair works and availability of large quantities of water for fire-fighting. Disadvantages include the need for large diameter and longer length pipes and the need for more cut off valves.

6.5.4 Radial system

In the radial system, the service area is divided into small distribution zones with individual distribution reservoirs for each zone. Pipelines are laid radially from the distribution reservoir to the service zones (Figure 6.6).

Advantages include high velocity and high pressure, small head loss and easy hydraulic analysis. Disadvantages include cost of distribution reservoirs and stagnation of water due to many dead ends.

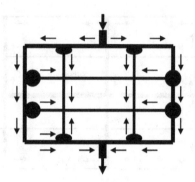

Figure 6.5 Ring distribution system (Circles indicate cut-off valves).

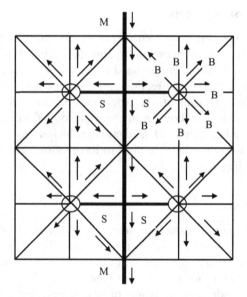

Figure 6.6 Radial distribution systems (The circles in the middle represent distribution reservoirs; M – Main pipe; S – Sub-main pipes; B – Branch pipes).

6.6 HOUSEHOLD WATER QUALITY

In rural areas, especially in low-income countries, public water supply is unavailable. Individual households improvise by various means how to get the daily water needs from various sources. The sources may be rainwater, which is not available all the time, shallow well water and water from a stream. Wells can be dug out at the premises but when the source is a stream, the water has to be carried to the consumers' ends and stored in some containers. This is usually a daily routine since water, which is not disinfected, cannot be stored for long periods of time. The task of carrying the water from a source to the house is usually carried out by women. Contamination can occur at the source, during transportation and in storage. In general, surface water or water from shallow wells should not be used as a potable source without treatment.

6.6.1 Contamination at the source

All natural sources of water contain inorganic and organic chemicals. Inorganic chemicals come from rocks and soil through which water flows or percolates whereas organic chemicals come from breakdown of vegetation, algae and other microorganisms that grow on water or on sediments. Before deciding on a potable source of water, it is important to decide whether or not water from a source needs treatment before use.

Rainwater is relatively clean but somewhat acidic. It lacks minerals essential for human health in appropriate concentrations. These include calcium, magnesium, iron and fluoride. The presence of *E. coli* in rainwater is quite common, especially if collected just after rainfall.

Sources of household water include tube wells (with a hand pump), spring and/or stream water, shallow wells and rainwater. Contamination can occur at each of these sources. For example, in the case of tube wells, ingress of contaminated water into the tube well directly or due to poor construction; in the case of shallow wells, contamination can occur due to poor construction and lining as well as from buckets used to take water out of the well; in the case of springs and/or streams, contamination can occur due to runoff from upstream, which may carry pollutants including human and animal excreta; in the case of rainwater, contaminants may include bird droppings and other pollutants in roof tops.

6.6.2 Contamination during transport

Transport from the source to the dwelling is normally done by women using containers, which they are able to carry. Contamination of the water can occur due to unhygienic handling of water during collection, transport and delivery as well as due to the container not being clean.

6.6.3 Contamination during storage

Contamination during storage can be due to the storage container not being clean, whether it is covered or left open, whether or not there is a safe and clean method of taking water from the container for domestic use, especially for drinking and washing uncooked food, lack of protection against contamination introduced by vectors (flies, cockroaches, rodents etc.) and inadequate cleaning of containers to prevent biofilm formation and accumulation of sediments and pathogens. Other factors that can contribute to microbial contamination include higher temperatures, long storage times and the presence of higher levels of airborne particulates. The containers also need to have facility to withdraw water in a sanitary manner, such as a tap, spigot, spout or other narrow orifice.

6.6.4 Treatment

Treatment of household water, in principle, is not different from the treatment of public water supplies. However, some treatment methods at household level may not be practical due to the associated costs as well as the inadequacy of skills required.

Treatment at the consumers' end can be by sedimentation to get rid of solid matter present in the water, filtration and disinfection using chemicals such as chlorine, heat treatment and/or solar radiation. They all have their negative effects too. For example, boiling will leave a flat taste, solar radiation will make the water hot and chemicals can leave a bad taste, all of which may not be acceptable to the consumers.

Straining the water using a cloth can block out fine suspended particles in the raw water. It is a simple process with no equipment or technological knowledge but that alone does not make the water safe for drinking. Sedimentation, which allows suspended solids to settle down to the bottom of the container either by gravity or assisted by adding chemicals (flocculation and coagulation), makes the water clearer but does not eliminate germs. It should be a pre-process before disinfection. Filtration in individual houses can be either by mini slow sand filtration or by ceramic filters. Mini slow sand filtration is the same as normal slow sand filtration except for the size of the filtration unit. Typically, a mini slow sand filter unit can be of the size shown in Figure 6.2. With slow sand filtration as a treatment process, it is important to ensure that the consumer is aware and familiar with the operation and maintenance of the filter unit. Ceramic filters are of the readymade type but expensive and do not have any residual effect.

6.6.4.1 Sedimentation

Sedimentation can reduce water turbidity but is not effective in reducing microbial contamination. Reduced levels of solids (turbidity) improve penetration of UV radiation, decrease oxidant demand (e.g. chlorine) and decrease solid pathogens. It is easy and cost effective and can settle solids such as sands, silts and larger microbes but clays and smaller microbes can only be settled with chemicals to form flocs. Chemically assisted sedimentation, which include coagulation and flocculation, is not recommended at the household level.

6.6.4.2 Filtration

Although many different types of filtration are available, the most widely and adaptable ones at household level are mini slow sand filter units and clay or ceramic filters. The basic principle and operation of a mini slow sand filter is the same as that used in public water supplies, but the scale is much smaller. Proper maintenance is important to achieve optimum results. Consumers should be educated about the operating principles of slow sand filters. With proper education and training, consumers at household level can also use Xylem filters (described in Section 4.8.3).

6.6.4.3 Aeration

Although aeration does not have microbiocidal effect, it introduces oxygen, which can cause chemical reactions that precipitate certain dissolved solutes in anaerobic water, thereby indirectly contributing to microbial reductions.

6.6.4.4 Disinfection

Boiling will kill all pathogens but re-contamination can occur because there is no residual effect. Once boiled, water should not be kept for long periods of time. Boiling requires some source of energy, which in most rural areas is firewood. Boiling makes the water to give a flat taste, which can be restored by adding a pinch of salt.

Solar disinfection can be done by exposing the water in a clear plastic or glass container to direct sunlight. The exposure time in tropics should be about 5 h around mid-day. Solar disinfection does not carry any residual effect, which means that the water should not be kept for long periods of time. It takes more time than other modes of disinfection but because of its simplicity, low cost and the need for only beverage bottles and sunlight, it is recommended as an appropriate technology for developing countries. Achieving a sufficiently high temperature (preferably 55°C or higher for several hours) is an important factor for microbial inactivation by solar disinfection systems. Inactivation of microbes is by the combined action of heat and UV radiation.

Chemical disinfection requires some understanding of the effects and safe use of chemicals. The widely used chemical is chlorine, which is available in tablets as well as in liquid form. Chlorine disinfection has a residual effect that minimizes re-contamination. It can leave some tastes, which may not agree with the consumers. In an emergency or for a short-term effect, lime and other citrus juices can be added to water to inactivate certain pathogens and to lower the pH value of the water to less than about 4.5.

Disinfection by vegetable matter has been practiced since ancient times. Two of the widely used methods are the use of seed cake (remaining after oil extraction) of *Moringa*, a vegetable tree known as *murunga* in Sinhalese language, *murugai* in Tamil language and drumstick in English with the botanical name *Moringa oleifera*, and ripe seeds of *nirmali* plant or *Strychnos potatorum* (also called the clearing nut) as a flocculent to purify water. The leaves and seeds of *moringa* tree have several healthy compounds such as vitamin A, vitamin B1 (thiamine), calcium, potassium, iron, magnesium, phosphorus and zinc. Both these plants are also considered as Ayurvedic medicinal plants.

Sludge removal is the same as for sedimentation.

6.7 ACCEPTABILITY

Water should be free of tastes and odours that would be objectionable to the majority of consumers. Tastes and odours can originate from natural inorganic and organic chemical contaminants, biological sources, synthetic chemicals, corrosion as well as from water treatment. Their presence can be indicative of some form of harmful pollution. Another indicator of acceptability is colour, which may be due to the presence of particulate matter and visible organisms.

Microbial, chemical and physical water constituents may affect the appearance, odour or taste of the water, and the consumer will evaluate the quality and acceptability of the water on the basis of these criteria. In extreme cases, consumers may avoid aesthetically unacceptable but otherwise safe drinking water in favour of more pleasant but potentially unsafe sources. Some characteristics not related to health, such as those with significant impacts on the acceptability of water, may also be of importance. Where water has unacceptable aesthetic characteristics (e.g., appearance, taste and odour), further investigation may be required to determine whether there are problems with significance for health.

REFERENCES

American Water Works Association (AWWA) (1995). *Water Treatment, Second Edition, Principles and Practices of Water Supply Operations*, Denver, CO, pp. 523.

Twort, A. C., Law, F. M., Crowley, F. W. and Ratnayaka, D. D. (1994). *Water Supply*, 4th edition, Edward Arnold, London, p. 511.

World Health Organization (2003). Hardness in drinking-water. Background document for preparation of WHO Guidelines for drinking-water quality. Geneva, World Health Organization (WHO/SDE/WSH/03.04/6).

Chapter 7

Water and health

7.1 INTRODUCTION

The health of a nation depends on the quality of its drinking water, which should be maintained at the highest possible level from a health point of view. An adult weighing about 60 kg needs about 2 litres of water per day, which goes down to about 1 litre per day for a body weight of about 10 kg and about 0.75 litres per day for bottle-fed infants weighing about 5 kg. Pregnant women need about 2.4 litres and breastfeeding women need about 3 litres since large amount of fluid is lost during nursing. Normally about 20% of water intake comes from food while the rest comes from drinking water and beverages. It is to be noted that too much water can cause hyperhydration, which can be fatal. The standards vary from country to country and sometimes from region to region within the same country. The World Health Organization (WHO) provides certain guidelines, which are followed by many countries, but some advanced countries maintain standards more stringent than WHO guidelines. The quality of drinking water primarily depends on the source of raw water, which may be rainwater (collected using rainwater harvesting), surface water from rivers, streams, lakes etc. where the water is conveyed on the surface of the land, and groundwater extracted from aquifers using deep wells as well as shallow wells from the water table. Groundwater from deep aquifers is relatively clean but may sometimes be polluted with saltwater intrusion and industrial discharge.

Drinking water can be contaminated at the source, during conveyancing, storage and distribution. Drinking water contaminated with pathogenic pollutants can cause many health problems, some of which can be fatal. The WHO has indicated that unsafe water is the leading killer causing about 3.4 million water related deaths a year (www.voanews.com/a/a-13-2005-03-17-voa34-67381152/274768.html). In this chapter, the different types of contaminants in water are identified and brief descriptions of water-borne diseases are given.

DOI: 10.1201/9781003329206-7

7.2 WATER CONTAMINATION

The common types of contamination include sediment and debris contamination, microbial contamination, chemical contamination and radiological contamination. The last type is very rare except in situations resulting from nuclear accidents. Even after the raw water is treated to overcome the above types of contamination, it is also important to ensure acceptability criteria by the consumers.

7.2.1 Sediment and debris contamination

Sediment and debris contamination include particles of soil, clay, minerals and colloids, some of which can be inert and some capable of biological and chemical degradation. They add colour and turbidity to water. Larger particles can be removed by sedimentation and smaller and colloidal particles can be made to become bigger particles or flocs, which can also be removed by straining.

7.2.2 Microbial contamination

7.2.2.1 Types of pathogens

(i) Viruses

Viruses are the smallest pathogens and hence are more difficult to remove by physical processes such as filtration. Most viruses have sizes ranging from 20 to 400 nm in diameter and can only be seen under an electron microscope. They are infectious agents of sub-microscopic size that can multiply only in living cells of animals, plants and bacteria. They can reproduce and carry on metabolic processes only within a host cell. They are not living cells. They all contain nucleic acid (either DNA (deoxyribonucleic acid) or RNA (ribonucleic acid)) enclosed within a protein shell or coat. The nucleic acid encodes the genetic information of the virus. The protein coat that encloses the genetic material is known as a capsid, which can have several shapes such as polyhedral, rod or helical. The capsids protect the genetic material from damage.

In the context of water, the viruses of significance are those that can infect the gastronomical tracts. They are excreted in the faeces of humans and are identified as enteric viruses, which can be transmitted via person-to-person contact through inhalation of aerosols and droplets, urine and faeces.

When infected in the host, the host cells reproduce copies of the original virus manifold. Viruses can infect animals, plants, bacteria, protozoa, fungi and archaea (single-celled microorganisms that lack cell nuclei and are therefore prokaryotes). There are many diseases caused by viruses. Some viruses cause an immune response in the host that kills the virus but viruses such as HIV (human immunodeficiency virus) do not. Antibiotics are ineffective in treating diseases caused by viruses, which can only be treated with antiviral drugs and vaccines. Among the many viruses causing diseases, rotaviruses cause most gastrointestinal infections in children, which can lead to severe consequences, including hospitalization and death, with the latter being far more frequent in low-income countries.

The presence of any enteric virus in drinking water is an indicator for the potential presence of other enteric viruses. It also is an indication of shortcomings in the disinfection process in water treatment. Viruses can persist for long periods in water and infective doses are typically low.

(ii) Bacteria

Bacteria are unicellular organisms (prokaryotic cells) that replicate by binary fission, considered as one of the oldest forms of life on earth. Their sizes are much larger than those of viruses and range from about 200 to 2,000 nm. Bacteria can infect animals, plants and fungi. Diseases caused by bacteria can be treated with antibiotics. Typical diseases caused by bacteria in humans include tuberculosis, food poisoning, anthrax and meningitis. A widely present bacteria in polluted waters is *E. coli*, which is considered as the most suitable indicator of faecal contamination.

Bacteria can be classified based on the shape, composition of the cell wall, mode of respiration and mode of nutrition. Shapes can vary from rod shaped (called bacillus), spiral shaped, spherical shaped and comma shaped. Cell wall composition can be of peptidoglycan (Gram-positive bacteria), or lipopolysaccharide (Gram-negative bacteria). Nutrition can be autotrophic (as in cyanobacteria) or heterotrophic (as in all disease-causing bacteria). The mode of respiration can be anaerobic or aerobic. Of all pathogens, bacteria are most sensitive to inactivation by disinfection. Enteric bacteria typically do not grow in water and survive for shorter periods than viruses or protozoa. Many types of bacteria that cause infections in humans are carried by animals.

Not all bacteria are disease causing. Some are useful. For example, lactic acid bacteria that convert milk into curd, bacillus that ferments food products, actinobacteria that helps to build up the immune system and soil bacteria that helps in the production of antibodies used in the treatment of bacterial diseases.

(iii) Protozoa

Protozoa are unicellular eukaryote parasitic organisms that feed on organic matter such as other microorganisms and debris. Replication is by binary fission. Infections caused by protozoans can range from being mild to life-threatening depending on the individual. Most protozoans in humans have sizes less than about 50 nm and they perform various tasks by different organs such as mouth, intestines, anus etc. of higher animals. They lack a cell wall and therefore do not have a fixed shape. Their habitat is in the water environment. Examples include Amoeba, Entamoeba etc. Malaria and amoebiasis are two of the widely known diseases caused by protozoa pathogen. Another painful infection of the cornea that can lead to blindness known as acanthamoebic keratitis can occur in individuals wearing contact lenses when tap water is used in preparing solutions for washing contact lenses. Protozoa are least sensitive to inactivation by chemical disinfection.

(iv) Helminths

Protozoa and helminths are among the most common causes of infection and disease in humans and animals. Helminth parasites infect humans and animals worldwide. The word 'helminth', which in Greek means 'worm' refers to all types of worms, both free-living and parasitic. Except for guinea worm (*Dracunculus medinensis*) and liver flukes (Fasciola spp. (*F. hepatica* and *F. gigantica*)), drinking water is not a significant route of transmission of helminths.

Microbial contamination includes bacteriological, viral, protozoan and other biological contaminants which are pathogens. The main microbial contaminant is animal or human faeces, which can be a source of pathogenic bacteria, viruses, protozoa and helminths (parasitic roundworms and flatworms). An indicator of faecal contamination is the *E. coli* concentration. Absence of *E. coli* does not necessarily indicate that the water is free of pathogenic microorganisms. Drinking water is the only mode of pathogen transmission by the faecal-to-oral route but other modes of transmission include contaminated food, hands, utensils and clothing, particularly when domestic sanitation and hygiene are poor. Transmission can also be by the ingestion of the eggs in food contaminated with faeces or soil contaminated with faeces. Other microbial contaminants include guinea worm, toxic cyanobacteria and legionella. Cyanobacteria, sometimes known as blue-green algae, derives its name from the blue-green colour. They derive their energy through photosynthesis. With the presence of excessive nutrients (nitrogen and phosphorous) combined with hot sunny dry weather, cyanobacteria can produce a range of toxins known as cyanotoxins that can be harmful to humans and animals. Cyanobacterial toxins are neurotoxic affecting the nervous system with symptoms such as high fever, muscle aches, cough, headache and diarrhoea, and hepatotoxic affecting the liver. Filtration can effectively remove cyanobacterial cells, thereby removing some of the cyanotoxins as well. Disinfection by chlorination and oxidation by ozone can also remove most cyanotoxins. Legionella bacteria are more prone to be present in warmer climates, especially in pipe distribution systems and in stagnant waters. Humans are exposed to legionella bacterium primarily through inhalation of water droplets (aerosols) into the respiratory system that may lead to legionnaires' disease, which is a type of pneumonia that can be fatal. Sources of legionella include cooling towers, mist machines, humidifiers, whirlpools and hot springs, all engineered.

Higher microbial contamination indicated by *E. coli* is quite common in rainwater collected shortly after rainfall although the presence of pathogens is generally lower than in unprotected surface water. The occurrence of pathogens in groundwater and surface water sources depends on the physical and chemical characteristics of the catchment area and the aquifer, and the level of human activities and animal sources that release pathogens to the environment. In surface waters, potential sources of pathogens include point sources such as sewage and storm-water overflow as well as non-point sources such as runoff from agricultural and livestock areas, septic tanks and latrines. Their concentrations usually decrease due to dilution, natural decay and settling. Groundwater is less susceptible to contamination if extracted from deep confined aquifers. However, it is not the same if extraction is from shallow unconfined aquifers.

7.2.3 Chemical contamination

Naturally occurring chemicals in drinking water can be of inorganic nature derived from rocks and soil through which water percolates or flows over, or of organic nature derived from the breakdown of plant material or from algae and other microorganisms that grow in the water or on sediments. The main naturally occurring chemicals in drinking water are fluorides and arsenic but may also include barium, boron, chromium, manganese, molybdenum, selenium and uranium. Exposure to excessive fluorides can lead to mottling of teeth and that due to arsenic can lead to cancer and kidney diseases. The presence of other naturally occurring chemicals such as uranium and selenium may also lead to health problems if they are present in excessive quantities. The only naturally occurring organic contaminant is microcystin-LR, which is a toxin produced by cyanobacteria or blue-green algae. Nitrates and nitrites from excessive application of fertilizers as well as from leaching of wastewater into surface and ground sources have been shown to be associated with ethaemoglobinaemia (potentially life-threatening health condition in which the oxygen-carrying capacity of circulating haemoglobin is significantly reduced), especially in bottle-fed infants. In addition to naturally occurring chemicals, emphasis should be placed on controlling the chemicals from industrial sources and human dwellings as well as additives used in the production of drinking water. Chemicals from industrial sources include cadmium, cyanide and mercury. The main chemical used in water treatment processes is chlorine, which in excessive quantities becomes a chemical contaminant. In the past, DDT (dichloro diphenyl trichloroethane), which had been used as an insecticide to control malaria, has been used for public health purposes, but this practice has now been abandoned. Contaminants from pipes and fittings in the water distribution system include antimony, benzo[a]pyrene (an aromatic hydrocarbon found in coal tar, $C_{20}H_{12}$), copper, lead, nickel and vinyl chloride. Table 7.1 gives a summary of WHO's provisional guideline values for various contaminants that are of health significance in drinking water (WHO, 2011). A guideline value normally represents the concentration of a constituent that does not cause any significant risk to health over a lifetime of consumption. They are designated as provisional when there is a high degree of uncertainty in the toxicology and health data. Such guidelines are based on limited studies on human populations as well as toxicity studies using laboratory

Table 7.1 Provisional guideline values for chemicals that are of health significance in drinking water

Contaminant	Source	Guideline concentration (mg per litre)
Antimony	Pipes and fittings	0.02
Arsenic	Naturally occurring	0.01
Barium	Naturally occurring	0.7
Benzene	Industrial sources	0.01
Boron	Naturally occurring	2.4
Bromate	Disinfectant by-product (DBP)	0.01
Cadmium	Industrial sources	0.003
Carbon tetrachloride	Industrial sources	0.004
Chlorine	Disinfectant	5
Chlorate	Disinfectant by-product (DBP)	0.7
Chlorite	Disinfectant by-product (DBP)	0.7
Chloroform	Disinfectant by-product (DBP)	0.3
Chromium	Naturally occurring	0.05
Copper	Pipes and fittings	2
Cyanide	Industrial sources	0.07
DDT	Pesticide	0.001
Fluoride	Naturally occurring	1.5
Lead	Pipes and fittings	0.01
Manganese	Naturally occurring	0.4
Mercury	Industrial sources	0.006
Molybdenum	Naturally occurring	0.07
Nickel	Pipes and fittings	0.07
Nitrate (NO_3)	Agricultural sources	50
Nitrite (NO_2)	Agricultural sources	3
Selenium	Naturally occurring	0.04
Sodium	Treatment by-product	50
Toluene	Industrial sources	0.7
Uranium	Naturally occurring	0.03
Vinyl chloride	Pipes and fittings	0.0003

Source: Extracted from WHO (2011).

animals and extrapolated to human populations. The limitations of the latter include the high doses of contaminants administered to small samples of animals and the uncertainty associated with the extrapolation techniques to human populations subjected to relatively low doses of corresponding contaminants.

WHO publishes a series of guidelines to assist national authorities to establish their standards appropriate for their national conditions. Table 7.1 gives a summary of guideline values for chemicals (naturally occurring, industrial and agricultural sources) extracted from a WHO publication (WHO, 2011).

7.2.4 Radiological contamination

Environmental radiation originates from a number of naturally occurring (e.g. uranium, thorium and potassium-40) and human-made sources (e.g. nuclear plants, nuclear weapons and accidents). The largest fraction of natural radiation exposure comes from radon, which exists almost everywhere but particularly in the air over land and in buildings. It is a gas due to decay of radium contained in rocks and soil. Underground rocks containing natural uranium continuously release radon into water in contact with it (groundwater). Background radiation exposures vary widely across the earth, but the average is about 2.4 mSv per year (Table 7.2), with the highest local levels being up to about 10 times without any detected increased health risks from population studies. The current guidelines are based on a recommended reference dose level (RDL) equal to 0.1 mSv from a year's consumption of radiologically contaminated drinking water. This is a very small dosage compared to the average background radiation across the earth. Radiological hazards from drinking water are therefore rarely of public health significance. However, there is

Table 7.2 Average radiation dose from natural sources

Source	Worldwide average annual effective dose mSv	Typical range (mSv)
External exposure		
Cosmic rays	0.4	0.3–1.0
Terrestrial Gamma rays[1]	0.5	0.3–0.6
Internal exposure		
Inhalation (mainly radon)	1.2	0.2–10[2]
Ingestion (food and drinking water)	0.3	0.2–0.8
Total	2.4	0.2–10

Source: UNSCEAR (2000).

[1] Terrestrial exposure is due to radionuclides in the soil and building materials.
[2] Dose from inhalation of radon may exceed 10 mSv per year in certain residential areas.

evidence from both human and animal studies that radiation exposure at low to moderate doses may increase the long-term incidence of cancer. It is also important to note that the guidelines may be exceeded in the case of accidental release of radioactive material into the environment as a result of accidents.

7.3 ACCEPTABILITY

Water should be free of tastes and odours, which may be objectionable to the majority of consumers. The presence of microbial, chemical and physical constituents in water can affect the appearance, odour or taste. Although the evaluation is subjective, acceptability is an important criterion in water supply schemes. Sometimes, consumers may avoid aesthetically unacceptable but perfectly safe drinking water in favour of pleasant but unsafe water. This is particularly important in the use of recycled water. The present-day wastewater treatment technology permits wastewaters to be treated to standards sometimes exceeding the guidelines recommended for safe drinking water. The stigma attached to the concept of 'toilet to tap' is difficult to be ignored by the consumers. An example in which the 'toilet to tap' concept is executed is the 'NEWater' of Singapore where the wastewater is treated to a very high standard and blended with water from other sources in impounding reservoirs and further treated using conventional water treatment methods.

7.4 WATER-BORNE DISEASES

Water-borne diseases are transmitted through contact or consumption of microbially contaminated water. Most are life-threatening and are sometimes referred to as intestinal or filthy diseases because the pathway is usually faecal-to-oral route by water or food contaminated with human or animal faeces, although some diseases may be caused by contact with polluted water. Food and water-borne diseases may be caused by pathogens such as protozoa (e.g. amoebiasis), bacteria (e.g. cholera, dysentery, typhoid fever), viruses (e.g. hepatitis A, polio), algae (e.g. desmodesmus infection through open wounds) and helminths (parasitic roundworms and flatworms) (e.g. guinea worm disease). Most water-borne diseases are infectious, but there are also some water-borne diseases caused by organic and inorganic chemical contaminations in water that are non-infectious. Lack of clean water supply, sanitation and hygiene (WASH) are the major causes for the spread of water-borne diseases in a community. If every person in the world practices safe sanitation and hygiene and have access to clean water, there will be no water-borne diseases. Although governments, NGOs and communities have made great strides in the past 20 years to end water-borne diseases, there is still much to be done. Globally, in 2017, 1.2 million people died prematurely due to unsafe water which is 2.2% of global deaths. In low-income countries, this percentage is about 6%. (https://ourworldindata.org/water-access#unsafe-water-is-a-leading-risk-factor-for-death). The basic information of some of the common water-borne diseases (arranged alphabetically) is given below.

7.4.1 Amoebiasis

Amoebiasis is caused by the protozoic pathogen *Entamoeba histolytica*. Normal symptoms include loss of weight, colonic ulcerations, abdominal pain, and diarrhea, which may be bloody. In complicated cases, inflammation and ulceration of the colon, tissue death and anaemia due to prolonged gastric bleeding can occur. Diagnosis is usually by stool examination, but a more reliable method is by finding specific antibodies in the blood. Prevention of amoebiasis is by improved sanitation, including separating food and water from faeces. Treatment may require a combination of medications. Infected persons can spread the parasite to others. It is estimated that over 40 million people worldwide are infected with about 50,000 deaths annually (Podolsky et al., 2015).

7.4.2 Arsenicosis

Arsenicosis is a disease caused by the highly poisonous metallic chemical element arsenic contained in groundwater, infected soil and rocks and arsenic-preserved wood. Natural arsenic salts are present in all waters but in very small concentrations of less than about 0.01 mg/litre, which is also the WHO guideline value for safe drinking water. Health problems, which are usually delayed, occur when dangerous amounts of arsenic enter the body. Immediate symptoms after swallowing arsenic include drowsiness, headache, confusion and severe diarrhoea whereas in severe cases the symptoms may include excess saliva, problems in swallowing, blood in the urine, cramping muscles, hair loss, stomach cramps, convulsions, excessive sweating and vomiting. In extreme cases, the disease can lead to cancer of the skin, bladder, kidney and lung and diseases of the blood vessels of the legs, diabetes and reproductive disorders. Diagnosis is done by testing the source of drinking water, testing samples of hair and nail, urine and blood. Treatment methods include bowel irrigation, medication and chelation therapy (a process in which potent medications are used to remove heavy metals from the human body). Prevention is by using arsenic free source of drinking water. WHO estimate that more than 200 million people worldwide are exposed to water that contain potentially unsafe levels of arsenic (www.medicalnewstoday.com/articles/241860#causes-and-compli cations). Some regions of Bangladesh are known to experience this problem on a large scale.

7.4.3 Ascariasis

Ascariasis is an infection of the small intestine caused by *Ascaris lumbricoides*, which is a species of roundworm. The roundworm infects the small intestine and acts as a parasite deriving nutrients from the host's intestinal tract. It is prevalent in developing countries that lack safe and modern sanitation. *A. lumbricoides*, together with whipworm and hookworm, belong to the soil-transmitted helminth (STH) group that causes a group of diseases called helminthiases. The roundworm lays eggs, which then exits with the person's stool. Transmission is by accidental ingestion of the eggs of the *A. lumbricoides* roundworm, which can be found in soil contaminated by human faeces or uncooked food contaminated by soil that contains roundworm eggs. It can also be transmitted from person to person. Children are particularly vulnerable as they often put their soil contaminated hands in their mouths. The infection usually causes no symptoms, but a high infestation can lead to problems in the lungs such as coughing, wheezing or shortness of breath, blood in mucous, chest discomfort or in the intestines such as abdominal swelling, abdominal pain and diarrhea. Diagnosis is done by examining stool samples for parasites and eggs. Once diagnosed, further tests such as X-ray, CT scan, ultrasound, magnetic resonance imaging (MRI) scan and endoscopy may be necessary. Treatment is with antiparasitic drugs such as albendazole (Albenza), ivermectin (Stromectol) and mebendazole (Vermox). Ascariasis can be prevented by avoiding unsafe food and water and observing good hygiene.

7.4.4 Botulism

Botulism is caused by the bacteria *Clostridium botulinum* that can enter an open wound from contaminated water or through the consumption of contaminated water or food. There are three types of botulism, namely foodborne botulism, wound botulism and infant botulism. Symptoms include dry mouth, blurred vision, difficulty in swallowing, difficulty in breathing, slurred speech, vomiting and diarrhea. Diagnosis is done usually by testing stool. The treatment is usually through botulism antitoxin. Respiratory failure may require mechanical ventilation. Preventive measures include food safety for foodborne botulism, avoiding feeding honey for infant botulism and use of antibiotics for wound botulism.

7.4.5 Campylobacteriosis

Campylobacteriosis is an infectious disease caused by Campylobacter bacteria. The usual symptoms include nausea, vomiting, diarrhoea (sometimes bloody) and abdominal pain whereas in complicated cases they may include bowel syndrome, temporary paralysis and arthritis. The disease can be life-threatening for people with weakened immune systems, such as those with a blood disorder, with AIDS or receiving chemotherapy, due to Campylobacter spreading to the bloodstream. Transmission happens by spreading Campylobacter from animals to people through contaminated food, especially raw or undercooked chicken, through contact with animals and their environments, by drinking raw milk, eating contaminated fresh produce or drinking contaminated water. There is no person-to-person transmission. Most people recover from the infection without antibiotic treatment but should drink extra fluids as long as diarrhoea lasts. Campylobacteriosis can be prevented by observing good personal hygiene.

7.4.6 Cholera

Cholera is caused by the bacterium *Vibrio cholerae* contained in contaminated drinking water. Symptoms include watery diarrhea, nausea, cramps, bleeding from the nose, vomiting and rapid pulse. In severe cases, death can occur in 12–18 hours. Transmission is linked to access to clean water and sanitation. Risk areas include slums, overcrowded camps and places where refugees and internally displaced people are made to live. Diagnosis is by identifying *V. cholerae* in stool samples from affected patients. Cholera is a curable disease. In early stages of infection, patients can be treated successfully through prompt administration of oral rehydration solutions. Severely dehydrated patients need to be treated with rapid administration of intravenous fluids and antibiotics. WHO has prequalified three oral vaccines: Dukoral, Shanchol and Euvichol-Plus as preventive measures. All three vaccines require two doses for full protection. According to WHO (www.who.int/news-room/fact-sheets/detail/cholera), there are about 1.3 to 4.0 million cases of cholera and 21,000 to 143,000 deaths worldwide annually. Mass prevention is accomplished by providing access to safe drinking water and adequate hygiene and sanitation (WASH). A global strategy called Global Task Force on Cholera Control (GTFCC) aims to reduce cholera deaths by 90% by 2030.

7.4.7 Cryptosporidiosis

Cryptosporidium infection (cryptosporidiosis) is a disease caused by a tiny unicelled cryptosporidium protozoan parasite. When it enters the body, it travels to the small intestine, burrows into the walls and exits out with the faeces. Common symptoms include watery diarrhoea, low-grade fever, crampness, dehydration, weight loss, fatigue, nausea and vomiting whereas less common symptoms include arthritis and jaundice (yellowing discoloration of the skin and sclera of the eyes, dark urine and pale stool). Primary transmission is via the faecal-to-oral route, often through contaminated water. It is one of the most common water-borne diseases in the world. The infection begins when humans consume food or water containing cysts of the Cryptosporidium organism. Diagnosis is done by testing stool and possibly a biopsy from the intestine. Most healthy people with cryptosporidiosis recover without treatment in a short period of time. For people with compromised immune system, the treatment goal is to relieve symptoms and improve the immune response. Options include antiparasitic drugs, anti-motility agents that slow the movement of the intestines and increase fluid absorption to relieve diarrhoea and fluid replacement. Cryptosporidiosis may turn life-threatening in small children, elderly and immunocompromised persons, especially in AIDS patients. This disease can be prevented by practicing good hygiene and avoiding swallowing water from swimming pools, recreational water parks, lakes and streams. Since Cryptosporidium is highly resistant to chlorine with high enough concentrations and contact time, it can be inactivated by chlorine dioxide and ozone treatment.

7.4.8 Cyanobacterial toxins

Cyanotoxins are produced by cyanobacteria (also known as blue-green algae), which are found almost everywhere, but particularly in lakes, rivers and in coastal waters with high concentration of nutrients. They derive their energy through photosynthesis. When conditions are favourable, they reproduce exponentially to form blooms popularly known as harmful algal blooms (HAB), sometimes referred to as 'red tides'. Favourable conditions in freshwater systems include light availability, water temperature, alteration of water flow, vertical mixing, pH changes, nutrient loading (both nitrogen and phosphorus) and trace metals. Several types of cyanobacteria have gas-filled cavities that allow them to float to the surface or near surface depending on light availability and nutrient levels resulting in the formation of a blue-green scum. In marine environments, HABs cause fish kills. Transmission to humans is by exposure to cyanobacteria mostly via accidental inhalation and ingestion through contaminated water, resulting in gastrointestinal symptoms such as abdominal pain, malaise, nausea, vomiting, and diarrhoea as well as hay fever symptoms, which in some cases can be fatal. It can be prevented by avoiding exposure to water systems contaminated with cyanobacteria. Traditional methods of water treatment can remove cyanobacterial cells but are not effective in removing dissolved toxins. Biological filtration (slow sand filtration) is an effective way of removing toxins.

7.4.9 Dracunculiasis

Dracunculiasis, also referred to as guinea worm disease, is a vector-borne parasitic disease caused by *Dracunculus medinensis* roundworm. Infection starts after drinking water that contains copepods (small water fleas) infected with *D. medinensis* larvae. Once ingested, the copepod dissolves by stomach acid-releasing the larvae, which migrate through the host intestine into the abdominal cavity. Symptoms include painful blisters, usually on the leg or foot accompanied by burning pain, swelling and vomiting, which burst releasing the adult worm. The adult female worm grows up to about 1 m in length and exits through the infected person's skin, causing extreme pain. To soothe the pain, infected persons usually immerse the affected parts in water. The worms then release larvae (baby worms) into water, thereby repeating the cycle. Transmission is by the consumption of contaminated water from ponds and wells. Prevention is by avoiding wading in contaminated waters, filtering water before drinking and ensuring wider access to safe drinking water.

7.4.10 Dysentery

Dysentery is caused mainly by bacterium *Shigella dysenteriae* but may also be caused by Shigella and Salmonella. The disease is also sometimes called bloody diarrhea. Symptoms include extreme abdominal pain, rectal pain, low-grade fever and bloody diarrhoea and sometimes vomiting blood. The pathogens typically enter the body orally through ingestion of contaminated water or food as well as through oral contact with contaminated objects or hands and finally reach the large intestine, damaging the intestinal linings. Treatment is by oral rehydration and in severe cases by intravenous fluid replacement and drugs to kill the parasite and antibiotics to treat bacterial infection. There are between 120 million and 165 million cases of Shigella infection, of which about 1 million are fatal. Over 60 percent of these fatalities are children under the age of five years in developing countries. (www.medicalnewstoday.com/articles/171193#treatment). In the long term, prevention is by having access to clean water, adequate hygiene and sanitation (WASH).

7.4.11 Escherichia coli *(E. coli)* infection

E. coli infection is caused by the bacteria *Escherichia coli*. *E. coli* bacteria normally live in the intestines of healthy people and animals and are harmless but a few strains such as *E. coli* 0157:H7 can cause bloody diarrhoea and vomiting and damage the lining of the small intestine. It can also cause kidney failure in children. Symptoms include diarrhea, stomach crampness, nausea and vomiting. Diagnosis is done by testing stool samples. The causes of infection include exposure to contaminated water and food, especially raw fruits and vegetables, partially cooked ground beef and unpasteurized milk. It can also be transmitted by a person-to-person contact. The infection can be prevented by avoiding exposure to contaminated water and food and personal contact with infected people.

7.4.12 Giardia

Giardia, or giardiasis, is caused by the microscopic parasite *Giardia duodenalis* (or 'Giardia' for short). Once infected, the parasite lives in the intestines and is passed out in stool. It can survive outside the body for weeks and sometimes for months. It is one of recreational water illnesses transmitted specially in swimming pools. Symptoms include diarrhea, stomach cramps, nausea, vomiting and dehydration. Transmission is by person-to-person contact or through contaminated water, food, surfaces or objects. Diagnosis is done by testing stool. Giardia is treated by prescription drugs. It can be prevented by avoiding exposure to contaminated water. Particular attention should be paid to avoid accidental swallowing of water in swimming pools.

7.4.13 Hepatitis A

Hepatitis A is a liver disease caused by the hepatitis A virus (HAV). There are five main strains of the hepatitis virus, referred to as types A, B, C, D and E. They all cause liver disease but differ in modes of transmission, severity of the illness, geographical distribution and prevention methods. Symptoms of HAV include fever, malaise, loss of appetite, diarrhoea, nausea, abdominal discomfort and jaundice. Transmission is mainly by the faecal-to-oral route but can also be via person-to-person contact including sexual contact. There is no specific treatment but therapy including adequate nutritional balance and replacement of fluids can provide comfort. Hepatitis A can be prevented by improved sanitation, food safety and immunization. Vaccination is highly recommended for international travellers visiting infected regions.

7.4.14 Hepatitis E

Hepatitis E is a liver disease caused by the hepatitis E virus (HEV). Symptoms are similar to those of other viral forms of hepatitis and include fever, reduced appetite, nausea, vomiting and jaundice. The normal mode of transmission is via the faecal-to-oral route including faecal contamination of drinking water and ingestion of undercooked meat, but in rare cases it can also be via transfusion of infected blood products, organ transplants and transmission from pregnant women to their foetus. Diagnosis is done by a laboratory blood test. There is no specific treatment for hepatitis E. It can be prevented by maintaining good personal hygiene, good food hygiene and good environmental hygiene. It has been reported that hepatitis E virus caused around 20 million infections a year worldwide and about 44,000 deaths during 2015 (www.who.int/news-room/fact-sheets/detail/hepatitis-e). A vaccine to prevent hepatitis E virus infection has been developed and is licensed in China, but not yet available elsewhere (www.who.int/news-room/fact-sheets/detail/hepatitis-e).

7.4.15 Hookworm infection

Hookworm infection is caused by an intestinal parasite known as hookworm. Initial symptoms include itching and rash at the infected site. Severe infections may cause coughing, chest pain, wheezing, fever, iron-deficiency anaemia, protein deficiency and cardiac failure. Protein and iron deficiency can occur as a result of blood loss at the site of the intestinal attachment of the adult worms, particularly when children are continuously infected by many worms, retarding their growth and development. Diagnosis is done by finding characteristic worm eggs on microscopic examination of the stools. Infective larvae can survive on damp dirt, particularly sandy and loamy soil but not on dry soil or clay. Treatment is by anthelminthic medications (drugs that rid the body of parasitic worms), such as albendazole and mebendazole. Hookworm infection can be prevented by observing good hygiene, avoiding walking bare feet in infected areas where there may be human faecal contamination of the soil and proper sewage disposal.

7.4.16 Japanese encephalitis

Japanese encephalitis is a mosquito-borne disease caused by the Japanese encephalitis virus. Symptoms include fever and headache in mild cases and high fever, stiff neck, impaired mental state coma, convulsions in children and paralysis in severe infections. Transmission is by the bites of infected mosquitoes. Blood transfusion and organ transplant are also considered to be potential modes of transmission. There is no cure for the disease. Treatment is focused on relieving severe clinical signs and supporting the patient to overcome the infection. This infection can be prevented by avoiding mosquito-infected areas, particularly in the night and vaccination. WHO reports a global estimate of 68,000 clinical cases and approximately 13,600 to 20,400 deaths annually (www.who.int/news-room/fact-sheets/det ail/japanese-encephalitis).

7.4.17 Lead poisoning

Lead poisoning, also known as plumbism and saturnism, is caused by the chemical lead (Pb) in the body. Exposure to lead can occur by contaminated air, water, food and consumer products. Diagnosis is done by measuring the lead level in blood, which according to the US Centers for Disease Control has a limit 10 µg per 100 g for adults and 5 µg per 100 g for children. Lead poisoning may occur from intense exposure for a short duration or repeated low-level exposure over long periods of time. Symptoms of organic lead poisoning include insomnia, delirium, cognitive deficits, tremor, hallucinations and convulsions, which predominantly originate in the nervous system. Lead from the atmosphere and soil ends up in surface and ground water. In water distribution systems, lead contamination can arise from plumbing and fixtures that are either made of lead or have lead solder. To prevent or to reduce break down of lead in plumbing systems, chemicals can be added to municipal water to increase the pH, thereby reducing the corrosivity of the public water supply system. Lead poisoning is preventable and the preventive strategy is by avoiding exposure to lead. Lead poisoning can also be an occupational hazard for people working in smelting, recycling, stripping leaded paint and using leaded gasoline and aviation fuel. Diagnosis is done by laboratory testing of blood lead level. Transmission happens through inhalation of lead particles generated by burning material containing lead and ingestion of lead-contaminated dust, water and food. Lead poisoning symptoms in children include developmental delay, learning difficulties, irritability, loss of appetite, weight loss and constipation. For adults, the symptoms include high blood pressure, joint and muscle pain, difficulties with memory or concentration, headache, abdominal pain and reduced sperm count and abnormal sperm. Organic lead poisoning can be treated by removing the lead compound from the skin. Chelation therapy that uses special drugs that bind to metals in the blood may be used for people with high blood lead concentrations.

7.4.18 Legionnaires' disease

Legionella is a bacterium that can cause Legionnaires' disease and Pontiac fever, collectively known as legionellosis. Legionella occurs naturally in freshwater environments like lakes and streams. It grows well in warm waters. Legionnaires' disease is a potentially deadly lung infection (pneumonia), and Pontiac fever is a less serious infection with milder symptoms similar to flu. Diagnosis is through tests to detect the bacteria in coughed-up mucus. Symptoms include fever, cough, shortness of breath, tiredness, headache, muscle and abdominal pain and diarrhea. In extreme cases, respiratory failure may appear, which may cause death. Transmission is by breathing contaminated droplets (aerosols) and mist generated from artificial water systems such as water tanks, cooling towers, humidifiers, water fountains etc. as well as in handling garden soils contaminated with Legionella bacteria. It is not transmitted by person-to-person contact or by eating or drinking. Prevention is by maintaining good household water systems. There is no vaccine at the present time.

7.4.19 Leptospirosis

Leptospirosis is caused by the bacteria Leptospira in water contaminated with urine from animals such as mice, rats, cattle, pigs and dogs. Symptoms include fever, severe headache, sore muscles, chills, vomiting and red eyes. If not treated, patients may develop complications leading to infections in kidneys, liver, brain, lungs and heart, which can be fatal. Diagnosis is done through a blood test. Transmission happens through contact with water, food or soil containing urine from infected animals. The disease can also be transmitted through rodent bites. Person-to-person transmission is rare. Treatment is through antibiotics. Leptospirosis can be prevented by avoiding contact with fresh water, soil and vegetation, which may have been contaminated with urine of infected animals, especially rodents, as well as by maintaining good personal hygiene and wearing appropriate protective clothing and footwear when in contact with potentially contaminated soil and water.

7.4.20 Lymphatic filariasis

Lymphatic filariasis, also known as elephantiasis, is caused by filarial parasites through mosquito infection. The three closely related parasitic nematodes (roundworms) of the family Filariodidea are *Wuchereria bancrofti*, *Brugia malayi* and *Brugia timori*. The disease impairs the lymphatic system and symptoms include swelling and disfigurement of the limbs (elephantiasis) and breasts and genitalia that cause social stigma. Infection is usually acquired during childhood. Transmission happens through infected mosquito bites. Lymphatic filariasis can be prevented by preventive chemotherapy with safe medication to stop the spread of the parasitic infection. Pest mosquitoes breed in polluted waters such as cesspools, cesspits, drains and septic tanks. Long-term prevention is achieved by having clean water supply and drainage system.

7.4.21 Methaemoglobinemia

Methaemoglobinemia is a condition of elevated methaemoglobin in the blood caused by the decreased ability of blood to carry vital oxygen around the body. Arterial blood with elevated methaemoglobin levels has a characteristic chocolate-brown colour compared with the normal bright red oxygen-containing arterial blood. Infants have lower levels of a key methaemoglobin reduction enzyme (NADH-cytochrome b5 reductase) in their red blood cells. This results in a major risk of methemoglobinemia caused by nitrates ingested in drinking water. The most common cause of methaemoglobin is the high level of nitrates (> 50 mg/litre) in drinking water, which may come from manures and fertilizers in agricultural land. Nitrates are also found in vegetables. Symptoms may include headache, dizziness, shortness of breath, nausea, poor muscle coordination and blue-coloured skin (cyanosis), commonly known as 'blue baby syndrome'. Of special concern is in bottle-fed infants and water from wells in rural areas. Diagnosis is done by complete blood count (CBC), checking enzymes, blood colour and DNA sequencing. Treatment is done through oxygen therapy and mythylene blue (methylthioninium chloride, a salt used as a medication and dye). Methaemoglobinemia can be prevented by controlling nitrate levels in drinking water sources to below around 50 mg/litre.

7.4.22 Onchocerciasis

Onchocerciasis, also known as river blindness, is an eye and skin tropical disease caused by infection with the parasitic worm *Onchocerca volvulus*. It is called river blindness because the blackfly of the genus Simulium that transmits the infection lives and breeds near streams and rivers. The infected blackfly introduces third-stage filarial larvae onto the host skin and penetrate into the bite wound. Symptoms include severe itching, bumps under the skin and blindness. It is one of the major infections that cause blindness. Diagnosis is done by biopsies of the skin to see the larva. Transmission happens by exposure to repeated bites of the infected blackflies. The infection is treated by ivermectin on a community basis. There is no vaccine against the disease at the present time. The infection can be prevented by using insect repellents and appropriate clothing to avoid being bitten by flies as well as by spraying insecticides to decrease or eliminate the fly population.

7.4.23 Polio

Polio (or poliomyelitis) is a potentially deadly disease caused by poliovirus. The virus enters the body through the mouth with water or food that has been contaminated with faecal matter from an infected person. It then multiplies in the throat and the intestine and is excreted in faeces, which can pass on to others. In areas with poor sanitation, the virus easily spreads from faeces into the water supply and food. Symptoms include fever, fatigue, headache, vomiting, stiffness in neck, pain in limbs and trouble with swallowing and breathing. Diagnosis is done by symptoms and by laboratory testing for poliovirus by examining throat secretions, stool samples or cerebrospinal fluid. Transmission happens through person-to-person contact and via the stool or droplets from a sneeze or cough of an infected person. There is no cure for polio. Polio can be prevented by polio vaccine, oral or by injection.

7.4.24 Ring Worm or tinea

Ringworm, also known as dermatophytosis, dermatophyte infection, or tinea, is a fungal infection of the skin. Although it is called ringworm, it is not caused by a worm but by a fungus. Three different types of fungi, namely, Trichophyton, Microsporum and Epidermophyton, which may live as spores in soil, can cause ringworm. Humans and animals can contract ringworm after direct contact with such soil. Infection is common among children who own pets such as cats and dogs. Animals can pass the infection to humans. Infection of the foot is also known as the athlete's foot, which is common among people who walk bare feet. Initial symptoms include red patches on affected areas of the skin, which later may spread to other parts of the body. It may affect the scalp, feet, nails, groin or beard. Diagnosis is done by examining the skin. Transmission happens through a person-to-person contact, animal-to-person contact, through indirect contact with objects and via showers and swimming pools. The infection is treated with antifungal creams, ointments, gels or sprays. If left untreated, ringworm can lead to hair loss and scarring and nail deformities. The infection can be prevented by observing good hygiene and avoiding contact with animals.

7.4.25 Rotavirus infection

Rotavirus infection is caused by Rotavirus A, the most common of 9 species referred to as A, B, C, D, F, G, H, I and J. The virus is transmitted by the faecal-to-oral route through ingestion of contaminated water or food, or contact with contaminated surfaces, hands, and objects. It infects and damages small intestine linings and causes gastroenteritis (which is often called 'stomach flu'). Symptoms include vomiting, severe fatigue, high fever, irritability, dehydration and abdominal pain. Symptomatic infection rates are highest in children under two years of age and decrease progressively towards 45 years of age. Diagnosis of infection with rotavirus A is done by testing stool by enzyme immunoassay. Rotaviruses are highly contagious and cannot be treated with antibiotics or other drugs. Rotavirus infection can be prevented by good personal, food and environmental hygiene and vaccination. WHO recommends vaccination of infants in all the countries. Two of the recommended oral vaccines are RotaTeq and Rotarix. Vaccination, however, does not guarantee 100% immunity but subsequent infections are less severe. According to WHO estimates, 453,000 child deaths occurred in 2008 due to rotavirus gastroenteritis worldwide (www.who.int/ith/diseases/rotavirus/en/).

7.4.26 Salmonella

Salmonella is caused by the salmonella bacteria that live in the intestines of birds, animals and humans. There are many types of Salmonella bacteria, which can cause a range of illnesses including typhoid fever and gastro-enteritis. Symptoms include diarrhoea, which can be bloody, fever, stomach cramps, nausea, vomiting and headache. Food poisoning is closely associated with salmonella. Diagnosis is by testing stool or blood. Transmission is via food including sprouts and other vegetables, eggs, chicken, pork, fruits and processed foods, which may look and smell normal. It can also be transmitted from animals to people and from people to people as well as from water contaminated with the bacteria in contact with the faeces of infected people or animals. Warmer weather makes an ideal condition for salmonella to grow. The infection can be prevented by observing cleanliness in food and water, avoiding contact with poultry and other furry animals.

7.4.27 Schistosomiasis

Schistosomiasis, also called bilharzia or bilharziasis, is caused by small parasitic flatworms (Schistosomatidae), also known as blood flukes. Symptoms include itchy skin, abdominal pain, diarrhoea, bloody stool or blood in the urine. In the long term, patients may also experience liver damage, kidney failure, infertility or bladder cancer. It is one of the most serious parasitic infections affecting some 200 million people yearly in Africa, Asia, South America and the Caribbean. It is mostly prevalent in low-income rural communities in tropical countries where the standard of hygiene is poor. Diagnosis is done by testing urine and faeces samples, blood tests, chest X-rays and ultrasound or CT/MRI scans. Transmission happens by contact with water that contain snails, which act as intermediary hosts. Treatment is done by an oral medicine called Praziquantel. The infection can be prevented by improving access to clean water and reducing the number of snails.

7.4.28 Trachoma

Trachoma is a leading infectious eye disease worldwide caused by the bacterium *Chlamydia trachomatis*. Blindness from trachoma is irreversible. Symptoms include irritation of the eyes with tearing, pain and vision loss, eye discharge, swollen eyelids, misdirected eyelashes, swelling of lymph nodes in front of the ears, sensitivity to bright lights, increased heart rate and ear, nose and throat complications. The infection is diagnosed by examining the eyes and eyelids of the patient. Transmission happens by direct and indirect contact with an infected person's eyes or nose and throat secretions. Indirect transmission may be via clothing and/or flies that have come into contact with such secretions from an infected person's eyes, nose and throat. Poor sanitation, crowded living conditions and lack of access to clean water and toilets increase the transmission rate. The infection can be prevented by improving access to clean water as well as by communal treatment with antibiotics. WHO has developed an elimination strategy summarized by the acronym 'SAFE', which means Surgery for advanced disease, Antibiotics to clear *C. trachomatis* infection, Facial cleanliness and Environmental improvement to reduce transmission. There is no vaccine at the present time but the disease is preventable.

7.4.29 Trichuriasis

Trichuriasis, also known as a whipworm infection, is an infection of the large intestine caused by a parasite called *Trichuris trichiura*. This parasite is known as a whipworm because it resembles a whip and belongs to the soil-transmitted helminths (parasitic worms) that account for a large number of diseases worldwide. It is a tropical disease affecting over 700 million people worldwide. Whipworms live in the large intestine of infected persons and their eggs are passed in the faeces. Infection is caused by ingesting eggs, which may be present in human faeces when used as fertilizers. This can happen when contaminated hands or fingers are put in mouth, or, by consuming vegetables or fruits that have not been carefully cooked, washed or peeled. Mild infections do not show any symptoms but severe infections show symptoms such as frequent watery and sometimes bloody and painful bowel movements. In extreme cases, rectal prolapse (when the rectum sags and comes out of the anus) can also occur. Diagnosis is by stool samples examined under a microscope for whipworm eggs. Transmission is via food and drinks that contain the eggs of the worm. Treatment can be individual using medication or by preventive mass drug administration, especially for children who are at a high risk of infection. Prevention is by observing good hygiene and avoiding contact with soil that may be contaminated with human faeces, including fertilizers containing human faecal matter.

7.4.30 Typhoid fever

Typhoid fever is a disease caused by Salmonella serotype Typhi bacteria. Paratyphoid fever is caused by Salmonella Paratyphi bacteria. Both are life-threatening. Symptoms vary from mild to severe fever accompanied by weakness, abdominal pain, constipation, headache and mild vomiting. The fever is diagnosed by culturing the bacteria or detecting their DNA in the blood, stool or bone marrow. Transmission happens by eating or drinking food or water contaminated with the faeces of an infected person. Humans are the only carriers of this disease. The infection can be prevented by having access to clean water, observing good hygiene and sanitation and as well as by vaccination. Vaccines used currently do not have long-lasting immunity although a longer lasting new typhoid conjugate vaccine, has been pre-qualified by WHO in December 2017 for use in children from the age of six months. Worldwide, typhoid fever affects an estimated 11–21 million people and paratyphoid fever affects an estimated 5 million people each year (www.cdc.gov/typhoid-fever/sources.html). WHO estimates the global annual deaths ranging from 128,000 to 161,000 (www.who.int/news-room/fact-sheets/detail/typhoid). Most casualties are from countries which lack access to safe drinking water and adequate hygiene and sanitation (WASH).

REFERENCES

Podolsky, D. K., Camilleri, M., Fitz, J. G., Kalloo, A. N., Shanahan, F. and Wang, T. C. (2015). *Yamada's Textbook of Gastroenterology*. John Wiley & Sons. p. 2323. ISBN 978-1-118-51215-9.

UNSCEAR (2000). Sources, effects and risks of ionizing radiation. UNSCEAR 2000 report to the General Assembly. United Nations Scientific Committee on the Effects of Atomic Radiation, New York, NY.

World Health Organization (2011). *Guidelines for Drinking Water Quality* (4th Edition) incorporating the first and second addenda, Vol 1, Recommendations, 3rd Edition, Geneva, pp 515.

Chapter 8

Water and food

8.1 INTRODUCTION

Globally, nearly one in nine people, or about 820 million people, are hungry or undernourished, and about 132 million people live with acute hunger that approach starvation. (Malnutrition Is the Leading Cause of Death Globally: Report 2020 (globalcitizen.org)). In sub-Saharan Africa, the percentage of population undernourished is much higher. Even as hunger rises around the world, more people are becoming overweight or obese, with nearly a third of the global population falling into this category, according to the report.

Malnutrition refers to deficiencies, excesses or imbalances in a person's intake of energy and/or nutrients. The different forms of malnutrition include undernutrition, micronutrition-related malnutrition and overweight, obesity and diet-related non-communicable diseases. Undernutrition causes low weight-to-height ratio known as wasting, low height for age known as stunting and low weight for age known as underweight. Micronutrition is caused by the lack of important vitamins and minerals or their excesses. Overweight and obesity cause heart diseases, stroke and diabetes. Undernutrition is usually associated with low income, poor socio-economic conditions, poor maternal nutrition and health. Micro-nutrients (vitamins and minerals) enable the body to produce enzymes, hormones and other substances that are essential for proper growth and development. Their deficiency is prevalent, particularly in children and pregnant women in low-income countries.

Overweight and obesity can be considered as the opposite effect of undernutrition, resulting from too much energy consumed and too little energy expended. It is measured by the body mass index (BMI), which is defined as a person's weight in kilograms divided by the square of his/her height in meters (kg per m^2). A BMI of 25 or more is considered as overweight and a BMI of 30 or more is considered as obesity.

DOI: 10.1201/9781003329206-8

According to WHO (Fact Sheets – Malnutrition (who.int)), there were approximately 462 million underweight adults worldwide and 1.9 billion either overweight or obese in 2014 and around 45% of deaths among children under five years of age were linked to undernutrition.

On April 1, 2016, the United Nations General Assembly proclaimed 2016–2025 the United Nations Decade of Action on Nutrition. Led by WHO and the FAO, the UN Decade of Action on Nutrition calls for policy actions across six key areas that include creating sustainable and resilient food systems for healthy diets, providing social protection and nutrition-related education for all, aligning health systems to nutrition needs, providing universal coverage of essential nutrition interventions, ensuring trade and investment policies to improve nutrition, building safe and supportive environments for nutrition at all ages, and strengthening and promoting nutrition governance and accountability. It also encompasses two sustainable development goals (SDGs), the SDG 2 to end hunger, achieve food security, improved nutrition and promote sustainable agriculture, and SDG 3 to ensure healthy lives and promote well-being for all at all ages. According to a Global Nutritional Report (Inequalities in the global burden of malnutrition – Global Nutrition Report), the mean salt intake in 2017 has been 5.6 g per day and that in 2015, 1.1 billion adults had high blood pressure.

The biological needs of drinking water for a human being are on average about 2–4 litres per day. It takes about 1,000 times as much water to produce the food for a normal diet for a person. Since food is traded across countries, the concept of virtual water has become important. For example, if a country imports 1 million tons of wheat, it is also importing 1 billion m^3 of virtual water.

Water footprint is defined as the total volume of freshwater that is used to produce the goods and services by an individual or a community. Water footprints of animal products are very much larger than those of crop products of equivalent nutritional value. It is reported that 29% of the total water footprint of agricultural products in the world is related to the generation of animal products and that one-third of the global water footprint of animal products is related to beef cattle (Mekonnen and Hoekstra, 2010). Table 8.1 gives the trend in per capita consumption of food in different regions of the world.

8.2 PHOTOSYNTHESIS

Photosynthesis is the biological process of manufacturing the 'food' in plants by combining water with CO_2 in the presence of sunlight. It takes place in the leaves of plants, which contain microscopic cellular organelles known as chloroplasts that contain green-coloured pigment known as chlorophyll

Table 8.1 Per capita food consumption from 1965 to 2030 in kcal per capita per day

	1965	1975	1985	1998	2015	2030
World	2,358	2,435	2,655	2,803	2,940	3,050
Developing countries	2,054	2,152	2,450	2,681	2,850	2,980
Sub-Saharan Africa	2,058	2,079	2,057	2,195	2,360	2,540
Near East/North Africa	2,290	2,591	2,953	3,006	3,090	3,170
Latin America/ Caribbean	2,393	2,546	2,689	2,826	2,980	3,140
South Asia	2,017	1,986	2,205	2,403	2,700	2,900
East Asia	1,957	2,105	2,559	2,921	3,060	3,190
Industrial countries	2,947	3,065	3,206	3,380	3,440	3,500
Transition countries	3,222	3,385	3,379	2,906	3,080	3,180

Source: FAO (2002).

(mainly chlorophyll 'a'), which absorbs energy from sunlight. The chemical reaction of photosynthesis is of the form

$$6CO_2 + 6H_2O \Rightarrow C_6H_{12}O_6 (Glucose) + 6O_2$$
$$(In\ the\ presence\ of\ sunlight)$$

(8.1)

Water is absorbed by the roots and conveyed to the leaves through the xylem whereas CO_2 is absorbed through the stomata located in the leaves. Oxygen is released into the atmosphere as a by-product. The product glucose is sent to the roots, stems, leaves, fruits, flowers and seeds of the plant via the phloem. Glucose is a source of food for plants that provide energy for growth and development. The glucose molecules then combine with each other to form more complex carbohydrates like cellulose and starch. Cellulose is the structural material in plant cell walls.

8.3 SOURCES OF FOOD FOR HUMAN CONSUMPTION

The main staple foods in the world are corn, rice, wheat, potatoes and cassava (tapioca or manioc) of which the cereals corn, wheat and rice are the most popular. Rice is produced and consumed mostly in Asia with China, India, Indonesia and Bangladesh taking the lead in that order. Wheat is popular in the western world while potato is popular in Europe. Cassava is popular in Nigeria.

Of about 50,000 edible plants in the world, only 15 of them provide about 90% of the world's food energy intake. Rice, corn and wheat provide two-thirds of this intake. Other staple food crops include millet and sorghum, potatoes, cassava (Manioc), yams and taro. Table 8.2 gives the most important staple foods in the world including their calorific intake contributions as a percentage of the total calorific intake from all sources.

Table 8.2 Most important staple foods in the world

Rank	Staple food	Share of global calorific intake from all sources
1	Maize corn	19.5%
2	Rice	16.5%
3	Wheat	15.0%
4	Cassava	2.6%
5	Soybeans	2.1%
6	Potatoes	1.7%
7	Sorghum	1.2%
8	Sweet Potato	0.6%
9	Yams	0.4%
10	Plantain	0.3%

Source: What Are the World's Most Important Staple Foods? – WorldAtlas.

8.3.1 Cereals

FAO projects that the annual production of cereals across the world will increase, but the relative importance of rice will decrease in favour of wheat. Only about 50% of the cereals is used as food for humans while about 44% is used as animal feed and the rest is used for other purposes including waste. Oil crops such as palm oil, soybean, sunflower, groundnut, sesame nuts and coconuts account for about 25% of the calorie intake. There is also an increasing contribution to calorie intake by livestock (beef, pork, poultry) and eggs as well as from fisheries (sea food as well as inland fish). Food trade is projected to increase in developing countries.

8.3.1.1 Rice

Worldwide, there are about 79 million hectares of irrigated lowlands with an average yield of about 5 metric tons per hectare, 54 million hectares of rain-fed lowlands with an average yield of about 2.3 metric tons per hectare, 14 million hectares of irrigated uplands with an average yield of 1 metric tons per hectare and 11 million hectares of flood-prone areas with an average yield of 1.5 metric tons per hectare (p. 516, IWMI, 2007). The world annual production of rice is around 550–600 million metric tons (before husking) and the production rate has in recent years been keeping up with the demand. However, with the current population growth, there is no guarantee that the production will meet the expected demand unless the productivity and affordability are increased. A balance needs to be struck between a reasonable return on the investments and efforts of the producers and the affordability of poor consumers. Factors such as droughts, reducing losses from flooded areas due to evaporation, percolation and drainage etc. should be considered in managing the limited water resources. Efficiency of irrigation needs to be increased. China and India are the major producers of rice in the world followed by Indonesia, but the highest per capita consumption of milled rice goes to Myanmar and Vietnam (Table 14.1, IWMI, 2007). Outside of Asia, Brazil is the largest rice producer. Rice grows in warm, wet climates and thrives in waterlogged soil, such as in the flood plains of Asian rivers like the Ganges and the Mekong. 'Deepwater rice' is a variety of rice that is adapted to deep flooding and is grown in eastern Pakistan, Vietnam and Myanmar.

Attempts to increase the productivity of rice with less water include replacing two crops by a single crop with water-efficient irrigation, introducing a pricing structure based on the volumetric consumption of water, improving the irrigation canal system by lining to minimize seepage losses, as well as reusing drainage water. New techniques such as adopting alternate wet and dry systems, which partially or totally suppress the need for ponding the field have been proven to be increasing the water productivity. The Zhangye Irrigation System in the middle reaches of the Yellow River basin has been practising the alternate wetting and drying method in the mid-1990s with decreasing water supply and increasing rice production (Box 14.8, IWMI, 2007). This technique has both advantages and disadvantages. An advantage is the water savings whereas the disadvantage is the high water cost and the high-quality service required as well as the need to have weed control. Whether or not such approaches are feasible depends very much on the community that would be served as well as the political, institutional, social and financial constraints.

Rice is the only cereal that can survive periods of submergence without damage. Rice plants have the capacity to elongate their stems to escape oxygen deficiency when water tables rise. They can also withstand severe droughts. Ecologists have identified five water-related categories of rice plant: rain-fed lowland, deep water, tidal wetland, rain-fed upland and irrigated rice. The water that is applied via irrigation and/or rainfall is used for evapo-transpiration, seepage, percolation, land formation, drainage as well as water needs for seed preparation. Terraced cultivation in mountainous areas helps in preventing soil erosion and landslides and allows the drainage water in upland terraces to be used in lowland terraces, thereby increasing the reuse of water.

Total crop water requirement for rice (paddy) ranges from about 1,100 mm to 1,250 mm with the daily consumptive use varying from about 6 to 10 mm depending upon the agro-climatic conditions. Approximately 3% of the total water requirement (about 40 mm) is used for the nursery, 16% (about 200 mm) for land preparation and 81% (about 1,000 mm) for field irrigation of the crop.

The water requirement at the time of transplanting is about 20 mm, which is maintained for about 7 days after which a submergence depth of about 50 mm is necessary for the development of new roots. This water level is maintained until flowering. Water requirement during ripening is less and water is not necessary after yellow ripening. Water can be drained 15–21 days before harvesting.

According to FAO (factsheet1.pdf (fao.org)), the consumptive use of water for irrigated rice fields ranges from a low of 900 mm to a high of 2,250 mm, which is used up for land preparation (150–250 mm), evapo-transpiration (500–1,200 mm), seepage and percolation (200–700 mm) and mid-season drainage (50–100 mm).

On average, it takes 1,432 litres of water to produce 1 kg of rice in an irrigated lowland production system. The total seasonal water input to rice fields varies from as little as 400 mm in heavy clay soils with shallow ground-water tables to more than 2,000 mm in coarse-textured (sandy or loamy) soils with deep groundwater tables. Around 1,300–1,500 mm is a typical amount of water needed for irrigated rice in Asia. Irrigated rice receives an estimated 34–43% of the total world's irrigation water, or about 24–30% of the entire world's developed freshwater resources. After transplanting, water levels should be around 30 mm initially, and gradually increase to 50–100 mm (with increasing plant height) and remain there until the field is drained 7–10 days before harvest.

(www.knowledgebank.irri.org/step-by-step-production/growth/water-management)

8.3.1.2 Corn

Corn (or maize in Spanish) has been a staple food in North and Central America since pre-historic times. It has now spread to many parts of the world as a grain as well as in other forms such as corn flour, corn syrup, cornflakes, cornmeal and corn oil. Corn remains the most widely grown crop in the United States with about 40% of the world production. Other countries growing large amounts of corn include China, Brazil, Mexico and Argentina. Since corn can be stored in a relatively easy way, it has become a staple food in several countries. Fermented corn is also the source of some alcohol such as whiskey. In addition to the use of corn as a food crop, it has recently been used to produce ethanol, a green fuel that can be used as a substitute for gasoline. In 2007, because of the increasing demand for corn for producing ethanol, a crisis known as 'tortilla crisis' occurred in Mexico where tortilla made from corn is a major staple food item.

Corn needs relatively little water, usually about 25 mm per week applied by drip irrigation. The yield for irrigated corn is generally higher than that for non-irrigated corn. Ideal soil type for growing corn is a well-drained organic soil with a pH of around 6.

8.3.1.3 *Wheat*

For wheat, barley and canola, the minimum water requirement is about 100 mm. From germination to reproductive growth stage where it can produce grain, the water requirement is about 125 mm. In cooler climates the crop water requirements are lower. For example, in Alberta, Canada, the crop water requirements for wheat in the spring varies from 420 to 480 mm whereas in winter, it varies from about 400 to 430 mm.

Wheat grows well in temperate climates, even those with a short growing season. Presently, China, India, United States, Russia and France are among the largest wheat producers in the world.

The majority of bread is made with wheat flour, which is also used in pasta, pastries, crackers, breakfast cereals and noodles. Wheat can be crushed into bulgur, which has a high nutritional value and is often used in soups and pastries in the Middle East. Bread made from wheat is a common staple food around the world. Malt made from wheat is also an ingredient for brewing beer.

8.3.2 Potatoes

Potatoes are native to cold climates and are now a staple food in Europe and parts of the Americas. The leading potato producers are China, Russia, India, United States and Ukraine. Potatoes are the fourth largest food crop after corn, wheat and rice.

In 2010, some 18.6 million hectares of land have been used for potato cultivation in the world with an average yield of about 17.4 tons per hectare. In the same year, the United States had the highest production with a nationwide average yield of 44.3 tons per hectare followed by the United Kingdom. China and India accounted for over a third of the world production with yields of 14.7 and 19.9 tons per hectare, respectively. The big gap in the yield in different economies are due to factors such as crop breed, seed age and quality, crop management practice and the plant environment (Potatoes are the staple food for millions | The Financial Daily)

According to FAO, for high yields, the crop water requirements for a 120- to 150-day crop are 500–700 mm, depending on climate (Potato Water Requirements and Irrigation Systems – Wikifarmer). Irrigation of potato crops includes overhead rain guns, boom irrigation, sprinkler irrigation and drip irrigation. Drip irrigation uses less water than sprinkler irrigation for comparable yield.

8.3.3 Other staple foods

In tropical regions, other staple foods include roots and tubers such as cassava (manioc), yams and taro. Cassava, which has its origin in the Amazon rainforest of South America, is a staple food for over 500 million people in Latin America and Africa. Yams are an important food item in the rainforests of West Africa and taro is a staple food in some of the Pacific islands such as Hawaii, Fiji and New Caledonia and also in West Africa.

8.3.3.1 Vegetables

The water requirements for vegetables have a high degree of variation. For example, for radish, it is about 150 mm, whereas for tomatoes, it is about 610 mm per season. Fruity vegetables such as cucumbers, melons, pumpkins, beans, peas, peppers, sweet corn and tomatoes are sensitive to drought stress during flowering as fruits and seeds develop. Adequate water supply (irrigation or rain) during the period of fruit enlargement is necessary to ensure there will be no fruit cracking and rotting of tomatoes.

8.4 AGRICULTURAL DEMAND FOR FOOD PRODUCTION

The main process of food production for human population in the world is agriculture, and this term also includes livestock husbandry, fisheries and forestry. According to FAO, 777 million people in developing countries do not have access to sufficient and adequate food because they do not have the resources to buy or in case of subsistent farmers to produce it (FAO, 2003).

FAO anticipates that the total irrigated land of some 93 developing countries would be 242 million hectares by 2030. Agricultural water withdrawals are expected to increase by 14% from 2000 to 2030.

It is important to realize that not all the water supplied as irrigation is used for food production. Wastages include evaporation, infiltration, soil evaporation and water used up by non-food crops such as weeds.

Agricultural demand can be considered as the second most important need for water after the biological requirements. It is necessary for survival and sustenance of life. Agricultural demand in future is likely to exceed all other demands.

Food for humans comes directly or indirectly from plants, and water is an essential ingredient in their life. It carries nutrients to cells and wastes from cells. In the natural environment, water requirement for plant life is provided by rainwater. But in places where the rainfall is low, improved crops can be made to grow with the provision of irrigation. Irrigation has been practiced from as far back as 4000 BC in countries like Egypt, Syria, Persia, India, Sri Lanka in the eastern hemisphere and about 2,000 years ago in countries such as Mexico and Peru in the western hemisphere.

Irrigation requires large amounts of water and, in most cases, it cannot be reused except through the natural processes of the hydrological cycle. Irrigation requirements vary from crop to crop. Some require flooding, for example, rice; others require only sprinkling. Rice needs flooding to about 150–200 mm. For an area of one hectare, flooding with 200 mm requires 2,000 m³. Typical requirements for some crops are as follows:

Sugar beet	1,000 tons per ton during the growing season
Wheat	1,500 tons per ton during the growing season
Rice	4,000 tons per ton during the growing season

Irrigation requirements excepting the gardening done in individual houses do not come under 'water supply engineering'. The quality of irrigation water need not be as high as for domestic water.

The highest consumption of agricultural water is in Asian and African countries whereas the highest consumption for industrial use is in Europe and America. Asian countries use about 2,556 km³ per year, which is the highest agricultural use in the world. On a percentage basis, water withdrawal for agriculture in Asia and Africa is 81%, Americas 48% and Europe 25% whereas for industrial use, Europe has the highest percentage at 54%, Americas 37%, Asia 10% and Africa 4% (Bhagwat, 2019).

8.5 SOURCES OF WATER FOR AGRICULTURE

8.5.1 Rainfall

At any instant of time, the atmosphere contains about 13×10^3 km³ of water as vapour, liquid or solid. Precipitation is the process by which this water is deposited on the earth's surface. The two main forms of precipitation are rainfall and snowfall. Other minor forms of precipitation include mist, hail, sleet, dew and frost. The effect of rainfall, which is liquid precipitation, is immediate whereas that of snowfall which is in solid form is slow and attenuated.

There are many stages of the precipitation process. Firstly, the air must contain the necessary water vapour. A water vapour particle undergoes various phases and physical changes before precipitation takes place. The water vapour must first be carried to upper levels where expansion and cooling take place. When the temperature has reached the dew point, condensation will take place releasing the latent heat of condensation to an otherwise adiabatic process. Cloud formation will take place with nucleation around impurities in water vapour. Droplets coalesce with other droplets forming raindrops, which are large enough to cause precipitation. The stages of precipitation consist of lifting, expansion and cooling, nucleation, condensation and raindrop formation. The details of these processes are described in several textbooks on hydrology including one by the author (Jayawardena, 2021).

Monsoons bring large amounts of rainfall. The world's highest recorded annual rainfall of 26,470 mm, and an average annual rainfall of about 12,000 mm, have been in Cherrapunji (25°15'N, 91°44'E) in Northeast India, which also has a monthly record of 9,300 mm. This rainfall is brought about by the South-West (SW) monsoon. In India, approximately 70% of the annual rainfall takes place during the SW monsoon (June to September). In temperate climates, rainfalls are much lower and in higher latitudes, precipitation is mainly in the form of snowfall.

Although rainfall is the most natural source of water for agriculture, it has a high degree of variability in space and time and therefore cannot be taken as a guaranteed source. In places where rainy season lasts for a reasonable period of time (at least three months) in the year, rain-fed agriculture is widespread. They include permanent crops such as rubber, tea and coffee as well as seasonal crops such as wheat, maize and rice. In rural communities in developing countries, rain-fed agriculture is the main, if not the only, means of production of their staple food. It accounts for more than 95% of farmed land in sub-Saharan Africa, 90% in Latin America, 75% in the Near East and North Africa, 65% in East Asia and 60% in South Asia (Summary – Rain-fed Agriculture: IWMI (cgiar.org)). Water productivity in rain-fed agriculture is generally low and the losses due to evaporation, infiltration and runoff are high. For staple foods, only one crop per year is the norm in rain-fed agriculture. It is also a fact that rain-fed agriculture is practiced mainly by poor farmers at a subsistence level. Crop failures can occur due to drought, floods, early or late arrival of monsoon in monsoon-based rain and inefficient use of water leading to food insecurity in poor communities. There are no inputs and/or incentives from governments, NGOs and research communities that could enhance the productivity from rain-fed farming. The only way the farmers and/or communities can expect an increase in yield is by increasing the farmed land area. No attempts have been made to increase the water productivity and the productivity of the land and/or the crop.

High yields have been observed in several temperate regions with reliable rainfall and productive soil and even in humid and sub-humid tropical regions with commercially managed rain-fed agriculture. For example, it has been reported that rain-fed agriculture production managed commercially exceeds 5–6 tons per hectare whereas in dry sub-humid and semi-arid regions, it has been between 0.5 and 2 tons per hectare with an average of 1 ton per hectare in sub-Saharan Africa, and 1–1.5 tons per hectare in South Asia, Central and West Asia and North Africa (Rockstrom and Falkenmark, 2000; Wani et al., 2003a, b).

8.5.2 Irrigation

Irrigation has a positive impact on alleviating poverty, which is the root cause of malnutrition. It has been reported that in India, 69% of people in unirrigated districts are poor whereas only 26% are poor in irrigated districts (World Bank, 1991). More details of irrigation are presented in Chapter 9.

8.6 WATER PRODUCTIVITY

The productivity of water in agriculture is generally quantified using the crop water productivity (CWP), which varies with the crop type, evapotranspiration as well as with climatic condition, soil type and several other factors. The main crops, which end up as staple food in the world, are rice, wheat and maize, which are all cereals. Of these, rice consumes more water than other cereals and thus the water productivity of rice is significantly lower than those of other cereals. Rice has a worldwide average growing season of 136 days from crop emergence to harvest, including mountain and lowland rice. Reported values of CWP vary from region to region and from source to source too. Based on data for the period 1998–2008, a global average value of CWP for rice is reported to be 0.98 kg per m^3 (Bastiaanssen and Steduto, 2016) whereas a high value of 2.2 kg per m^3 has been reported for rice crop in the Zhangye irrigation system in China (Dong et al., 2001 as cited by Bastiaanssen and Steduto, 2016). The lowest CWP is in sub-Saharan Africa with values ranging from 0.10 to 0.25 kg per m^3. In the United States, values range from 0.9 to 1.9 kg per m^3 with higher values in the north than in the south and the highest in the north-western regions (Cai and Rosegrant, 2003). The projected values of water productivity and water use for the period 2021–2025 for rice in developing countries have been reported to be 0.53 kg per m^3 and 8,445 m^3 per hectare. The corresponding figures in developed countries are 0.57 kg per m^3 and 9,730 m^3 per hectare, respectively (Cai and Rosegrant, 2003).

Other indicators of water productivity in agriculture include the yield per unit area and the water consumption per unit area. In sub-Saharan Africa, which has the lowest values in the world, the average yield for rice is 1.4 tons per hectare and the water consumption is close to 9,500 m^3 per hectare. In the developed world, the yield and water consumption for rice are 4.7 tons per hectare and 10,000 m^3 per hectare, respectively, whereas in the developing world the corresponding figures are 3.3 tons per hectare and 8,600 m^3 per hectare (Cai and Rosegrant, 2003).

In a study by Cai and Rosegrant (2003), the projected values of three indicators (basin efficiency, water withdrawals and irrigation consumptive use) for the period 2021–2025 under different scenarios (baseline, high efficiency and high efficiency with low water withdrawal) have been reported. Of these, the basin efficiencies in developing countries are 0.59 as baseline, 0.77 for high basin efficiency and 0.77 with high basin efficiency and low water withdrawal, respectively, whereas the corresponding figures in developed countries are 0.69, 0.81 and 0.81, respectively. The water withdrawals in developing countries are 3,486 km^3, 3,347 km^3 and 3,043 km^3, respectively, as baseline, high basin efficiency and with high basin efficiency and low water withdrawal, respectively, whereas the corresponding figures in developed countries are 1,277, 1,228 and 1,183 km^3, respectively. The irrigation consumptive use in developing countries are 1,214, 1,135 and 283 km^3, respectively, whereas in developed countries, the corresponding figures are 274, 250 and 227 km^3, respectively (Cai and Rosegrant, 2003).

The challenge therefore is to increase water productivity in food production. Any saving in the water needs for food production could be used for other water needs. It is reported that a 1% increase in water productivity in food production generates a potential of water use of 24 litres per capita per day (www.fao.org/3/y4525e/y4525e06.htm). Increasing the water productivity in the agriculture sector appears to be the best way of freeing water for other purposes. This can be achieved by minimizing the outflow from paddy fields, reuse of any outflows and drainage from fields, adopting aerobic irrigation in place of flooded irrigation, reducing evaporation from bare soil and reducing seepage and percolation. Aerobic irrigation reduces the yield but is compensated by less water requirement compared with flooded irrigation. Wet seeding is another method of increasing water productivity. In this method, farmers soak the rice seeds for 24 hours and then sow them in puddled or muddy fields. By this method, which is popular in parts of Thailand, Vietnam and the Philippines, the water requirements are about 20–25% less than for traditional methods. Intermittent irrigation, where the fields are flooded and allowed to dry and flooded again, is a practice adopted in China where the per capita freshwater availability is among the lowest in Asia. Land levelling where the slopes are reduced to improve uniform field conditions, thereby requiring less water to flood, is another approach.

Using pricing policies to increase water productivity requires governmental intervention and even so it does not seem feasible at least in the Asian context as it would be difficult to recover the true cost of providing water for agriculture. In the Indian subcontinent, it has been suggested that the charge required to affect demand significantly would be about ten times the cost of operation and maintenance of irrigation systems. It is also not feasible to charge agricultural water users on a volumetric basis because of the high cost of installing water meters as well as maintaining them. This applies to surface waters as well as groundwater, which require energy for pumping.

In some regions, crop substitution has been used as a means of increasing water productivity. Farmers tend to grow crops that require less water but have higher market values such as flowers and vegetables, which can also be exported. Such practices can conflict with national policies, which aim at self-sufficiency in staple foods. Aiming at long-term sustainability of water and food security should be the way forward.

Methods of improving water productivity include shifting from surface irrigation to drip irrigation, reducing evaporation, which is high in rain-fed systems with low plant densities, water delivery management as well as using biotechnology crop breeding targeting rapid early growth, breeding drought-resistant varieties, breeding varieties resistant to disease, pest and salinity and reducing agrochemical inputs.

8.7 ARABLE LAND AVAILABILITY FOR FOOD PRODUCTION

The earth's land surface area is only 29% of which 71% is habitable area. Of the habitable area, 50% is used for agriculture. Of this 50% agriculture area, 37% is forest, 11% shrubs and grassland, 1% water surface area and the remaining 1% built-up urban area (Land Use – Our World in Data). The arable land in developing countries is projected to increase by 13% from 1998 to 2030, but the per capita arable land availability has decreased and will decrease in the future too because of the population explosion. However, it has been compensated by the increased yield. Worldwide, the arable land needed to produce a fixed quantity of crop in 2014 has decreased by 70% relative to 1961 (Land Use – Our World in Data). The world average agricultural land as a percentage of the total land area in 2018 has been 36.9% with the highest of 80% in Kazakhstan and lowest of 0.5% in Suriname (Agricultural land (% of land area) | Data (worldbank.org)).

FAO estimates that the total irrigated land of some 93 developing countries would be 242 million hectares in 2030 and that agricultural water withdrawals would increase 11% from 2000 to 2030 (www.fao.org/3/1927 8EN/i9278en.pdf).

8.8 CROP WATER EVAPO-TRANSPIRATION

8.8.1 Transpiration

Transpiration consists of the vaporization of liquid water contained in plant tissues and the vapour removal to the atmosphere. Crops predominately lose their water through stomata. These are small openings on the plant leaf through which gases and water vapour pass. The water, together with some nutrients, is taken up by the roots and transported through the plant. The vaporization occurs within the leaf, namely in the intercellular spaces, and the vapour exchange with the atmosphere is controlled by the stomatal aperture. Nearly all water taken up is lost by transpiration and only a tiny fraction is used within the plant.

Transpiration, like direct evaporation, depends on the energy supply, vapour pressure gradient and wind. Hence, radiation, air temperature, air humidity and wind terms should be considered when assessing transpiration.

The soil water content and the ability of the soil to conduct water to the roots also determine the transpiration rate as do water-logging and soil water salinity. The transpiration rate is also influenced by crop characteristics, environmental aspects and cultivation practices. Different kinds of plants have different transpiration rates. Not only the type of crop, but also the crop development, environment and management should be considered when assessing transpiration.

8.8.2 Evapo-transpiration (ET)

Evaporation and transpiration occur simultaneously and there is no easy way of distinguishing between the two processes. Apart from the water availability in the topsoil, the evaporation from a cropped soil is mainly determined by the fraction of the solar radiation reaching the soil surface. This fraction decreases over the growing period as the crop develops and the crop canopy shades more and more of the ground area. When the crop is small, water is predominately lost by soil evaporation, but once the crop is well developed and completely covers the soil, transpiration becomes the main process. At sowing, nearly 100% of ET comes from evaporation, while at full crop cover more than 90% of ET comes from transpiration.

8.8.3 Reference evapo-transpiration (ET_0)

The reference evapo-transpiration is the evaporating power of the atmosphere at a specific location and time. It does not consider soil factors. Although there are several empirical equations to estimate ET_0 such as the Blaney–Criddle equation, radiation method, Penman equation etc., the FAO recommends the Penman–Monteith equation as the sole method of estimating ET_0. This is based on careful comparative analysis of the different methods of estimating reference evapo-transpiration by the Committee on Irrigation Water Requirements of the American Society of Civil Engineers and a consortium of European research institutes. This method explicitly incorporates physical and aerodynamic parameters. The calculation requires radiation, air temperature, air humidity and wind speed data. ET_0 can also be estimated from pan evaporation using an empirical pan coefficient. Two types of pans have been widely used, the class A pans and the sunken pans.

In this context, the reference surface is a hypothetical grass reference crop with an assumed crop height of 0.12 m, a fixed surface resistance of 70 s m^{-1} and an albedo of 0.23. The reference surface closely resembles an extensive surface of green, well-watered grass of uniform height, actively growing and completely shading the ground. This is considered as standard conditions. The fixed surface resistance of 70 s m^{-1} implies a moderately dry soil surface resulting from about a weekly irrigation frequency.

The FAO Penman–Monteith equation, which is derived from the original Penman–Monteith equation, and the equations of the aerodynamic and surface resistance take the forms (www.fao.org/3/X0490E/x0490e06.htm)

$$ET_0 = \frac{0.408\Delta(R_n - G) + \gamma\dfrac{900}{T_a + 273}u_2(e_s - e_a)}{\Delta + \gamma(1 + 0.34u_2)} \quad \text{mm per day} \tag{8.2}$$

where R_n is the net radiation at the crop surface (MJ per m² per day), G is the soil heat flux density (G is positive when the soil is warming and negative when the soil is cooling; MJ per m² per day), e_s is the saturation vapour pressure (kPa), e_a is the actual vapour pressure (kPa), Δ is the slope of the saturation vapour pressure deficit versus temperature (kPa per °C), T_a is the mean daily air temperature (°C), γ is a psychrometric constant (kPa per °C) and u_2 is the wind speed at 2 m above the surface (m per s). The input variables are averaged over a 24-hour time scale. The factor 0.408 converts energy in MJ per m² per day to equivalent evapo-transpiration in mm per day. The original Penman–Monteith equation and the resistance equations are of the form

$$\lambda ET = \frac{\Delta(R_n - G) + \rho_a c_p\left(\dfrac{e_s - e_a}{r_a}\right)}{\Delta + \gamma\left(1 + \dfrac{r_s}{r_a}\right)} \tag{8.3}$$

where λ is the latent heat of vaporization, ρ_a is the mean air density at constant pressure, c_p is the specific heat of the air, r_s is the surface resistance and r_a is the aerodynamic resistance. The aerodynamic resistance r_a is given as

$$r_a = \frac{ln\left[\dfrac{z_m - d}{z_{om}}\right]ln\left[\dfrac{z_h - d}{z_{oh}}\right]}{k^2 u_z} \quad \text{s per m} \tag{8.4}$$

where z_m is the height of wind measurements [m], z_h is the height of humidity measurements [m], d is the zero-plane displacement height [m], z_{om} is the roughness length governing momentum transfer [m], z_{oh} is the roughness length governing transfer of heat and vapour [m], k is the von Karman's constant (= 0.41), and u_z wind speed at height z [m per s]. The surface (bulk) resistance r_s is given as

$$r_s = \frac{r_1}{LAI_{active}} \quad \text{s per m} \tag{8.5}$$

where r_l is the bulk stomatal resistance of the well-illuminated leaf [s per m], and LAI_{active} is the active (sunlit) leaf area index [m^2 (leaf area) per m (soil surface)].

The crop evapo-transpiration under standard conditions is given as

$$ET(crop) = k_c ET_0 \tag{8.6}$$

where k_c is an empirical crop coefficient.

8.8.4 Crop water requirements

FAO defines crop water requirements as the depth of water needed to meet the water loss through evapo-transpiration (ET_{crop}) of a disease-free crop, growing in large fields under non-restricting soil conditions including soil water and fertility and achieving full production potential under the given growing environment. US Soil Conservation Service (USDA) defines it as the amount of water required to compensate the evapo-transpiration loss from the cropped field (USDA Soil Conservation Service, 1993). It is the amount of water that needs to be supplied through irrigation and rain and is equal to the crop evapo-transpiration. Crop water requirement is the sum total of

- transpiration losses through leaves,
- evaporation loss through soil surface in cropped area,
- amount of water used by plants for its metabolic activities,
- conveyance losses, percolation losses, runoff losses etc. and
- water required for puddling operations, ploughing operations, land preparation, leaching, weeding etc.

As it is difficult to measure crop evapo-transpiration (ET_{crop}) for different types of crops, it is estimated by relating to the reference evapo-transpiration values (ET_0) obtained from the FAO Penman–Montheith equation (Eq. 8.2) and a single crop coefficient (k_c) as in Eq. 8.7a, or two coefficients for each type of crop as in Eq. 8.7b:

$$ET(crop) = k_c ET_0 \tag{8.7a}$$

or,

$$ET(crop) = (k_{cb} + k_e)ET_0 \tag{8.7b}$$

In Eq. (8.7b), k_{cb} is a basal crop coefficient to represent plant transpiration and k_e is a soil evaporation coefficient.

Equations (8.7a, 8.7b) are considered as under standard surface conditions referred to above. Under non-standard conditions, this equation is modified by introducing a water stress coefficient k_s or by adjusting the k_c values for all kinds of other stresses and environmental constraints on crop evapo-transpiration as

$$ET(crop_{adjusted}) = k_s k_c ET_0 \qquad (8.7c)$$

The water stress coefficient k_s is estimated from

$$k_s = 1 - \frac{b}{100k_y}(EC_e - EC_t) \qquad (8.8)$$

where EC_e is the mean electrical conductivity of the saturation extract for the root zone (dS per m), EC_t is the electrical conductivity of the saturation extract at the threshold of EC_e when crop yield first reduces below Ym (deci Siemens per m, dS per m), b is the reduction in yield per increase in EC_e (% per (dS per m)); (for most natural waters, 1 dS per m is equivalent to about 640 mg per litre salts), k_y is a yield response factor, and Ym is the maximum expected yield when $EC_e < EC_t$. Both k_s and k_c are dimensionless and ET_{crop} is expressed in mm per day. The values of k_c and k_s differ depending on the salinity conditions in the water.

The amount of water required to compensate the loss by evapo-transpiration from a particular crop is known as the crop water requirement for that crop. The crop coefficient values relate to evapo-transpiration of a disease-free crop grown in large fields under optimum soil water and fertility conditions and achieving full production potential under the given growing environment.

It is difficult to separate transpiration, evaporation losses through soil surface and water used for metabolic activities. Therefore, crop water requirement is taken as the evapo-transpiration loss, which depends upon the location, crop type, climate and agronomical management practice. Location factors include soil type, soil texture, topography and soil chemical composition. Crop factors include the type of crop, stage of growth, crop population and the crop-growing season. Climatic factors include temperature, sunshine hours, relative humidity, wind velocity and rainfall. Agronomical management factors include method, frequency and efficiency of irrigation, tillage, weeding, mulching etc.

Typical values of seasonal crop evapo-transpiration values for some selected crops are given in Table 8.3.

The values of crop coefficients depend upon the season, weather conditions such as humidity, wind as well as the stage of growth of the crop. For rice, it ranges from a value of 1.1 to 1.35 depending upon the climatic condition, wind and stage of maturity of the crop.

Table 8.3 Typical values of seasonal crop evapo-transpiration for some selected crops

Type of crop	Range of values of crop evapo-transpiration (mm)
Rice	500–950
Potatoes	350–625
Maize	400–750
Grains (small)	300–450
Onions	350–600
Sugar cane	1,000–1,500
Vegetables	250–500
Soybeans	450–825
Alfalfa	600–1,500

Source: FAO (1996). Crop water requirements, FAO Irrigation and drainage paper 24, M-56 ISBN 92-5-100279-7.

The water requirements for each crop are calculated on the basis of satisfying the crop evapo-transpiration rate of a disease-free crop growing in large fields under optimal soil conditions, which depend upon climate, growing season, stage of growth of the crop and agricultural and irrigation practices. Although the quantitative values of the two are the same, their physical interpretation is different. Crop water requirement refers to the amount of water that needs to be supplied by rainfall, soil water, groundwater and irrigation. The irrigation water also includes additional water required for leaching of salts and other losses in the application of irrigation.

The total growing period of a crop consists of an initial stage that spans from sowing or transplanting until the crop covers about 10% of the ground, a development stage spanning from the end of the initial stage until full grown cover has been reached, which is about 70–80% ground cover, a mid-season stage spanning from the end of development stage to until maturity and a late season stage spanning from the end of the mid-season stage to until the end of the harvesting time. For some crops such as lettuce and cabbage, the crop water requirements remain the same until harvesting stage since such vegetables need to be fresh at the time of harvesting. On the other hand, dry harvested crops such as cotton and maize, the water needs during the late season are minimal. No irrigation is needed during the late stage.

More details including sample calculations of crop water requirements for different type of crops under specific conditions can be found in: www.fao.org/3/U3160E/u3160e04.htm

8.9 WATER FOR LIVESTOCK

Water is essential for life and is a major factor in livestock rearing and management. Water is necessary for the digestion and absorption of nutrients such as carbohydrates, proteins and fats as well as for the removal of body wastes. As most animals are made up of 60–70% water, they need large quantities of safe drinking water every day. In the case of dairy cows, water accounts for about 80% of the milk produced. A dairy cow needs about 15–25 gallons of water per day. It is also a fast and easy medium for administering medication for sick animals. Water requirements for livestock depend upon the type of animal, age, production level, moisture, energy and protein contents of the feed and the ambient temperature. With the changing dietary habits in the developing world, the demand for meat and meat products is likely to increase in the future. With the dwindling fresh-water resources, the competition for water for livestock industry is likely to increase.

The total water requirements for livestock farming include water consumed by the animals, the irrigated or rainwater needed for grazing land and feed preparation, the water needed for cleaning up the animals and their living habitats. For drinking water, the quality is also important although it can be lower than that for human consumption. It needs to be relatively clean and free from toxic materials, which may include high salinity (salts dissolved in water), inorganic elements (calcium, magnesium, arsenic, barium, fluoride, sulphates, nitrates etc.), organic wastes, pathogens, herbicides and pesticides. Water with high salinity is not suitable for animal feed. The limits of salinity as measured by the electrical conductivity (EC) should ideally be less than 1.5 dS per m (deci Siemens per m) and tolerable upto about 5 dS per m. Water with EC values greater than 15 dS per m is unfit for drinking by animals. To prevent health problems in animals, it is important to ensure that the drinking water qualities are within the toxicity guidelines for various toxic substances (Water quality for agriculture (fao.org)). Under natural conditions, the concentrations of these substances pose no danger to animals, but human activities can increase the concentrations of some substances that can become toxic to animals. Tables 8.4 and 8.5 give some suggested limits for magnesium and guideline values for toxic substances in drinking water.

The virtual water content of animal products is calculated based on the virtual water content of their feed and the volumes of drinking and service

Table 8.4 Suggested limits for magnesium in drinking water for livestock

Livestock	Suggested limits for magnesium in drinking water for livestock (mg per litre)
Poultry	<250
Beef cattle	400
Milk cows	250
Sheep (dry feed)	500

Source: Water quality for agriculture (fao.org).

Table 8.5 Guideline values for toxic substances in livestock drinking water

Toxic substance	Upper limit (mg per litre)
Arsenic	0.2
Boron	5.0
Cadmium	0.05
Chromium	1.0
Copper	0.5
Fluoride	2.0
Lead	0.1
Manganese	0.05
Mercury	0.01
Nitrates + Nitrites ($NO_3 + NO_2$)	100
Selenium	0.05
Vanadium	0.1
Zinc	24

Source: Water quality for agriculture (fao.org). Adapted from National Academy of Sciences and National Academy of Engineering. 1972.

Table 8.6 Average virtual water contents of selected meat products in developed country production systems

Type of meat	World average virtual water content (litres per kg)
Beef	15,497
Pork	4,856
Goat meat	4,043
Sheep meat	6,143
Chicken meat	3,918
Eggs	3,340
Milk	990

Source: Adapted from Hoekstra and Chapagain (2007).

waters consumed during their lifetime. The average virtual water contents in some meat products are given in Table 8.6.

8.10 WATER FOR FISHERIES

Fish is an important source of animal protein. Although the main habitat for fish is the ocean, communities living in landlocked countries or regions resort to fresh water fish by aquaculture in rivers, streams, reservoirs and even small ponds. Fish production in such areas is in small scale and mainly for local consumption. Many factors, some of which interact with others, determine the quality of water in aquaculture for the growth and survival of fish. They include temperature, nitrogenous waste (excreted by fish),

Table 8.7 Preferred ranges of concentrations for water in fish culture

Water quality parameter	Preferred ranges for fish culture (mg per litre)
Temperature	Species dependent
Dissolved oxygen	< 4
Total ammonia nitrogen	< 0.02
Nitrite	< 1; < 0.1 in soft water
pH	6–8
Alkalinity	50–300 Calcium Carbonate
Hardness	< 50 Calcium Carbonate
Iron	< 0.5
Chlorine	< 0.02
Hydrogen sulphide	No detectable level
Clarity	Species dependent

Source: Buttner et.al. (1993).

alkalinity, hardness, pH, carbon dioxide, salinity, iron, chlorine, hydrogen sulphide and water clarity. Table 8.7 gives the preferred ranges for water in fish culture.

8.11 WATER FOR FOOD PROCESSING

Food processing industry uses a large amount of water either as an ingredient, or for cooking ingredients and for cooling cooked food. In addition, water is used for cleaning purposes, dilution of detergents, sanitizers, detergents and also for rinsing off cleaning chemicals. High-quality potable water is essential in the food processing and catering industry. As water is a vehicle of transmission of diseases, it is important to ensure that the quality of water used in the food processing and catering industry is of highest value. Although water authorities and private companies provide water of good quality, poorly managed water systems can become polluted with microorganisms including faecal pathogens. In addition, water supplies can be subject to chemical contamination too. Contaminants in water are finally passed on to food that people consume.

Food and catering industry are spread worldwide. Different countries have different standards of potable water quality. A widely accepted guideline is that which meets the requirements of the WHO Guidelines for Drinking Water Quality. In the context of water in food processing, potable water may be defined as water that is wholesome or will not affect the wholesomeness of the food.

The Private Water Supplies Regulations 2009 apply in England and came into force on January 1, 2010. These regulations apply in relation to private supplies of water intended for human consumption; and for these purposes, water intended for human consumption means all water either in its original state or after treatment, intended for drinking, cooking, food preparation or other domestic purposes, regardless of its origin and whether it is supplied from a distribution network, from a tanker, or in bottles or containers; all water used in any food-production undertaking for the manufacture, processing, preservation or marketing of products or substances intended for human consumption.

Since water is a universal solvent, in the context of food handling, it is used to remove dirt and associated non-food material using water and detergent chemicals. The first step is flushing with water to remove visible dirt and non-food material. Cleaning also includes washing of equipment including personnel who handle food.

Diverse uses of water in food production and processing such as irrigation and livestock watering, as an ingredient in food products, washing and cleaning of food, heating, cooling and cleaning have different water quality requirements. Treatment processes to reduce microbial contamination may sometimes have residual effects that may impact water quality in food production and processing facilities. This effect should be taken into consideration when assessing the deterioration of drinking water quality in food production and processing.

Table 8.8 gives the water requirements per unit of some types of food.

8.12 REUSE OF AGRICULTURAL WATER

Water is a finite and precious resource. Wherever possible, it should be reused over and over to minimize the need to look for new resources although drainage water is normally of inferior quality compared to the

Table 8.8 Water consumption in food and agriculture

Product	Unit	Water required in m^3 per unit
Cereals	kg	1.5
Pulses	kg	1.0
Citrus fruit	kg	1.0
Palm oil	kg	1.0
Fresh poultry	kg	6.0
Fresh lamb	kg	10.0
Fresh beef	kg	15.0
Sheep and goat	Per head	500.0
Cattle	Per head	4,000.0

Source: Adapted from FAO (2003). Agriculture, Food and Water. Food and Agriculture. Organization of the United Nations, Rome, Italy www.fao.org/docrep/006/Y4683E/y4683e07.htm#P0_0 (accessed 8.11.18.).

original irrigation water. In agricultural areas where irrigation water is scarce, drainage water is an important supplementary resource. In addition to supplementing irrigation water, reuse can also help to alleviate drainage disposal problems by reducing the volume of disposable drainage water involved, thereby minimizing the overall problems of water pollution. Drainage water can also be used to grow salt-tolerant crops, in wildlife habitats and wetlands and for initial reclamation of salt-affected lands.

Salinity in drainage water is an important water quality parameter that needs to be kept to a minimum when reused for traditional agricultural purposes. To maximize the crop yield, soil salinity is reduced by leaching. Drainage water from reused water is usually highly concentrated and should be carefully disposed. Although salinity in water or soil reduces crop growth but low salinity, especially among crop species such as cotton or the halophytic sugar beet, may actually improve crop production. As salinity increases beyond some threshold tolerance, decline in yield is inevitable. Visible symptoms of the presence of high salinity in soil water include burning or scorching of leaves or shoots of crops as well as longer time to crop maturity. Sugar beet, sugar cane, dates, cotton and barley are among the most salt tolerant whereas beans, carrots, onions, strawberries and almonds are considered sensitive to salinity.

The type of irrigation also affects the effect of salinity on crops. For example, in the case of sprinkler irrigation, the effects come from the soil as well as from salts absorbed by wetted leaves. Drip irrigation, on the other hand, avoids wetting the leaves. Also, since drip irrigation is applied frequently, there is a continuous leaching of soil from which the plants absorb water. Thus, drip irrigation gives the best advantage when using saline water.

Another approach of using drainage water is by blending drainage water with good-quality irrigation water. It is considered as most economical provided that the blended water is sufficiently low in total salinity and toxicity although it has the potential to increase the salinity of groundwater in the long term.

Case studies where drainage water has been used include the reuse of over 4,000 million m³ of agricultural drainage water from the upper Nile delta in Egypt used to supplement irrigation in the lower delta and the agriculture-forestry system in which drainage water is used and disposed into a solar evaporator in California in the United States. The agriculture-forestry system consists of the sequential flow of water from salt-sensitive crops to salt-tolerant trees to more tolerant halophytes and finally to a solar evaporator.

Guidelines for the reuse of agricultural drainage water have been compiled by several institutes including the US Food and Drug Administration (USFDA), US Environmental Protection Agency (EPA), World Health Organization (WHO), the European Union (EU) and the United Nations (UN).

8.13 FOOD SECURITY

Food security, as defined by the United Nations' Committee on World Food Security, means that all people, at all times, have physical, social, and economic access to sufficient, safe and nutritious food that meets their food preferences and dietary needs for an active and healthy life. With the world population projected to reach about 10 billion by 2050, the demand for food is expected to be about 60% greater than it is today. The United Nations Sustainable Development Goal 2 (SDG2) aims at ending hunger, achieving food security and improved nutrition, and promoting sustainable agriculture by the year 2030. Food in sufficient quantity and quality is a basic human right just as water. It is estimated that over 420,000 people die and some 600 million people fall ill after eating contaminated food (www.fao.org/3/ca4289en/CA4289EN.pdf).

Food insecurity is caused by several factors including lack of physical availability, weather and climatic factors, economic affordability and utilization such as selection, preparation and storage of food. There can also be food insecurity due to conflicts. Internally displaced people due to political conflicts are the most vulnerable to food insecurity. Consequences of food insecurity can lead to preventable deaths due to diseases, mental disorders and malnutrition.

Possible ways of addressing food insecurity include increasing the food productivity, which can be accomplished by increasing the arable land, increasing the crop frequency, increasing the yield and increasing food trade. In rural communities, the only practical option is to increase the extent of arable land. In addition, researchers, governments, NGOs and international institutions can and should provide advice and assistance to food producers, especially farmers, on how the yield per unit arable area can be increased. It is also important to introduce new species of crops based on biotechnology that are resistant to weather and other negative influences.

REFERENCES

Bastiaanssen, W. G. M. and Steduto, P. (2016). The water productivity score (WPS) at global and regional level: Methodology and first results from remote sensing measurements of wheat, rice and maize, *Sci. Total Environ.*, http://dx.doi.org/10.1016/j.scitotenv.2016.09.032

Bhagwat, V. R. (2019). Safety of water used in food production. In: *Food Safety Human Health*. Chapter 9, 219–247. Academic Press. Published online 2019, August 9. doi: 10.1016/B978-0-12-816333-7.00009-6

Buttner, J. K., Soderberg, R. W. and Terlizi, D. E. (1993). An Introduction to Water Chemistry in Freshwater Aquaculture, NRAC Fact Sheet No. 170-1993, 4 pp.

Cai, Ximing and Rosegrant, Mark W. (2003). World water productivity: current situation and future options. In *Water Productivity in Agriculture: Limits and Opportunities for Improvement* (J. W. Kijne, R. Barker and D. Molden, eds). Wallingford, UK: CAB International, pp. 163–178.

Dong, B., Loeve, R., Li, Y. H., Chen, C. D., Deng, L., Molden, D. (2001). Water productivity in the Zhangye Irrigation System: issue of scale. In: Barker, et al. (Eds.), *Water Saving for Rice*, pp. 97–115. Proceedings of an international workshop held in Wuhan, China, March 23–25, 2001.

FAO (1996). Crop water requirements, FAO Irrigation and drainage paper 24, M-56 ISBN 92-5-100279-7) www.fao.org/3/s8376e/s8376e.pdf

FAO (2002). *World Agriculture: Towards 2015/2030, A FAO Study*. Rome: FAO.

FAO (2003). *Agriculture, Food and Water*. A Contribution to the World Water Development Report, ISBN 92-5-104943-2, pp 61.

Hoekstra, A. Y. and Chapagain, A. K. (2007). Water footprints of nations: water use by people as a function of their consumption pattern. *Water Resources Management*, 21: 35–48.

International Water Management Institute (2007). In *Water for Food and Water for Live: A Comprehensive Assessment of Water Management in Agriculture* (David Molden , ed.). London: Earthscan. pp. 645.

Jayawardena, A. W. (2021). *Fluid Mechanics, Hydraulics, Hydrology and Water Resources for Civil Engineers*, London: CRC Press, Taylor and Francis Group, pp. 894.

Mekonnen, M. M. and Hoekstra, A. Y. (2010). A global and high-resolution assessment of the green, blue and grey water footprint of wheat. *Hydrol Earth Syst. Sci.* 14(7): 1259–1276. CrossRefGoogle Scholar

National Academy of Sciences and National Academy of Engineering (1972). *Water Quality Criteria*. Washington DC: United States Environmental Protection Agency, Report No. EPA-R373-033. 592 p.

Rockstrom, J. and Falkenmark, M. (2000). Semiarid crop production from a hydrological perspective: Gap between potential and actual yields. *Critical Rev. Plant Sci.*, 19(4): 319–346.Google Scholar

USDA Soil Conservation Service (1993). Irrigation water requirements. *National Engineering Handbook*, Part 623, chapter 2, National Technical Information Service.

Wani, S. P., Pathak, P., Jangawad, L. S., Eswaran, H. and Singh, P. (2003a). Improved management of Vertisols in the semi-arid tropics for increased productivity and soil carbon sequestration. *Soil Use Mgmt* 19: 217–222.

Wani, S. P., Pathak, P., Sreedevi, T. K., Singh, H. P. and Singh, P. (2003b). Efficient management of rainwater for increased crop productivity and ground-water recharge in Asia. In *Water Productivity in Agriculture: Limits and Opportunities for Improvement* (J. W. Kijne, R. Barker and D. Molden, Eds.). Wallingford, UK: CAB International, Colombo, Sri Lanka: International Water Management Institute (IWMI), pp. 199–215.

World Bank (1991). *India Irrigation Sector Review*, vols 1 and 2, Washington DC: World Bank.

Chapter 9

Water and irrigation

9.1 INTRODUCTION

Irrigation played an important role in human civilization. Egypt, Mesopotamia, India and China have been the cradles of the so-called hydraulic civilization, which flourished along the banks of rivers. An irrigation system consists of an intake structure or a pumping station, which directs water from the source, a conveyance system that transports the water from the main intake structure to the field, a distribution system that transports the water to the irrigated fields, a field application system that assures the transport of water within the fields and a drainage system to remove excess water from the field. Watering plants in the garden is the simplest form of irrigation.

The canals that transport water may be rectangular, triangular, trapezoidal, circular and irregular in cross-section and may be earthen or lined. They may have structures such as drops and chutes when there are abrupt changes of slope in canals. Flumes, culverts and siphons are used when water has to be carried across gullies, ravines, roads and/or other natural depressions. Weirs (rectangular, triangular, trapezoidal) and Parshall flumes are used to measure flow rates. A drainage system is necessary in any irrigation system because plants (except rice) cannot withstand saturated soil for long periods. Plant roots need air and water for their health.

Irrigation has a positive impact on alleviating poverty, which is the root cause of malnutrition. It has been reported that in India, 69% of people in unirrigated districts are poor whereas only 26% are poor in irrigated districts (World Bank, 1991).

9.2 TYPES OF IRRIGATION

There are many types of irrigation systems. The most widely used type is surface irrigation where the water is distributed to the land by gravity. Other types include localized irrigation where water is distributed under low pressure to each plant through a pipe network; drip irrigation, which is also localized,

DOI: 10.1201/9781003329206-9

where water is distributed at or near the plant roots from a higher level; sprinkler irrigation where the water is distributed under high pressure from overhead sprinklers; center pivot irrigation where the water is distributed by a system of sprinklers that move in a circular pattern; lateral move irrigation where water is distributed by a series of pipes connected to wheels and sprinklers, which move across the field in short distances; sub-irrigation where water is distributed by raising the water table through pumping stations; lift irrigation in which water is pumped from a lower elevation and manual irrigation where water is distributed across the land manually.

An indicator of irrigation efficiency is the water use efficiency (WUE), which is defined as

$$WUE = \frac{Dry\ matter\ or\ crop\ yield}{Water\ consumed\ in\ evapotranspiration}$$

9.2.1 Ponded irrigation

Ponding, or flooding, has several advantages. It can contribute towards flood control since the bunds that border the field have sufficient capacity to store water during times of heavy rain. It also recharges the groundwater through percolation. Ponding also prevents weed development.

9.2.2 Basin irrigation

A basin in this context is a field enclosed by bunds (dikes) to contain the water. Channels and/or pipelines convey the water into the fields.

9.2.3 Border irrigation

In border irrigation, there are no bunds (dikes) to contain the water. Instead, water is allowed to drain from the high elevation end to the low elevation end. This method is suitable for sloping fields.

9.2.4 Furrow irrigation

In furrow irrigation, the water is directed along channels within the field. By controlling the flow in each channel, the farmer can control the amount of water in different parts of the field.

9.2.5 Overhead irrigation

In this type, water under pressure is applied from an overhead sprinkler system. This method is suitable for irrigating gardens but not suitable for irrigating rice fields.

9.2.6 Sub-surface irrigation

In this method, a sub-surface water distribution system provides the water to the root zone through upward water movement. It is suitable during the early stages of growth of the plant.

9.2.7 Drip irrigation

In this method, water is applied near the base of the plant leaving the upper part of the foliage dry. It is one of the efficient methods of micro-irrigation as it has the advantage of lower evaporation. Water is applied to the roots and stems from above or from below through holes in pipes.

9.2.8 Lift irrigation

In normal irrigation, the flow of water takes place by gravity. In situations where the source of water is at an elevation below that of the destination, the flow has to be assisted by some external force. The external force is provided by pumps, which lift the water from a lower elevation to a higher elevation. It is the only way groundwater can be used for irrigation, but there are cases where surface waters have also been carried to higher elevations to be used as lift irrigation. Multi-stage submerged pumps are at the heart of most lift irrigation systems.

The advantages of lift irrigation include minimal land acquisition problems and low water losses. During drought years, groundwater lift irrigation may be the only means of providing crop water requirements. The disadvantage is the external power requirement. Lift irrigation is practised in many countries. Currently, the Kaleshwaram Lift Irrigation Project in Telangana state in India has the world's largest multi-stage lift irrigation system where the Godavari River water is lifted some 618 meters from its source at Medigadda to the last reservoir. The system consists of 19 pumping stations, 1,531 km of gravity canals, 203 km of underground canals, 98 km of pressure mains and 20 reservoirs designed to irrigate 1.82×10^6 hectares and provide 70% of state drinking water in addition to providing the industrial needs. It has the capacity to store 141×10^9 m^3 by lifting water at the rate of about 0.057×10^9 m^3 per day for 90 flood days (www.kaleshwaram project.com/).

The San Joaquin Valley in California is the largest agricultural region in the United States and a major contributor to the food supply in the country. Early irrigation in the valley, starting at the end of the 19th century, was confined to gravity diversions from the San Joaquin River, which later on, from the 1920s, resorted to lift irrigation from groundwater. With severe competition for water, over exploitation of the shallow unconfined or partially confined aquifers has led to several undesirable consequences such as land subsidence and soil salinization. The water table in some places has declined by more than 30 meters and the maximum land subsidence near Mendota has been reported to be more than 7 meters. The Central Valley of California, which includes the San Joaquin Valley, the Sacramento Valley and the Sacramento-San Joaquin Delta, produces about 25% of the nation's table food in only 1% of the country's farmland (SanJoaquin Valley (usgs. gov)).

9.2.9 Micro-irrigation

Micro-irrigation is the slow application of water on, above or below the soil by surface drip, subsurface drip, bubbler and micro-sprinkler systems. The following features are common characteristics of micro-irrigation systems:

- Water is applied at a low rate.
- Water is applied over a long period.
- Water is applied at frequent intervals.
- Water is applied near or into the root zone.
- Water is applied by a low-pressure delivery system.
- Water is routinely used to transport fertilizers and other agricultural chemicals (Lamm et al., 2007).

Among many other benefits, the capability to save water has been a major reason for many countries to select micro-irrigation as a promising modern irrigation technology for several decades. According to the census conducted by the International Commission on Irrigation and Drainage (ICID, 2018), the total area of micro-irrigation in the world has increased 25 times reaching 14.4 million hectares, equivalent to 6.2% of the total irrigated area (231.9 million hectares), during the past 30 years.

9.3 IRRIGATION REQUIREMENTS

The net irrigation requirements are calculated using a simple water balance equation such as:

Irrigation requirement = ET_{crop} – Rainfall – Groundwater contribution – Soil water storage at the beginning of the chosen time period

The chosen time period can be seasonal, monthly or any other time period. The irrigation requirements are expressed as a depth of water (mm). If more than one crop is cultivated in the land, the sum of the individual irrigation requirements for each crop would be the total irrigation requirement. It is also important to make allowance for leaching requirements, which is the portion of the irrigation requirement that must be applied to the active root zone to remove accumulated salts. Crops grown in tropical regions with windy and sunny conditions need high crop water requirements, whereas

Table 9.1 Lengths of growing season and crop water needs during the growing season

Crop	Typical growing season (days)	Crop water needs during the growing season (mm)
Banana	300–365	1,200–2,200
Citrus	240–365	900–1,200
Sugarcane	270–365	1,500–2,500
Barley/Oats/Wheat	120–150	450–650
Maize	125–180	500–800
Potato	105–145	500–700
Alfalfa	100–365	800–1,600
Rice	90–150	450–700
Onion	75–95	350–550
Radish	35–45	

Source: Extracted from Tables 6 and 14 of Chapter 3: Crop Water Needs (fao.org).

those grown in cool humid areas with little or no wind need lower crop water requirements. Thus, the same crop grown in different climatic zones will have different crop water requirements. In cooler climates, the crop water requirements are lower. Table 9.1 gives lengths of typical growing season and the crop water needs during the growing season for some selected crops.

Irrigation requirement can therefore be taken as the difference between the crop water requirement and the sum of effective rainfall, soil moisture and groundwater contributions. Quantitatively, it is the depth of water (usually in mm) that should be applied to a crop to satisfy the specific crop water requirement. Allowance must also be made for irrigation losses since all the water supplied does not end up as crop water requirement.

Irrigation applied for wheat in the Nile delta, Middle East and Upper Egypt regions ranges from about 500 mm to about 800 mm and for maize in the same regions range from about 700 mm to about 1,100 mm (www.researchgate.net/profile/Khaled-Abdellatif-2/publication/326698071_Water_Requirements_for_Major_Crops/links/5b600f79458515c4b2545 26d/Water-Requirements-for-Major-Crops.pdf?origin=publication_detail).

9.4 RANKING OF IRRIGATED LAND IN THE WORLD

Irrigation is practised almost all over the world. Even providing water to potted plants in household gardens is a type of irrigation, but the main purpose of irrigation is to grow food crops. In this respect, China, the most populous country in the world, has the largest irrigated area in the world followed by India, the United States, Pakistan and Iran. The ranking of irrigated land in the world is shown in Table 9.2. The total irrigated land in

Table 9.2 Ranking of irrigated land in the world

Rank	Country/Region	Irrigated area (km²)	Date of information
1	China	690,070	2012
2	India	667,000	2012
3	USA	264,000	2012
4	Pakistan	202,000	2012
-	European Union	154,540	2011
5	Iran	95,530	2012
6	Indonesia	67,220	2012
7	Mexico	65,000	2012
8	Thailand	64,150	2012
9	Brazil	54,000	2012
10	Bangladesh	53,000	2012
11	Turkey	52,150	2012
12	Vietnam	46,000	2012
13	Russia	43,000	2012
14	Uzbekistan	42,150	2012
15	Italy	39,500	2012
16	Spain	38,000	2012
17	Egypt	36,500	2012
18	Iraq	35,250	2012
19	Afghanistan	32,080	2012
20	Poland	32,000	2012
20	Rumania	31,490	2012
21	France	26,420	2012
22	Peru	25,800	2012
23	Australia	25,500	2012
24	Japan	24,690	2012
25	Argentina	23,600	2012
26	Myanmar	22,950	2012
27	Ukraine	21,670	2012
28	Kazakhstan	20,660	2012
29	Turkmenistan	19,950	2012
30	Sudan	18,900	2012
31	South Africa	16,700	2012
32	Philippines	16,270	2012
33	Saudi Arabia	16,200	2012
34	Greece	15,550	2012
35	Ecuador	15,000	2012
36	Morocco	14,850	2012
37	North Korea	14,600	2012
38	Syria	14,280	2012
39	Azerbaijan	14,277	2012
40	Nepal	13,320	2012
41	Chile	11,100	2012
42	Colombia	10,900	2012
43	Madagascar	10,860	2012
44	Venezuela	10,550	2012
45	Kyrgyzstan	10,233	2012
46	Canada	8,700	2012
46	Cuba	8,700	2012

Table 9.2 Cont.

Rank	Country/Region	Irrigated area (km²)	Date of information
47	South Korea	7,780	2012
48	Tajikistan	7,420	2012
49	New Zealand	7,210	2012
50	Yemen	6,800	2012
52	Algeria	5,700	2012
52	Sri Lanka	5,700	2012
53	Portugal	5,400	2012
54	Netherland	4,860	2012
55	Libya	4,700	2012
56	Tunisia	4,590	2012
57	Denmark	4,350	2012
58	Georgia	4,330	2012
59	Taiwan	3,820	2012
60	Malaysia	3,800	2012
61	Mali	3,780	2012
62	Cambodia	3,540	2012
63	Guatemala	3,375	2012
64	Israel	3,350	2012
65	Albania	3,310	2012
66	Laos	3,100	2012
67	Dominican Republic	3,070	2012
68	Bolivia	3,000	2012
69	Nigeria	2,930	2012
70	Ethiopia	2,900	2012
71	Armenia	2,740	2012
72	Uruguay	2,380	2012
73	Moldova	2,283	2012
74	Somalia	2,000	2012
75	Nicaragua	1,900	2012
76	Tanzania	1,840	2012
77	Zimbabwe	1,740	2012
78	Hungary	1,721	2012
79	Sweden	1,640	2012
80	Zambia	2,900	2012
81	Guyana	1,430	2012
82	Paraguay	1,362	2012
83	Macedonia	1,280	2012
84	Senegal	1,200	2012
85	Mozambique	1,180	2012
86	Austria	1,170	2012
87	Belarus	1,140	2012
88	Lebanon	1,040	2012
89	Kenya	1,030	2012
90	Bulgaria	1,020	2012
91	Costa Rica	1,015	2012
92	South Sudan	1,000	2012
92	Niger	1,000	2012

(continued)

Table 9.2 Cont.

Rank	Country/Region	Irrigated area (km²)	Date of information
93	Germany	965	2012
94	Jordan	964	2012
95	Guinea	950	2012
95	Serbia	950	2012
95	United Kingdom	950	2012
96	United Arab Emirates	923	2012
97	Honduras	900	2012
97	Norway	900	2012
98	Slovakia	869	2012
99	Angola	860	2012
100	Mongolia	840	2012
101	Malawi	740	2012
102	Cote d'Ivoire	730	2012
103	Finland	690	2012
104	Switzerland	630	2012
105	Oman	590	2012
106	Surinam	570	2012
107	Burkina Faso	550	2012
108	Swaziland	500	2012
109	Cyprus	460	2012
110	El Salvador	452	2012
111	Mauritania	450	2012
112	Timor-Leste	350	2012
113	Ghana	340	2012
114	Panama	321	2012
115	Bhutan	320	2012
115	Czech Republic	320	2012
116	Chad	300	2012
116	Sierra Leone	300	2012
117	Cameroon	290	2012
118	Guinea Bissau	250	2012
118	Jamaica	250	2012
119	Croatia	240	2012
119	Palestine	240	2012
120	Belgium	230	2012
120	Benin	230	2012
120	Burundi	230	2012
120	Puerto Rico	220	2012
121	Eritrea	210	2012
122	Mauritius	190	2012
123	Uganda	140	2012
124	Qatar	130	2012
125	Democratic Republic of Congo	110	2012
126	Kuwait	105	2012
127	New Caledonia	100	2012
127	Sao Tome and Principe	100	2012
128	Rwanda	96	2012
129	Namibia	80	2012
130	Togo	70	2012

Table 9.2 Cont.

Rank	Country/Region	Irrigated area (km²)	Date of information
130	Trinidad and Tobago	70	2012
131	Slovenia	60	2012
132	Barbados	50	2012
132	Gambia	50	2012
133	Lithuania	44	2012
134	Bahrain	40	2012
134	Estonia	40	2012
134	Fiji	40	2012
134	Gabon	40	2012
135	Belize	35	2012
135	Cape Verde	35	2012
135	Malta	35	2012
136	Bosnia and Herzegovina	30	2012
136	Lesotho	30	2012
136	Liberia	30	2012
136	Saint Lucia	30	2012
137	Montenegro	24	2012
138	Botswana	20	2012
138	Republic of Congo	20	2012
138	Grenada	20	2012
139	Latvia	12	2012
140	Bahamas	10	2012
140	Brunei	10	2012
140	Central African Republic	10	2012
140	Djibouti	10	2012
140	French Polynesia	10	2012
140	Hong Kong	10	2012
140	St. Vincent and the Grenadines	10	2012
141	St. Kitts and Nevis	8	2012
142	Seychelles	3	2012
143	Guam	2	2012
144	Antigua and Barbuda	1.3	2012
144	Comoros	1.3	2012
145	Northern Marianna Islands	1	2012
145	US Virgin Islands	1	2012
	World	3,242,917	2012

Source: List of countries by irrigated land area – Wikipedia.

the world as of 2012 is about 3,242,917 km², which is approximately 2% of the land area of the world.

9.5 MAJOR IRRIGATION PROJECTS IN THE WORLD

Irrigation schemes are classified as major, medium and minor. A major scheme is one which has a culturalble command area (CCA) greater than 10,000 hectares, a medium scheme is if the CCA is in the range 2,000–10,000 hectares and minor is if the CCA is less than 2,000 hectares. In this context, CCA is the area in which a crop is grown at a particular time or crop season. Gross command area (GCA) is the total area that can be economically irrigated from an irrigation scheme without considering limitations of water. It includes cultivating land as well as roads, wastelands, forests, barren lands etc. In other words,

GCA = CCA + Uncultivable area

Irrigation schemes can also be classified as gravity type where the source of water to irrigation fields is at a higher elevation and therefore the flow is under gravity and lift irrigation where the source of water is at a lower elevation and therefore has to be lifted by pumping. The latter applies mainly to irrigation using groundwater from aquifers.

Since the capital expenditure for constructing major irrigation projects is quite high, they are usually combined with power generation, flood control, domestic water supply etc. Most large-scale projects are multi-purpose projects.

The largest irrigation project in the world, known as the Great Man-Made River (GMR), has been constructed in Libya where fossil water trapped in the Nubian Sandstone Aquifer system with an estimated 373×10^{12} m^3 of groundwater is pumped to provide irrigation and other water needs for most of Libya's urban centres. The aquifer system, which covers some 2×10^6 km^2, is the largest fossil freshwater reservoir in the world. This 'man-made river' covers a distance of up to 1,600 km and provides 70% of all freshwater used in Libya (Great Man-Made River – Wikipedia). The project, which started in 1985, is expected to be completed by 2030. When completed, it will irrigate more than 141,640 hectares of arable land. The system consists of 2,820 km of underground pipes, over 1,300 wells most of which are over 500 m deep supplying 6.5×10^6 m^3 freshwater per day to the urban cities including Tripoli and Benghazi and described as the 'Eighth Wonder of the World' by the late Libyan leader Muammar Gaddafi (Great Man-Made River – Wikipedia). Since the aquifer system lies beneath a desert, it is not rechargeable but believed to be so vast that it will take hundreds of years to deplete it.

Brief descriptions of some of the major irrigation projects in the world are given below.

9.5.1 China

Ancient agricultural developments in China started during the Neolithic period (circa 8,000 BC) in the Yellow and Yangtze River regions, followed by North-Eastern and North-Western regions including Gansu and Xinjiang provinces and the South-Eastern and South-Western regions. The Dujiangyan

Irrigation System in Sichuan province in Southwest China dating back some 2,200 years is the oldest irrigation system that is still providing water for over half a million hectares of land, drains flood waters and provides water for more than 50 cities in the province.

Although China's agricultural output is the largest in the world, only 10% of its total land area can be cultivated, which is approximately 10% of the arable land in the world despite China supporting over 20% of the world population. Most of the rice, which is the staple food in China, is grown in Huai River basin, Zhu Jiang delta (Pearl River) and in Yunnan, Guizhou and Sichuan provinces. Wheat, the second most prevalent grain crop, is grown on North China Plain, the Wei and Fen River valleys on the Loess plateau and Jiangsu, Hubei and Sichuan provinces whereas corn and millet are grown in North and Northeast China. Soybeans, from which tofu (bean curd) is made, are grown in North and Northeast China whereas the popular mandarin oranges are grown along and south of Yangtze River. Other food crops include potatoes, fruits and vegetables. Tropical fruits are grown in Hainan Island and apples and pears in Liaoning and Shandong provinces, and white potatoes in the north (China is the largest producer of potatoes in the world). Non-food crops such as cotton are grown predominantly in Xinjiang Uygur Autonomous Region using drip irrigation.

China's irrigated land in 2015 (over 66×10^6 hectares) has been greater than three times the irrigated land area in the United States. In 2013, the water use for irrigation has been 340×10^9 m^3, which is equivalent to the average annual flow in Pearl River (Agriculture in China – Wikipedia). It is also projected that an additional 18×10^6 hectares of land will adopt water saving irrigation technology by 2030.

In world ranking, China in 2018 was the largest producer of the staple foods rice, wheat, potato and the second largest producer of maize (second only to the United States) as well as the largest producer of several types of vegetables and fruits. It was also the third largest producer of cotton, after India and the United States.

Since water is becoming a scarce resource, drip irrigation and micro-irrigation have been adopted in several countries in recent times. It is the slow application of water on, above or below the soil by surface drip, sub-surface drip, bubbler and micro-sprinkler systems over a long period of time at a low pressure and applied near or onto the root zone. The main advantage of micro-irrigation is saving water. In China too, there has been a rapid expansion of micro-irrigation during the recent decades from 1970.

According to the data published by the ICID in 2016, the micro-irrigation area in China ranked first in the world. According to the more recent data published by the Year Book China Water Resources in 2016, micro-irrigation has been used in all provinces in China, while 62%, 28% and 10% of the total micro-irrigation area are distributed in the arid, semi-arid to sub-humid and humid regions, respectively.

Rapid advances in micro-irrigation systems and practices have occurred since drip irrigation was adopted from the middle of the 1970s in China. Micro-irrigation has been playing an increasingly important role in China.

Several weirs, canals and polder embankments in China have been recognized by the ICID as World Heritage Irrigation Structures.

9.5.2 India

In India, there are many major, medium and minor irrigation projects spread over the country, many of which are multi-purpose projects. A partial list of major irrigation projects is shown in Table 9.3. Their distributions, state-wise and basin-wise, are shown in Tables 9.4 and 9.5, respectively.

In India as a whole, the percentage of cultivated land under irrigation is about 48% with the highest in Punjab state with 98%. The average productivity in India is 1.9 tons per hectare, whereas in Punjab, it is 4.2 tons per ha (https://en.wikipedia.org/wiki/Irrigation_in_India).

Table 9.3 Major irrigation projects in India

Project name	River	State	Culturable command area (ha)	Year of completion
Bhakra Nangal Project	Sutlej	Punjab and Himachal Pradesh	4,000,000	1963
Beas Project	Beas	Punjab, Haryana and Rajasthan	2,100,000	1974
Indira Gandhi Canal	Harike (Sutlej and Beas)	Punjab	528,000	1965
Kosi Project	Kosi	Bihar and Nepal	848,000	1954
Hirakkud Project	Mahanadi	Odisha	1,000,000	1957
Tungabhadra project	Tungabhadra–Krishna	Andhra Pradesh–Karnataka	574,000	1953
Nagarjuna Sagar Project	Krishna	Andhra Pradesh	1,313,000	1960
Chambal Project	Chambal	Rajasthan and Madhya Pradesh	515,000	1960
Damodar valley project	Damodar	Jharkhand, West Bengal	823,700	1948
Gandak project	Gandak	Bihar–Uttar Pradesh	1,651,700	1970
Kakrapar project	Tapti	Gujarat	151,180	1954
Koyna Project	Koyna–Krishna	Maharashtra	172,470	1964
Malprabha project	Malprabha	Karnataka	218,191	1972
Mayurakshi Project	Mayurakshi	West Bengal	240,000	1956
Kangsabati project	Kangsabati and Kumari	West Bengal	348,477	1956

Source: http://ecoursesonline.iasri.res.in/mod/page/view.php?id=1950

Table 9.4 State-wise major and medium irrigation projects

State	Number of major and medium irrigation projects
Andhra Pradesh	95
Assam	20
Bihar	58
Chhattisgarh	49
Goa	3
Gujarat	139
Haryana	14
Himachal Pradesh	8
Jammu and Kashmir	34
Jharkhand	127
Karnataka	101
Kerala	23
Madhya Pradesh	150
Maharashtra	389
Manipur	9
Nagaland	1
Odisha	83
Punjab	17
Rajasthan	136
Tamil Nadu	73
Telangana	61
Tripura	3
Uttar Pradesh	106
Uttarakhand	12
West Bengal	33

Source: https://edurev.in/studytube/Major-Irrigation-and-Power-Projects-Geography/23401d6b-7087-45b3-a687-aa9423a7d3ac_t

The Bhakra-Nangal multi-purpose river valley project, which is a joint venture of the states of Haryana, Punjab and Rajasthan, is also the largest irrigation project in India. It comprises of a gravity dam across Sutlej River at Bhakra, Nangal dam downstream of Bhakra dam, a lined Nangal Hydel channel, two power houses at Bhakra dam and two power houses at Nangal dam at Ganguwal and Kotla and the Bhakra irrigation canal. The reservoir formed by the auxiliary Nangal dam serves for balancing daily fluctuations from Bhakra dam. The Bhakra irrigation canal system commands a gross area of about 2.7×10^6 hectares and provides irrigation to about 1.5×10^6 hectares of which 37.7% is in Punjab, 46.7% in Haryana and the remaining 15.6% in Rajasthan. In a period of 25 years from 1963 to 1988, the reservoir capacity has decreased from $7,438 \times 10^6$ m^3 to $6,784 \times 10^6$ m^3 due to silting (Bhakra Nangal Project of India (with interesting facts) (yourarticlelibrary.com)).

Table 9.5 Basin wise irrigation projects

Basin	Number of major and medium irrigation projects
Indus (Up to border)	69
Ganges (Ganga)	480
Brahmaputra	22
Barak and others	6
Godavari	286
Krishna	211
Cauvery	57
Subarnarekha	38
Brahmani and Baitarni	44
Narmada	44
Mahanadi	76
Mahi	39
Pennar	21
Tapi	81
West flowing rivers from Tapi to Tadri	28
West flowing rivers from Tadri to Kanyakumari	31
East flowing rivers between Mahanadi and Pennar	55
East flowing rivers between Pennar and Kanyakumari	47
West flowing rivers of Kutch and Saurashtra including Luni	108
Area of inland drainage in Rajasthan	12
Minor rivers draining into Myanmar (Burma and Bangladesh)	7

Source: https://edurev.in/studytube/Major-Irrigation-and-Power-Projects-Geography/23401d6b-7087-45b3-a687-aa9423a7d3ac_t

The Kaleshwaram multi-purpose project built across Godavari River in Telangana state, India, that lifts water to a height of over 500 m is considered as the largest lift irrigation project in the world. It is designed to irrigate 1.82×10^6 hectares for two crops in a year and provide 70% of domestic and industrial needs of the state of Telangana by lifting the waters of Godavari River. The project, which comprises 1,832 km of water supply routes, 1,531 km of gravity canals, 330 km of tunnel routes, 20 lifts, 19 pump houses and 20 reservoirs, with a total capacity of 141×10^9 m^3 started in 2016 and was completed in 2019. It is claimed to be the biggest of its kind (Explained: What is Telangana's Kaleshwaram water project? | Explained News, *The Indian Express*).

9.5.3 The United States

Although the United States is a relatively new country, traces of early irrigation go back as far as 1200 BC in the desert and plains of modern-day

Table 9.6 Percentages of irrigated areas in 13 leading states in the United States

State	Percentage of area irrigated
Nebraska	14.9
California	14.1
Arkansas	8.6
Texas	8.0
Idaho	6.0
Kansas	5.2
Colorado	4.5
Montana	3.4
Mississippi	3.0
Washington	2.9
Oregon	2.9
Florida	2.7
Wyoming	2.6
All other states	21.2

Source: USDA, Economics Research Services using USDA National Agricultural statistics service, 2012, Census of Agriculture, State data.

Arizona, Colorado and New Mexico. Archaeologists believe that a small community of Indigenous people that lived near the present-day Tucson had a network of canals providing irrigation water to small fields.

In 2015, a total of 25.698×10^6 hectares have been irrigated of which 14.043×10^6 hectares (55%) were with sprinkler irrigation, 9.429×10^6 hectares with surface irrigation and 2.222×10^6 hectares with micro-irrigation systems. In the same year, the total irrigation withdrawals were 447×10^6 m^3 per day that accounted for 42% of total freshwater withdrawals. Of this, withdrawals from surface water sources were 230×10^6 m^3 per day (52% of total withdrawals) whereas those from groundwater sources were 216×10^6 m^3 per day (48%) (Dieter et al., 2018).

In terms of land irrigated in 2012, Nebraska occupied the top place with 3.36×10^6 hectares, followed by California with 3.2×10^6 hectares, Arkansas with 1.94×10^6 hectares and Texas with 1.82×10^6 hectares. Table 9.6 gives the percentages of land irrigated in the 13 leading irrigation states.

The major food crops in the United States are corn, which accounted for roughly 25%, and soybeans, which accounted for about 29% of total US irrigated land harvested in 2012. Hay and other forage productions accounted for up to 18% of harvested irrigated land. Wheat, soybeans, orchards and vegetables are the second largest group of irrigated crops in the West, with crop land shares ranging from 7.0% to 9.8%. Nationally, other crops accounting for a significant share of total harvested irrigated areas include soybeans (14%), vegetables and orchard crops each (8%), cotton (7%), wheat (7%) and rice (5%).

9.5.4 Pakistan

Of a total land area of some 80 million hectares, the cultivated area is about 22 million hectares, of which 19 million hectares are under irrigation. Under the Indus Basin Irrigation System (IBIS), Pakistan has the largest contiguous irrigation system in the world consisting of three major reservoirs formed by Mangla dam, Tarbela dam and Chasma dam, 19 barrages, 44 canal systems with a total length of some 56,000 km, 107,000 water courses with a total length of some 1.6 million km and over 550,000 tube wells abstracting over 51 billion m^3 of groundwater annually (Shah (pdf)).

About 77% of the irrigated land is in Punjab, about 20% in Sindh and Baluchistan and about 3% in the Khyber Pakhtunkhwa (KPK) in the North-West Frontier along with Afghanistan–Pakistan border. Within the Indus Basin in Punjab, there are 13 barrages, 23 major canal systems, some 34,500 km of canals serving an area of 8.58 million hectares and 135 surface drainage systems with a total length of 6,600 km within the 23 major canal systems draining an area of some 5.8 million hectares.

The Indus River, which originates from the Himalayan mountain, runs through four countries and empties into the Arabian Sea near the port city of Karachi. Of a total basin area of 1.12 million km^2, 47% is within Pakistan, 39% within India, 8% within China and 6% within Afghanistan. Since the partition of India and Pakistan in 1947, the Indus River has been a bone of contention among the four countries through which it runs. The Indus Water Treaty brokered by the World Bank (then known as the International Bank for Reconstruction and Development) and signed by the then prime minister of India Jawaharlal Nehru and the then president of Pakistan Ayub Khan on September 19, 1960, paved the way for an equitable sharing of the waters of Indus River. Under this agreement, India got control over the three eastern rivers Ravi, Beas and Sutlej while Pakistan got control over three western rivers Indus, Chenab and Jhelum. The treaty gives 20% of Indus water to India and the remaining 80% to Pakistan. However, the implementation of the treaty was not without problems.

9.5.5 Iran

Iran with 89,930 km^2 of irrigated land ranks as the fifth largest country in terms of irrigated area. It has a long history of irrigation dating back to about 3,000 years. The well-known 'qanat' system of underground canals that convey water from aquifers in highlands to valleys below originated in Iran (Persia) can also be found in several other countries nowadays. Notable among them is the qanat system in Grape Valley, a cool and humid oasis situated near Turpan city in Xinjiang Province, which lies below the Flaming Mountain, which is the hottest place in China with average temperatures in the region of 50°C and surface temperature of 70°C where several types of

Table 9.7 Climate conditions in Iran

Main Climate	Area (km²)	Percentage
Extra arid	573,884	35.54
Desert arid	472,562	29.15
Semi arid	325,109	20.08
Mediterranean	80,007	4.90
Semi humid	55,097	3.37
Humid	58,006	3.36
Very humid	55,386	3.39
Total	1,620,051	100

Source: Microsoft PowerPoint – Poverty2011_3 Farahani.pptx (icidonline.org).

fruits including grapes are grown. Other countries where the qanat system is used include Pakistan, Iraq, Syria, Yemen, Spain and North America. The qanat system enabled the Persian farmers to provide irrigation to grow their crops successfully despite the prevailing dry weather in the region.

Qanat tunnels running from the mountainous alluvial fans to the valleys below were constructed manually. Vertical shafts just big enough for a person to work have been hand-dug at intervals of 20–30 m to remove excavated material and to provide ventilation. In some cases, there are underground reservoirs that store water during nighttime for use in the daytime. Qanat system prevented surface pollution and evaporation and provided water not only for irrigation but also for domestic consumption supplying about 75% of the water used in the country. The system consists of some 22,000 qanat units extending to over 273,500 km of underground channels delivering at the rate of about 500 m³/s to the plateau of Iran including Baluchistan and Afghanistan. In 2016, UNESCO inscribed the Persian Qanat as a World Heritage Site, listing 11 qanats.

Iran is a country with arid climatic conditions and very little rainfall. Water is therefore a precious resource distributed unequally in space. Rainfall is relatively high near the Caspian Sea (average about 1,280 mm per year) but low in the Central Plateau and in the lowlands to the south (seldom exceeding 100 mm per year) (*Source*: Ministry of Energy, Iran Water Resources Management Company). Tables 9.7 and 9.8 illustrate the climatic conditions and the rainfall distribution in the country.

9.5.6 Indonesia

Irrigation has been practiced in Indonesia since ancient times. Of particular importance is the Subak system in Bali Island, which is still being practiced. It is a system that has been established to meet the farmers' demands through a togetherness approach among its members. The system has been

Table 9.8 Rainfall distribution

Annual rainfall (mm)	Country area (%)
> 1000	0.40
500–1,000	9.14
300–500	29.23
150–300	32.20
< 150	29.30

Source: Microsoft PowerPoint – Poverty2011_3 Farahani.pptx (icidonline.org).

Table 9.9 Distribution of irrigation systems in Indonesia

Authority	Area (hectares)	Number of systems
Central government	2,682,897.8	244
Provincial governments	556,780.30	349
District governments	442,888.62	3,338

Source: Ministry of Public Works (2010).

designed to irrigate small areas of the order of 100 hectares for paddy cultivation. Subak system has been the way of life in Bali for hundreds of years
and has been developed to meet the farmers' evolutionary demands and
circumstances through a long-range practice. Through the 'harmonious-
togetherness' principle, the system has sustained from generation to generation without major changes.

Following the construction of the first canal called 'Oosterlokkans' by
the colonial Dutch government in the 18th century, the modern irrigation
practice started in the 19th century to harness the waters of Brantas River
in East Java. Of about 17,000 islands of which only about 1,000 are hab-
itable, Java and Bali islands have well-developed irrigation whereas other
islands have not developed to the same extent. Practically all the irrigation
works are designed to supply water to the rice fields. The three types of
irrigation work constructed are designated as 'technical', 'semi-technical'
and 'people's irrigation'. 'Technical' irrigation schemes are large works of
a permanent nature constructed and operated by a government agency,
'Semi-technical' irrigation schemes are minor works, either permanent or
temporary, constructed by the government and operated by the farmers
themselves, and 'People's irrigation schemes' are minor works with temporary or no weirs, constructed and maintained by the farmers themselves.
The distribution of irrigation systems in Indonesia is shown in Table 9.9 and
the major irrigation projects are shown in Table 9.10.

Swampy lands are found in several islands of Indonesia, particularly in
Kalimantan and Sumatra. The land in these regions is flat and almost at
sea level. People living in these areas have got used to digging channels

Table 9.10 Major irrigation projects in Indonesia

Project name	Area irrigated (hectares)
Jreu and Krueng Aceh Rivers Improvement Projects	20,000
River Ular Improvement Project	18,000
Teluk Lada Irrigation Project	31,000
Karawang Diversion Project	78,000
Cimanuk Diversion Project	37,000
Citanduy River Project	24,400
Cacaban Storage Project	41,000
Bali Irrigation Project	11,630
Jatiluhur in West Java multi-purpose	240,000
Karangkates in East Java multi-purpose	84,000

Source: Microsoft Word – Indonesia.doc (icid.org) Microsoft Word – Indonesia.doc (icid.org).

connecting the land with the tidal river. During high tide, water flows into the fields and inundates the rice fields. During low tide, the water in the channels flows back to the river carrying oxides, which are harmful to plants. This drainage and reclamation processes have transformed the marshy land to fertile land.

REFERENCES

Dieter, C. A., Maupin, M. A., Caldwell, R. R., Harris, M. A., Ivahnenko, T. I., Lovelace, J. K., Barber, N. L., and Linsey, K. S. (2018). Estimated use of water in the United States in 2015: U.S. Geological Survey Circular 1441, 65 p., https://doi.org/10.3133/cir1441.

International Commission on Irrigation and Drainage (ICID) (2018). Sprinkler and micro irrigated area in the world. (www.icid.org/sprinklerandmircro.pdf. download on November 28, 2018).

Lamm, F. R., Stone, L. R. and O'Brien, D. M. (2007). Crop production and economics in Northwest Kansas as related to irrigation capacity. Appl. Engr in Agric. 23(6):737–745.

Shah, Syed (pdf). Irrigation system in Pakistan, key facts, Academia.edu www.academia.edu/3813975/IRRIGATION_SYSTEM_IN_PAKISTAN_KEY_FACTS?email_work_card=view-paper&li=0

World Bank (1991). Report No. 9518-IN India Irrigation Sector Review (in Two Volumes) Volume 1: Main Report December 20, 1991 (World Bank Document).

Chapter 10

Water and soil

10.1 INTRODUCTION

All land plants absorb water and nutrients from the soil and therefore soil water plays a very important role in the life of plants. Plants, in turn, produce the food that humans and animals need for their lives. Soil is a porous medium that can retain as well as allow movement of water it receives from rainfall and irrigation. When the pores in the soil are completely filled up with water, the condition is known as saturated soil whereas when only a part of the pores is filled with water the condition is known as partially saturated soil. The water that infiltrates into the sub-soil from rainfall and/or irrigation usually saturates the soil immediately below the surface, but in a short time, the water will percolate to deeper layers of soil. The saturated condition does not last for a long time. Most plants cannot survive in saturated soils for a long time. Rice is the only exception. Plants need both water and air and therefore can only survive in a partially saturated soil. Directly or indirectly, plants produce the food necessary for the survival of all living things by a process called photosynthesis. The growth and yield of cultivated plants depend heavily on soil properties and on oxygen in soil water. The amount of oxygen needed by roots increases with the biological activity of the plant.

10.2 BASIC SOIL PROPERTIES

The two soil-based properties are the soil texture, as gravel, sand, silt and clay, and soil structure that refers to the arrangement and organization of particles in the soil. The latter is dependent on changes in climate, biological activity, soil management practice as well as forces acting on the soil, which are all time and space dependent. As a result, it is difficult to define soil structure quantitatively but qualitatively soil structure is classified into three categories identified as single grained, massive and aggregated. Coarse granular soils belong to the first category whereas tightly packed soils such as clays are identified as massive. In between is the aggregated soil in which

DOI: 10.1201/9781003329206-10

the soil particles are associated in quasi-stable small clods. The last type is the most desirable type for plant growth.

Most of the mineral particles in soil originate from the degradation of rocks. Those that originate from residues of plants or animals (rotting leaves, pieces of bone etc.) constitute organic particles (or organic matter). Although the soil particles seem to touch each other, there are always some spaces in between, which are called pores. When the soil is 'dry', the pores are mainly filled with air. After irrigation or rainfall, the pores are mainly filled with water. Living material found in the soil such as beetles, worms, larvae etc. help to aerate the soil and thus create favourable growing conditions for the plant roots. Pores act as conduits that allow water to infiltrate and percolate as well as storage compartment for water. Soil acts as a sponge to take up and retain water and the movement of water into soil is called infiltration, while the downward movement of water within the soil is called percolation. Water that percolates deep in the soil may reach the water table or groundwater aquifer. If the percolating water carries chemicals such as nitrates or pesticides, the aquifers will become contaminated.

Soils can be generally classified as sandy soils, loamy soils and clayey soils. The largest component of particles in soil is sand, followed by silt and clay in that order. Clay particles adsorb water, causing swelling upon wetting and shrinking upon drying. The relatively inert sand and silt constitute the 'soil skeleton' while the clay is considered as the 'flesh' of the soil. All three fractions constitute the matrix of the soil. Coarse-textured soils generally have lower porosities compared to those of fine-textured soils. Loam, which is considered to be the optimal soil for plant growth, is a balanced mixture of coarse and fine particles. Water is attracted to clay surfaces by electro-static attraction as well as by hydrogen bonding to oxygen atom in the clay crystal. A healthy soil is generally dark in colour, has high organic content with the soil surface covered with dead organic matter and allows infiltration.

10.2.1 Structure of the soil

Soil structure refers to the arrangement of soil particles (sand, silt, and clay) and their aggregates into certain defined patterns. Single particles, when assembled, appear as larger particles known as aggregates. Soil structure is not permanent. The basic types of aggregate arrangements are granular, blocky, prismatic and massive. Granular structures have individual particles of sand, silt and clay, which have rounded surfaces grouped into small grains. Water circulation is easy through such soils. Blocky structures have the soil particles cling together in rectilinear blocks having sharp edges. When the faces are flat and the edges sharp angular, they are called angular blocky whereas when the faces and edges are predominantly rounded, they are called sub-angular blocky. Large blocks impede the movement of water. Prismatic structures have soil particles formed into vertical columns separated by vertical cracks. Water circulation and drainage are poor. Platy (or massive) structure consists of soil particles aggregated in thin plates or sheets piled horizontally on top of one another with poor water circulation.

Soil structure may consist of particles unattached to each other such as in coarse granular soils sometimes referred to as single-grained structures, tightly packed such as in clay, and in between where soil particles are associated in quasi-stable conditions. The latter structure is desirable for plant growth.

10.2.2 Soil texture

Soil is composed of three categories of particles, sand, silt and clay, which are called soil separates. Of these, sand has the largest size (of the order of 0.005–2.00 mm), silt the next (of the order 0.002–0.05 mm) and clay the smallest (less than 0.002 mm). The proportion of soil separates in a soil defines its texture. There are 12 classes of soil texture (Figure 10.1) as can be seen in the triangular classification of soil texture that has been obtained by mechanical analysis (USDA – NRCS). They are

- clay,
- sandy clay,
- silty clay,

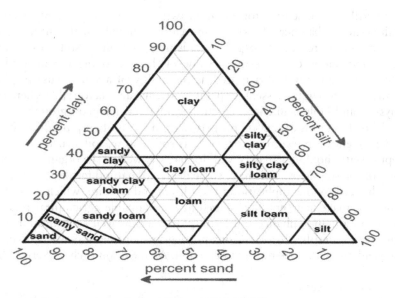

Figure 10.1 Soil texture classification (Source: USDA–NRCS).

- clay loam,
- sandy clay loam,
- silty clay loam,
- loam,
- sandy loam,
- silty loam,
- silt,
- loamy sand and
- sand.

In general, it is possible to identify soils as fine-textured that has a dominance of clay and coarse-textured that has a dominance of sand. Medium-textured soils have a dominance of silt. These three groups are divided by their particle size.

Medium-textured soils (fine sandy loam, silt loam and silty clay loam) have the highest water availability whereas coarse soils (sand, loamy sand and sandy loam) have the lowest water availability. The former provides a large number of pores that can hold water against gravity. The latter, on the other hand, has little aggregation and few small pores that could hold water against gravity. Fine-textured soils have a large number of small pores that can hold water against gravity, but plants find it difficult to adsorb the water because the water is held tightly in small pores.

10.3 SOIL FERTILITY

Soil fertility is essential for plant growth. Functions of plants include maintaining a balance of oxygen and carbon dioxide in the atmosphere, as a major source of atmospheric moisture, controlling soil erosion and as a contributor to the aesthetic value of the environment. Soil fertility, which can be natural or acquired, is the ability of a soil to sustain plant growth by providing essential plant nutrients and favourable chemical, physical and biological characteristics as a habitat for plant growth. It can be defined in terms of the soil properties or in terms of crop productivity. It is dependent on physical factors such as soil texture, soil structure, profile depth, water-holding capacity and drainage as well as the organic composition of the soil. Most soils have some degree of natural fertility, which can be enhanced by the application of fertilizers and manure, which have their limits. Management practices aimed at a sustainable agricultural practice can maintain and enhance soil fertility. Farmers as well as institutions can improve soil fertility by optimizing nutrient management so as to maximize the yield and minimize nutrient depletion. An ideal soil is characterized by

- a loamy texture,
- organic matter content to sustain microbial population,

- soil structure that promotes roots to grow into the soil mass and drainage and air exchange,
- sufficient clay and organic colloids to hold nutrients and soil moisture,
- a deep soil profile with a permeable sub-soil that allows root penetration and water drainage and
- a pH level that promotes root growth.

Soil fertility is crucial for agricultural productivity and therefore for food security. Factors that affect soil fertility include the mineral composition, soil pH, soil texture, organic matter and nutrient replenishment by adding manure and fertilizers and leguminous crops that contain nitrogen-fixing bacteria such as Rhizobium in the root nodules. These bacteria trap atmospheric nitrogen and make it available to the plants in the form of nitrogen compounds, which helps to reduce the use of commercial nitrogen fertilizers. Soil pH helps in maintaining nutrient availability of the soil. A pH range between 5.5 and 7 is considered to be optimum for soil fertility. For pH < 5.5, the availability of essential elements phosphorous and magnesium decline while the concentration of aluminium, manganese and for some soils copper begin to advance into toxic range. A pH > 6.5 causes the availability of copper, manganese, iron and zinc to decline (Jones, 2012, p. 60). Clayey soil can retain more nutrients and hence acts as a nutrient reservoir. Organic matter is a source of nitrogen and phosphorus. They also prevent tiny particles of clay from cementing themselves into solid masses. Organic matter promotes the growth of microorganisms, which helps to condition the soil. Once nutrients get into the soil, it is impossible to remove them and therefore over fertilization should be avoided.

10.3.1 Fertilizers

Fertilizers are chemical or natural substances that are used to provide nutrients to plants, usually via application to the soil. Nutrient sources include chemical and mineral fertilizers, organic fertilizers such as livestock manures and composts and recycled nutrients. A well-balanced supply of basic plant nutrients should include nitrogen, phosphorus, potassium, magnesium and calcium, which are essential for the normal growth and yield of crops. Most commercial fertilizers contain the first three in different proportions, which are the most common fertilizers required by the soil. Micronutrients essential for plant growth include boron, chlorine, copper, iron, manganese, molybdenum and zinc.

Of the inorganic fertilizers, nitrogen applied is either as the ammonium cation (NH_4^+) or nitrate anion (NO_3^-), or their combination NH_4NO_3. Urea ($CO(NH_2)_2$), which contains neither, is quickly hydrolysed when in contact with soil. Phosphorous fertilizer is applied in the form of orthophosphate anion (PO_4^{3-}), di-phosphate anion ($H_2 PO_4^{3-}$) or mono-phosphate anion

(H PO$_4^{3-}$). Potassium in fertilizers is present as potassium cation (K$^+$), which when in solution participates in cation exchange of the soil by mass flow. Details of the inorganic fertilizer sources can be found in Tables 19.1 and 19.2 of the book by Jones (2012).

Organic fertilizers are mainly from composted animal manures. Composted animal manures have better physical and chemical properties than 'fresh' animal manure. The element compositions of animal manure are nitrogen, phosphorous and potassium with potassium and phosphorous in the form of their oxides (K$_2$O and P$_2$O$_5$). Details of the organic fertilizer sources can be found in pages 170–171 of the book by Jones (2012). Organic materials, which originate from living organisms, are broken down in the soil by bacteria into inorganic, water-soluble forms. Inorganic materials are mineral salts that are water-soluble. They do not need bacteria to make them available to plants. Commercial organic fertilizers include dried manures, bone and blood meal, and cottonseed and soybean meals.

10.3.2 Plant nutrients

Plant nutrient elements can be broadly classified into macronutrients and micronutrients. Macronutrient elements include hydrogen, oxygen, carbon, nitrogen, potassium, calcium, magnesium, iron and sulphur whereas the micronutrient elements include chlorine, boron, iron, manganese, zinc, copper and molybdenum. Other than carbon, hydrogen and oxygen, which are used in photosynthesis, plants get their nutrients through the root system from the soil water solution. Modes of bringing the ions of the nutrient elements to the neighbourhood of roots are by mass flow driven by hydraulic gradients, diffusion of ions driven by concentration gradients and by root interception, the process of expansion of roots into soil mass. Normally, there is very little contact between roots and the soil mass.

10.3.3 Water in plants

Approximately 60–90% of plant mass is water, which is essential for cell functioning and photosynthesis. When the cells are full of water, the condition is known as turgid. Loss of turgidity results in wilting. Transpiration from the leaves helps to cool the leaves. Plants absorb their nutrients from soil water solution and water carries the carbohydrates in phloem. Too little water causes wilting and too much water leaves no space for air, thereby depriving oxygen.

Soil water includes gravitational water that drains from the voids due to gravity, capillary water which is retained by surface tension after the gravitational water has been drained, and hygroscopic water that is adsorbed by dry soil when exposed to a moist environment and held by soil at a very high tension (negative pressure). Of these three components, only the capillary water is available for plants. The other two components are not available for plants.

10.4 SOIL EROSION

Soil erosion that involves detachment, movement and deposition is the process in which the topsoil of a field is carried away by physical forces such as wind, water and tillage. The topsoil, which is high in organic matter, fertility and soil life, is carried away and deposited downstream of the source, which usually ends up in rivers, streams and lakes. The result is loss of crop productivity as well as contribution to pollution of watercourses, wetlands and lakes. All soils undergo erosion, but some are more vulnerable than others depending on the soil type and vegetative cover. Types of rain-induced erosion include rill erosion, gulley erosion, bank erosion and splash erosion.

Rill erosion results when overland flow is concentrated, forming small channels called rills where the soil has been washed away. Due to tillage, such rills usually get filled with time. Gully erosion occurs when running water erodes soil to form channels, which can be considered as an advanced stage of rill erosion. Gully erosion results in significant loss of productive land and can also create a hazardous condition for the operators of farm machinery. Bank erosion refers to the progressive undercutting, scouring and slumping of natural and artificial drainage channels. Splash erosion refers to erosion caused by the impact of raindrops on the soil surface.

The main causes that trigger soil erosion are rainfall, wind and farming practices such as tillage. Rainfall has kinetic energy, which can dislodge soil particles and the runoff caused by rainfall assisted by gravity, will carry the dislodged particles in the downstream direction until they get deposited in a stream, river or lake. High winds can contribute to soil erosion, particularly in dry weather and in arid and semi-arid regions. Wind forces can carry away loose soil particles to far-away places. Farming, grazing, mining, construction and recreational activities also contribute to soil erosion.

An indicator of the ability of soils to resist erosion is the soil erodibility, which is a physical characteristic of each soil. It depends on the soil texture, soil structure, organic matter content and permeability. Generally, soils with high infiltration capacities, high levels of organic matter and improved soil structure have a greater resistance to erosion. Sand and loam-textured soils are less liable to erosion compared to silt, fine sand and clay textured soils. Tillage and cropping practices that reduce soil organic matter can also contribute to increases in soil erodibility. However, compacted subsurface soil layers or a soil crust at the surface that decreases infiltration and increases runoff can have opposing effects on erodibility. Soil crust can decrease the soil loss from raindrop impact and splash but can increase the overland flow, thereby contributing more to soil erosion. Erosion caused by tillage results in a progressive down-slope movement of soil, causing soil loss on upper-slope and accumulation in lower-slope positions. It has a greatest potential for 'on-site' movement of soil and in many cases can cause more erosion than by wind and water. Another obvious factor that contributes to soil erodibility is the slope angle. The steeper and longer the slope, the higher the likelihood of erosion.

The consequences of soil erosion are reduced agricultural productivity, ecological damages, soil degradation, and the possibility of desertification. It can also carry pesticides, insecticides and other chemicals contained in the soil to the receiving waters, thereby polluting the hydro-environment. Sediment accumulation in streams and rivers can clog the waterways, leading to flooding. Wind-induced soil erosion can cause air pollution. Continued soil erosion can lead to desertification of arable land.

Soil erosion is a serious issue as it significantly affects food production. Measures to curb soil erosion include planting trees in barren lands and adding mulch to reduce erosion. Mulch is a layer added to the surface of the soil to suppress weeds and prevent water loss through evaporation. Examples of mulches include fabric sheet mulches, organic loose mulches such as chipped bark and inorganic loose mulches such as gravel and pebbles.

10.5 SLOPE FAILURE

A slope failure can be defined as an incident in which a slope fails as the outcome of the weak bearing capacity of the soil because of heavy rainfall, earthquake, landslide and other natural disasters or factors. Slope failures can be classified into four types:

- Rotational failure, which occurs by the rotation on the slip circle, which may be classified as
  - Base failure in which the failure plane passes through of base of the slope
  - Toe failure in which the failure plane passes through the toe of the slope. This is the most common mode of failure
  - Face failure in which the failure occurs on the surface of the slope passing through the toe of the slope which occurs when the soil above the toe contains weak strata
- Translational failure, which occurs in an infinite slope (a slope which has no boundaries) in which the movement of the soil is along the level surface
- Compound failure, which is a combination of rotation and translation
- Wedge failure, which is a block failure or plane slope failure that occurs along an infinite plane.

10.5.1 Causes and effects

Slope failures are caused by several human activities as well as by natural factors. Some of the major causes are listed below:

- Earthquake is a major cause of slope failure and instabilities in soil. Earthquakes induce vertical and horizontal forces, which are dynamic that result in stresses in the slope. They decrease the shear strength of the soil thereby making the soil liable to fail.
- Rainfall is also a major cause of slope failure as it softens the soil, making it liable to fail. Rainwater enters the cracks and reduces the strength (bearing capacity) of soil.
- Erosion by which soil particles get eroded by wind and rainwater causes changes in the geometry of the slope that can result in slope failure.
- Construction activities when carried out near the toe of the slope reduce or sometimes eliminate the resistance to slope failure.
- Geological characteristics prevailing in the soil mass can cause landslides (slope failure).
- External loading imposed on the surface (slope) add additional load to the gravitational load that can causes slope failure.
- Increase in shear force because of a sudden increase in the dynamic force of the soil mass like during an earthquake, or when the slope of the soil is increased or increase in weight of the soil due to absorption of rainwater.
- Decrease in shear force of soil due to the bond between soil particles getting weak and losing its bearing capacity.

Effects of slope failures resulting from reduced bearing capacity of soil include human casualties and destruction of property and infrastructure.

10.5.2 Prevention

Preventive measures for slope failures include providing adequate drainage to prevent water logging of soil, which reduces the bearing capacity, vegetation cover to stabilize the soil with the root system, constructing retaining walls that will provide lateral support to the soil mass and providing soil and rock anchors (Figure 10.2) and soil nails. Soil and rock anchors are drilled into the soil mass and stressed to provide a high load capacity even in poor ground conditions. They are simple in construction and lengths of anchors can be increased to accommodate the load required.

Soil nailing is an in-situ soil reinforcement technique involving the installation of closely spaced soil nails, which are structural elements to enhance the stability of slopes, retaining walls and excavations. In

Figure 10.2 Rock anchors in Hong Kong.

Hong Kong, it is the most widely used slope stabilizing method. A soil-nail reinforcement is the main element of a soil-nailed system. Its primary function is to provide tensile resistance. Installation may be done by the drill and grout method, self-drilling method or driven method using percussion or compressed air. Comprehensive details of the soil nailing system can be found in a document prepared by the Geotechnical Engineering Office of the Hong Kong SAR Government (Geoguide 7 (cedd.gov.hk)).

10.5.3 Early warning

Early warning can be based on measurements made by various types of instruments that can measure translation, rotation and settlement embedded in the soil mass. Monitoring data can be transmitted to a control centre that issues warnings of impending slope failures via the internet and other means. In regions where slope failures are primarily triggered by rainfall, warnings can be issued based on the intensity and duration of rainfall.

10.5.4 Case studies

The Geotechnical Engineering Office of the Civil Engineering and Development Department of the Government of Hong Kong Special Administrative Region has published a review of slope-specific early warning systems for rain-induced landslides (Kwan et al., 1979), giving details of site description, establishment of threshold values for issue of early warning, response plan and preparedness of stakeholders and monitoring parameters and instrumentation used citing case studies of slope failures in Australia, Canada, China, Italy, New Zealand, Norway, Scotland and South Korea.

10.6 PHYSICAL PROPERTIES OF SOIL

The key measurable physical properties of a soil are the uniformity coefficient, the grain size distribution, porosity, void ratio, bulk density and pore volume which are defined as follows:

The effective particle size is the 10% finer than value, d_{10}. The distribution of particle sizes is also identified by the uniformity coefficient, u_c, which is defined by Hazen as follows.

10.6.1 Uniformity coefficient

$$u_c = \frac{d_{60}}{d_{10}}$$

(10.1a)

There are other definitions too. For example, Kramer defined u_c as

$$u_c = \frac{\sum_0^{50} \Delta p_i d_i}{\sum_{50}^{100} \Delta p_i d_i}$$

(10.1b)

where Δp_i is the fraction of the particle size d_i.

A low uniformity coefficient implies a uniform material whereas a high uniformity coefficient implies a well-graded material.

10.6.2 Grain size distribution

A typical grain size distribution is shown in Figure 10.3. The ordinate shows the percentage of material finer than that of a given size on a dry weight basis. Organic components are not considered when grain size distribution is determined experimentally.

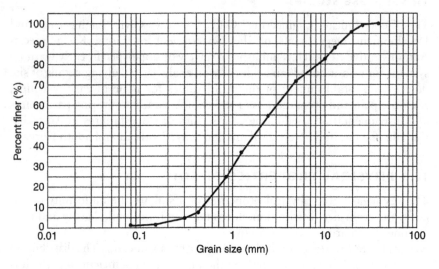

Figure 10.3 A typical grain size distribution.

10.6.3 Porosity, n

$$n = \frac{\text{Volume of interstices}}{\text{Total volume}} = \frac{V_i}{V} \tag{10.2}$$

Typical values are 0.35–0.45 for sandy soils and 0.40–0.60 for clays and peats.

10.6.4 Void ratio

$$\text{Void ratio} = \frac{\text{Volume of voids}}{\text{Volume of solids}}, \text{which can be greater than unity} \tag{10.3}$$

10.6.5 Bulk density

$$\text{Bulk density} = \frac{\text{Mass of soil}}{\text{Volume of soil}} \tag{10.4}$$

Bulk density usually refers to the dry bulk density corresponding to soil dried at 105°C.

Table 10.1 Typical values of bulk density, pore volume and void ratio of some soils

Soil type	Bulk density (g per cm³)	Pore volume (%)	Void ratio
Sandy soils	1.67–1.19	37–55	0.58–1.22
Loams	1.96–1.19	26–55	0.25–1.22
Silty loams	1.53–1.19	42–55	0.72–1.22
Clay-rich soils	1.32–0.92	50–65	1.00–1.85
Organic-rich soils	0.48–0.12	60–90	1.50–9.00

Source: Table 2.1, Horton et al. (2016).

10.6.6 Pore volume

$$\text{Pore volume} = \frac{\text{Total volume} - \text{Volume of solids}}{\text{Total volume}} \tag{10.5}$$

which is usually expressed as a percentage of the total volume. Typical values of these quantities as given by Horton et al (2016) are shown in Table 10.1.

10.7 HYDRAULIC PROPERTIES OF SOIL

In the context of the soil water system, the soil hydraulic parameters include the specific retention, specific yield, field capacity and permanent wilting point.

10.7.1 Specific retention, S_r

$$S_r = \frac{\text{Volume of water retained in pores against gravity}}{\text{Bulk volume of soil}} = \frac{W_r}{V} \tag{10.6}$$

10.7.2 Specific yield, S_y (or, effective porosity)

$$S_y = \frac{\text{Volume of water that can be drained by gravity}}{\text{Bulk volume of soil}} = \frac{W_y}{V} \tag{10.7}$$

Specific values range from about 0.05 to about 0.40. Table 10.2 show some typical values for different types of soils. More details about specific yields of various materials can be found in a USGS Water Supply Paper (Johnson, 1967). It can be seen that

$$W_r + W_y = \alpha = \text{volume of water in pores} \tag{10.8}$$

Table 10.2 Typical values of soil hydraulic parameters of some soils

Material	Porosity (%)	Specific yield (%)	Specific retention (%)
Soil	55	40	15
Clay	50	2	48
Sand	25	22	3
Gravel	20	19	1

Source: Heath (1983).

Note: Volume of water in pores is not necessarily equal to the volume of pores (i.e. $\alpha \neq n$).

10.7.3 Field capacity

The water remaining in the soil after the drainage by gravity has stopped is known as the field capacity. Usually, the water in large pores drains first, leaving the smaller pores still full of water. The soil at field capacity is considered as ideal for plant growth. It is synonymous with water-holding capacity and water retention capacity. It is the amount of water the soil can hold against gravity. The water content between saturation and field capacity is known as the drainable porosity of a soil, which controls the transient water table dynamics. Many factors affect field capacity. For example, the field capacity of a drying soil can be different from that of a wetting soil due to hysteresis effect. Also, finer soils such as clay can hold more water, resulting in a higher field capacity. Field capacity decreases as temperature increases. Despite its use in many field problems, field capacity is not an exact physical soil property. There is some confusion in the use of the term field capacity. Since water drainage is a continuous process, the term field capacity is vague because, the drainage does not stop at a particular value of soil water content. Also, field capacity does not apply to potted plants since potted plants do not have a physical underlying soil that pulls water deep down into the soil profile by capillarity. An analogous soil aeration index, which may be called 'field air-capacity', refers to the volume of air in a soil at the field capacity water content.

10.7.4 Permanent wilting point

The water stored in the soil is taken up by the plant roots or evaporated from the topsoil into the atmosphere. The dryer the soil becomes, the more tightly the remaining water is retained and the more difficult it is for the plant roots to extract it. At a certain stage, the uptake of water is not sufficient to meet the plant's needs. The plant loses freshness and the leaves change colour from green to yellow. The plant begins to wilt, which is the stage of the plant losing its rigidity. Wilting is the loss of rigidity of non-woody parts of plants. This occurs when the turgor pressure (pressure exerted by stored water against a cell wall) in non-lignified plant cells falls towards zero as a result of diminished water in the cells. This happens when the rate of loss of water from the plant is greater than the rate of absorption of water by the plant.

The soil water content at the stage when the plants begin to wilt during daytime but recovers in the night time or when water is added is known as the temporary wilting point. Permanent (or ultimate) wilting point refers to the case when the plants will wilt irreversibly. At this water content, plants are not able to extract water from the soil and they will not recover even when placed in a humid environment. The difference in the water content between the field capacity and the permanent wilting point is known as the available water content of the soil.

10.8 THE SOIL WATER SYSTEM

The soil water system is composed of three major components, air, water and solids mixed heterogeneously. The solid component is a mixture of organic matter and minerals. Mineral matter consists of sand, silt and clay whereas the organic matter consists of decaying plant and animal matter as well as bacteria and fungi that contain the elements carbon, nitrogen, phosphorous and sulphur. The void space around soil particles holds water and air and their relative composition is important for plant growth.

The water in the soil can be represented by three zones as shown in Figure 10.4. At the bottom is the saturated zone in which all the voids in the soil are filled with water. The upper limit of this zone is the water table at which the pressure is atmospheric. The pressure in the saturated zone is above atmospheric and is normally assumed to be hydrostatic.

The intermediate capillary water zone is also saturated because of the capillary action, which pulls water from the water table upwards against gravity. The water in this zone is held by capillary forces between the soil particles and is at less than atmospheric pressure. The depth of the capillary zone depends on the type and compaction of the soil. It can range from a few centimetres in a coarse sandy soil to several metres in clay and silt.

In a soil water system, the voids are filled with air and water. The water particles are held in contact with the soil particles by the capillary forces. If the system is allowed to drain under gravity, the drainage will take place until the water content of the soil is reduced to the field capacity.

10.8.1 Soil moisture content

The state of a soil water system is measured by two parameters, namely the volumetric soil moisture content, θ, and the soil suction, ψ, which is a negative pressure. The former is measured best by gravimetric methods, but in the field, it is now measured by a number of methods including neutron scattering techniques, which are non-destructive. The volumetric soil moisture content is defined as

$$\theta = \frac{\text{Volume of moisture}}{\text{Bulk volume of soil}} \qquad (10.9)$$

Thus, the maximum value of θ occurs when the soil is fully saturated in which case it is equal to the porosity of the soil. The soil moisture content can be measured in the laboratory by a gravimetric method as

$$\text{Gravimetric wetness} = \frac{\text{Wet weight} - \text{Dry weight}}{\text{Dry weight}} \qquad (10.10)$$

where the dry weight is obtained by oven drying at 105°C for 24 hours. It can also be measured by using electrical resistance, which depend on soil moisture content as well as on the soil structure and salt content. This method has limited accuracy. In the field, it can be measured using neutron scattering devices, as well as by Gamma ray absorption technique.

A neutron probe is an instrument to measure the moisture content in soil. It contains a probe and a gauge. The probe typically contains a pellet of americium-241 and beryllium. The alpha particles emitted by the decay of the americium collide with the light beryllium nuclei, producing fast neutrons. When these fast neutrons collide with hydrogen nuclei of water present in the soil, they lose much of their energy. The returning slow neutrons are detected by the gauge that allows an estimate of the amount of hydrogen present, which in turn allows an estimation of the amount of water present in the soil.

Measurements are made by lowering the probe in an access tube to the desired depth and counting the backscattered slow neutrons by the gauge. Calibration of the gauge is by gravimetric methods in which a relationship is established between the slow neutron counts and the soil moisture content. Since the neutron scattering is occurring over a spherical domain, the neutron probe gives the soil moisture content in a volume sample of soil rather than at a point. Because of the large volume of influence, the neutron probe is prone to errors when measuring moisture content at or near the ground surface since the adjoining air is also sampled. The method is non-destructive and relatively easy to use. On the negative side are the cost and the safety issues in handling the probe since it contains radioactive material.

Gamma ray absorption consists of a source which is usually a pellet of radioactive caesium (^{137}Cs) that emits gamma radiation and a detector usually consisting of a scintillation counter such as sodium iodide crystal connected to a photomultiplier and preamplifier.

During the flow of water through the partially saturated porous medium, three zones (Figure 10.5) can be identified:

- Saturated zone $\theta = \theta_{sat}$ (only a few mm. deep)
- Transmission zone ($\theta = \theta_{const}$ uniform θ, extends to a fair depth)
- Wetting zone (θ decreases with depth at an extremely high rate down to a wetting front)

The sharp boundary between the wet soil and the relatively dry soil is called the wetting front. At the wetting front, the moisture gradient is very steep. It is because the hydraulic conductivity $K(\theta)$ decreases as θ decreases, and therefore, to have any flux, the hydraulic gradient must be very large. The length of the wetted zone increases as the water penetrates deeper into the soil. As a result, the average suction gradient decreases (being divided by a larger number). This trend continues until the suction gradient in the upper part of the profile becomes negligible, leaving the constant gravitational gradient as the only driving force, that is,

$$q = -K \frac{\partial \phi}{\partial z} \tag{10.11a}$$

$$= -K \frac{d}{dz}(\psi - z) \tag{10.11b}$$

$$= K \tag{10.11c}$$

Therefore, the limiting value of flux = hydraulic conductivity.

10.8.2 Soil water potential

Energy in a soil water system can be in two principal forms – kinetic and potential. Since water movement in soils is very slow, the kinetic energy component is negligible, leaving the only form of energy as potential energy. The movement of water within the soil is therefore driven by the potential gradient between two points from a higher potential to a lower potential.

As defined by the International Society of Soil Science (1963), the total potential of soil-water is

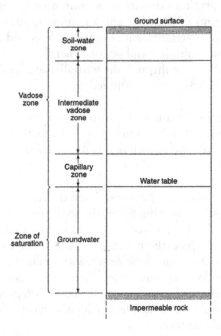

Figure 10.4 Soil water zones.

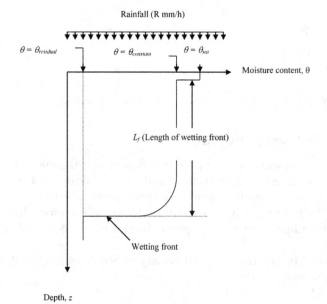

Figure 10.5 Soil water system.

the amount of work that must be done per unit quantity of pure water in order to transport reversibly and isothermally an infinitesimal quantity of water from a pool of pure water at a specified elevation at atmospheric pressure to the soil-water at the point under consideration.

This total water potential (ψ_t) can be divided into components to distinguish between the actions of different force fields. The algebraic sum of these component potentials is equal to the total water potential which consists of the following four components:

- The matric or capillary potential (ψ_m), which results from the interaction of soil particle surfaces with water
- The osmotic potential (ψ_0), which results from the solutes dissolved in the soil-water
- The gravitational potential (ψ_g), which results from elevation with respect to a reference level
- The pressure potential (ψ_p), which results from external pressure on the soil-water.

These potentials can be expressed in different ways. For example,

- Potential per unit mass (μ); μ = potential per mass (Nm per kg)
- Potential per unit volume (ψ); ψ = potential per volume = (N per m^3),
- Potential per unit weight (h); h = potential per weight (m, head unit)

As a result, it is not necessary to compute the soil-water potential directly by computing the amount of work needed. Instead, it can be indirectly measured from pressure or water height measurements.

Matric potential is due to adhesion of water molecules on mineral surfaces and cohesion between water molecules. Meniscus is formed by cohesion. Matric potential is measured by a tensiometer. It can also be defined as the amount of work needed to extract a given unit of soil water from a given location in a soil volume at a given pressure. Matric potential counteracts gravitational potential and has a negative sign. Osmotic potential exists only at the root zone where there is a semi-permeable membrane between the root and the soil water with dissolved solutes. Osmosis occurs when solutions of different concentrations are separated by a semi-permeable membrane. Random motion of water and solute molecules create a net movement of water by diffusion to the compartment with higher concentration, until equilibrium is reached. If the soil is unsaturated, the soil water matric potential is negative ($h < 0$) whereas it is positive if the soil is saturated ($h > 0$). Hydraulic potential is therefore the sum of matric potential and the gravitational potential. Below the water table, matric potential is zero.

In the field, soil suction is measured by a tensiometer, which has a ceramic porous cup attached to a tube and connected to a manometer. When the porous cup gets in touch with the soil, the negative pressure in the soil sucks water from the tensiometer through the pores in the porous cup until pressure equilibrium is attained. The drop in the manometer attached to the tensiometer tube gives the soil suction at the depth where the porous cup is placed. Standard tensiometers can measure soil suction up to about 850 kPa, but it is reported that a tensiometer developed by Imperial College is able to measure suction up to 1,500 kPa. Other devices for measuring soil suction include but not limited to transistor psychrometers capable of measuring suctions up to about 10,000 kPa and chilled mirror hygrometers capable of measuring up to about 30,000 kPa (Fondjo et al., 2020).

10.9 SOIL HYDRAULIC CHARACTERISTICS

There are three basic soil hydraulic characteristics that are necessary to define the flow of moisture through partially saturated porous media. They are as follows.

10.9.1 Soil suction – soil moisture content characteristic

Soil suction – soil moisture content characteristic is sometimes referred to as the retention curve (Figure 10.6). It gives the variation of the soil suction as a function of the moisture content. At the saturated end, the moisture content has its maximum value and the suction has its minimum value, which is

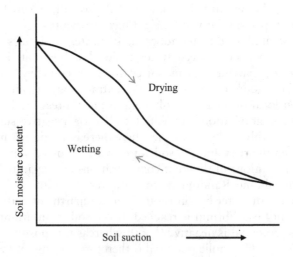

Figure 10.6 Typical soil water retention characteristics.

zero. At the other extreme is the dry end, which is not achievable in the field and can only be attained in the laboratory after oven drying the soil for at least 24 hours. The soil suction at the field dry end depends on the type of soil. The range of suctions corresponding to the range of applicable moisture contents can vary over several orders of magnitude.

An important property of this characteristic is that it depends on the history of the wetting and drying of the soil. The property known as hysteresis is analogous to the same property in magnetism. It can be obtained by observing the ψ–θ relationship of an initially saturated sample (desorption), or by observing the ψ–θ relationship of an initially dry sample (sorption). The two curves will be continuous but different. The moisture content on the drying curve is greater than the moisture content on the wetting curve for a given suction.

The shape of the ψ–θ relationship depends on the soil structure as well as on the soil texture. In clayey soils, the slope is more gradual indicating an increase in suction causes a gradual decrease in moisture content. In the case of sandy soils, once the large pores are emptied at a given suction, only a small amount of water remains. The structure of the soil affects the ψ–θ relationship particularly in the low suction range. For compacted soils, the suction decreases very rapidly to zero at saturation whereas for aggregated soils it is more gradual. At the high end of suction, there is hardly any difference in the shape of the ψ–θ relationship for both compacted soils and aggregated soils (Figure 10.7). An important property of soil water system in terms of soil water availability to plants is the slope of the soil moisture

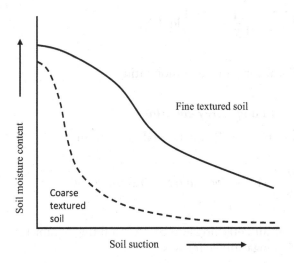

Figure 10.7 Typical soil water retention characteristic for different textured soils.

characteristic curve $d\theta/d\psi$ referred to as the differential water capacity or specific water capacity.

There are several empirical equations to describe the soil suction–soil moisture content characteristic. Some equations are described below:

10.9.1.1 Brutsaert equation

Brutsaert (1966) equation is of the form

$$S_e = \frac{\alpha}{\alpha + |\psi|^\beta} \text{ for } \psi \le 0, \text{ and } S_e = 1.0 \text{ for } \psi > 0 \tag{10.12}$$

where S_e ($0 \le S_e \le 1$), is the effective degree of saturation, defined as

$$S_e = \frac{\theta - \theta_r}{\theta_s - \theta_r} \tag{10.13}$$

in which θ_r and θ_s are the residual and saturated soil moisture contents, and θ and ψ are the soil moisture content and soil suction, respectively. In Eq. 10.12, α and β are the parameters that should be determined.

By taking logarithms, Eq. 10.12 can be written as

$$
\begin{aligned}
\beta \log(\psi) &= \log\left(\frac{1}{S_e} - 1\right) + \log(\alpha) \\
\log(\psi) &= \frac{1}{\beta}\log\left(\frac{1}{S_e} - 1\right) + \frac{1}{\beta}\log(\alpha)
\end{aligned}
\tag{10.14}
$$

which is a linear form for regression fitting.

10.9.1.2 Brooks and Corey equation

Brooks and Corey (1964) equation takes the form

$$S_e = \left(\frac{\psi_e}{\psi}\right)^\lambda \text{ for } \psi \le \psi_e, \text{ and } S_e = 1.0 \text{ for } \psi > \psi \tag{10.15}$$

where ψ_e is the air entry pressure, and λ is a fitting parameter.

Eq. 10.15 in logarithmic form is

$$\log(\psi) = \frac{1}{\lambda}\log(S_e) + \log(\psi_e) \tag{10.16}$$

10.9.1.3 Vauclin equation

Vauclin et al. equation (1979) is of the form

$$S_e = \frac{\alpha}{\alpha + (\ln|\psi|)^\beta} \text{ for } \psi \le -1 \text{ cm, and } S_e = 1.0 \text{ for } \psi > -1 \text{ cm} \qquad (10.17)$$

where α and β are the fitting parameters.
Eq. 10.17 in logarithmic form is

$$\ln\{\ln(\psi)\} = \frac{1}{\beta}\ln\left(\frac{1}{S_e} - 1\right) + \frac{1}{\beta}\ln(\alpha) \qquad (10.18)$$

10.9.1.4 van Genuchten equation

van Genuchten (1978, 1980) equation is of the form

$$S_e = \left[\frac{1}{1 + (\alpha|\psi|)^\beta}\right]^{\left(1 - \frac{1}{\beta}\right)} \text{ for } \psi \le 0, \text{ and } S_e = 1.0 \text{ for } \psi > 0 \qquad (10.19)$$

Eq. 10.19 cannot be transformed into an explicit linear form by taking logarithms. Estimates of α and β are obtained by an iterative procedure starting from an initial set.

10.9.1.5 Other equations

Other forms of representation include equations of the form (Campbell, 1974; Clapp & Hornberger, 1978):

$$\psi = \psi_e \left[\frac{\theta}{\theta_s}\right]^{-b} \qquad (10.20)$$

where b is a constant.

10.9.2 Soil hydraulic conductivity – soil moisture content characteristic

Soil hydraulic conductivity determines the capacity of the soil to conduct moisture. It is also sometimes referred to as the soil permeability. For partially saturated soils, the hydraulic conductivity depends on the degree of saturation (or the moisture content). At saturation, the hydraulic conductivity

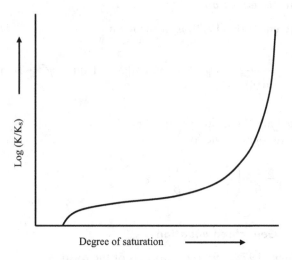

Figure 10.8 Typical soil hydraulic characteristics.

attains its maximum value asymptotically (Figure 10.8). The saturated hydraulic conductivity can be measured either in situ by field permeameters or in the laboratory using soil samples. In the latter case, the falling head permeameter and the constant head permeameter methods are commonly used. It is difficult to measure the unsaturated hydraulic conductivity and often various methods of estimation using probability concepts are used. A simple relationship used by Campbell (1974) and Clapp and Hornberger (1978) is of the form

$$K = K_s \left[\frac{\theta}{\theta_s} \right]^B \tag{10.21}$$

where B in Eq. 10.21 has been related to b in Eq. 10.20 as
$B = 2b+3$ (Clapp and Hornberger, 1978), and,
$B = 2b+2$ (Campbell, 1974)

10.9.3 Soil diffusivity – soil moisture content characteristic

Soil diffusivity is a derived parameter. It is obtained from the soil suction–moisture content and the hydraulic conductivity–moisture content relationships and defined as

$$D(\theta) = K(\theta)\frac{\partial \psi}{\partial \theta} \tag{10.22a}$$

Because of the hysteresis effect of the soil suction–moisture content characteristic, the diffusivity is not a single-valued function of the moisture content. At saturation, it tends to very large values. The exact representation
depends on the relationship between B in Eq. 10.21 and b in Eq. 10.20:

$$D = bK(\theta)\left[\frac{\psi_e}{\theta_s}\right]\left[\frac{\theta_s}{\theta}\right]^{b+1} \tag{10.22b}$$

10.10 CONSTITUTIVE EQUATIONS

The governing equations for flow through partially saturated porous media
are obtained by assuming that Darcy's law, which is defined for the flow
of water through saturated porous media is applicable to flow of moisture
through unsaturated porous media as well with some modifications. The
driving force comes from the hydraulic gradients due to gravity and pressure
forces. There are other forces such as osmotic and electrochemical, but they
are small in comparison with the two major forces in the soil water system.
The governing equations are obtained by combining the continuity equation
with Darcy's law.

Considering a control volume of soil of unit cross-sectional area, the continuity equation can be written as

$$-\frac{\partial q}{\partial z} = \frac{\partial \theta}{\partial t} \tag{10.23}$$

where q represents the volume flux of moisture through the control volume
per unit area.

Darcy's law can be written as

$$q = -K(\theta)\frac{\partial \phi}{\partial z} \tag{10.24}$$

where $K(\theta)$ is the unsaturated hydraulic conductivity, which is a function
of the moisture content, thus making the relationship non-linear, and ϕ is
the total potential, which is the sum of the pressure head and the gravitational head:

$$\phi = \psi + z \tag{10.25}$$

Therefore, combining Eqs. (10.23) and (10.24),

$$\frac{\partial \theta}{\partial t} = \frac{\partial}{\partial z}\left(K(\theta)\frac{\partial \phi}{\partial z}\right) \tag{10.26a}$$

$$= \frac{\partial}{\partial z}\left[K(\theta)\frac{\partial}{\partial z}(\psi + z)\right] \qquad (10.26b)$$

$$= \frac{\partial}{\partial z}\left[K(\theta)\frac{\partial\psi}{\partial z}\right] + \frac{\partial K}{\partial z} \qquad (10.26c)$$

This is the basic constitutive equation of moisture movement through unsaturated porous media.

10.10.1 Horizontal infiltration

For horizontal infiltration, the gravity term does not exist. The only driving force is due to pressure (suction) gradient. Therefore, Eq. (10.26a) simplifies to

$$\frac{\partial\theta}{\partial t} = \frac{\partial}{\partial x}[K(\theta)\frac{\partial\phi}{\partial x}] = \frac{\partial}{\partial x}\left[K(\theta)\frac{\partial\psi}{\partial x}\right] \qquad (10.27)$$

In this case, the water moves in the horizontal direction and the soil absorbs water by matric suction gradients only.

For homogeneous soils, replacing $K(\theta)\dfrac{\partial\psi}{\partial\theta}$ by $D(\theta)$, Eq. (10.27) becomes

$$\frac{\partial\theta}{\partial t} = \frac{\partial}{\partial x}\left[D(\theta)\frac{\partial\theta}{\partial x}\right] \qquad (10.28)$$

The usual initial and boundary conditions are

$$\theta = \theta_i \quad \text{for } x \geq 0, \quad t = 0$$

$$\theta = \theta_0 \quad \text{for } x = 0, \quad t > 0$$

By using the Boltzmann (1894) transformation, which is given as

$$\lambda(\theta) = xt^{-1/2} \qquad (10.29)$$

where x is the distance from the surface to the wetting front, the partial differential equation, Eq. (10.28) can be transformed into an ordinary differential equation:

$$-\frac{\lambda}{2}\frac{d\theta}{d\lambda} = \frac{d}{d\lambda}\left[D(\theta)\frac{d\theta}{d\lambda}\right] \qquad (10.30)$$

The boundary conditions then change to

$$\theta = \theta_i, \lambda \rightarrow \infty; \theta = \theta_0, \lambda = 0.$$

Since $D(\theta)$ is a non-linear function of θ, Eq. (10.30) has to be solved numerically. If θ is assumed to remain constant in the transmission zone, then, from Eq. (10.29), x would be proportional to $\sqrt{t}$, and a plot of x versus $\sqrt{t}$ would give a straight line of slope λ.

The cumulative infiltration, I, is given by

$$I = \int_{\theta_i}^{\theta_0} x d\theta \tag{10.31}$$

where θ_i is the initial moisture content and θ_0 the final moisture content. Substituting for x from Eq. (10.29)

$$I = \int_{\theta_i}^{\theta_0} \lambda(\theta)t^{1/2}d\theta = st^{1/2} \tag{10.32}$$

where

$$s = \int_{\theta_i}^{\theta_0} \lambda(\theta)d\theta = \frac{I}{t^{1/2}} \tag{10.33}$$

This is a constant called the sorptivity (Philip, 1969) and depends on θ_i and θ_0, and has dimensions of $LT^{-1/2}$. By plotting I versus $t^{1/2}$, a straight line with slope s can be obtained.

10.10.2 Vertical infiltration

For vertical infiltration, both gravity and suction terms apply. For the case of infiltration into a homogeneous semi-infinite medium, it is convenient to measure z positive in the downward direction. Then (by substituting $z = -z$), Eq. (10.26c) takes the form

$$\frac{\partial \theta}{\partial t} = \frac{\partial}{\partial z}\left[K(\theta)\frac{\partial \psi}{\partial z}\right] - \frac{\partial K(\theta)}{\partial z} \tag{10.34}$$

By introducing the soil diffusivity D, which is defined as

$$D(\theta) = K(\theta)\frac{\partial \psi}{\partial \theta} \tag{10.35}$$

Eq. (10.34) can be transformed into the following form:

$$\frac{\partial \theta}{\partial t} = \frac{\partial}{\partial z}\left(D(\theta)\frac{\partial \theta}{\partial z}\right) - \frac{\partial K(\theta)}{\partial z} \qquad (10.36)$$

Eq. (10.34) can also be written as

$$C(\psi)\frac{\partial \psi}{\partial t} = \frac{\partial}{\partial z}\left[K(\psi)\frac{\partial \psi}{\partial z}\right] - \frac{\partial K(\psi)}{\partial z} \qquad (10.37)$$

where

$$C(\psi) = \frac{\partial \theta}{\partial \psi}(= \text{specific moisture capacity}) \qquad (10.38)$$

Eqs. (10.36) and (10.37) form the constitutive equations that govern the one-dimensional vertical movement of moisture through a porous medium and are usually identified as the θ-based and the ψ-based equations respectively. Although θ and ψ are interdependent, D and K in Eq. (10.36) are usually expressed as functions of θ while C and K in Eq. (10.37) are expressed as functions of ψ.

In both Eqs. (10.36) and (10.37), the driving force consists of the suction gradient, which arises out of wetness, and the gravitational gradient. Their relative magnitudes differ depending upon initial and boundary conditions.

For example, when the soil is initially dry the suction gradient is very much greater than the gravitational gradient. Then the vertical infiltration rates approach the horizontal infiltration rates.

When $\theta \to \theta_{sat}$, $D \to \infty$, and hence the θ-based equation cannot be used. Instead, ψ-based equation is used.

As $\theta \to \theta_{sat}$, $K(\theta) \to K$ (constant), and Eq. (10.37) simplifies to

$$K\frac{\partial^2 \psi}{\partial z^2} = 0 \qquad (10.39)$$

which is the Laplace equation.

For initially wet soils, suction gradients are small.

10.10.3 Boundary and initial conditions

There are three different boundary and initial conditions that can be applied to Eqs. (10.36) and (10.37) when describing infiltration. They are briefly defined in the following equations.

10.10.3.1 Ponded condition

$$\theta (z,0) = \theta_i \text{ for } z \geq 0 \text{ } t = 0 \tag{10.40a}$$

$$\theta (0,t) = \theta_0 \text{ for } z = 0; t \geq 0 \tag{10.40b}$$

where θ_i and θ_0 are the initial and surface moisture contents respectively (usually $\theta_0 > \theta_i$). They may be constants or functions of z and/or t. The most common condition in infiltration is when there is a thin layer of water available at the surface. Then, the surface moisture content is the saturated value θ_{sat} and is called the ponded infiltration condition. Then

$$\theta (0,t) = \theta_{sat} \text{ for } z = 0; \text{ } t \geq 0 \tag{10.40c}$$

10.10.3.2 Rain condition – low rainfall intensities

$$\theta (z,0) = \theta_i \text{ for } z \geq 0; t = 0 \tag{10.41a}$$

$$\theta (\infty,t) = \theta_i \text{ for } z \to \infty; t \geq 0 \tag{10.41b}$$

$$\text{Flux} = -K(\theta)\left[\frac{\partial \psi}{\partial z} - 1\right] = R \text{ for } z = 0; \text{ } t > 0 \tag{10.41c}$$

where R is the rainfall intensity.

The flux boundary condition given by Eq. (10.41c) is obtained as follows:

$$q = -K(\theta)\frac{\partial \phi}{\partial z} = -K(\theta)\left[\frac{\partial \psi}{\partial z} + \frac{\partial z}{\partial z}\right]$$

Rainfall is an inward flux. Therefore, $R = -q$. Substituting for q and using the coordinate transformation $z = -z$,

$$-R = -K(\theta)\left[-\frac{\partial \psi}{\partial z} + 1\right] \tag{10.41a}*$$

$$R = -K(\theta)\left[\frac{\partial \psi}{\partial z} - 1\right] \tag{10.41b}*$$

Because Eq. (10.41a)* can be written in the form

$$-R = K(\theta)\frac{\partial \psi}{\partial \theta}\frac{\partial \theta}{\partial z} - K(\theta) \tag{10.41c}*$$

The condition (10.41c) can also be written as

$$\frac{\partial \theta}{\partial z} = -\frac{R - K(\theta)}{D(\theta)}$$

(10.41d)

This condition corresponds to rain infiltration and is applicable from the beginning of rainfall to the time of occurrence of incipient ponding. For low rainfall intensities $[R < K(\theta_{sat})]$, rain infiltration can continue without giving rise to ponding. As time passes, the surface moisture content approaches a limiting value θ_L.

10.10.3.3 Rain condition – high rain intensities

For high rainfall intensities $[R > K(\theta)_{sat}]$, the ψ-based equation is preferred because the $D(\theta)$ term in the θ-based equation tends to very large values near saturation. Then

$$\psi(z,0) = \psi_i \qquad \text{for } z \geq 0; \quad t = 0$$

(10.42a)

$$\psi(\infty, t) = \psi_i \qquad \text{for } z \to \infty; \quad t \geq 0$$

(10.42b)

$$\psi(0, t) = \psi_f \geq 0 \quad \text{for } z = 0; \quad t \geq t_p$$

(10.42c)

$$\text{Flux} = -K(\psi)\left[\frac{\partial \psi}{\partial z} - 1\right] = R \quad \text{for } z = 0; \quad 0 \leq t \leq t_p$$

(10.42d)

where ψ_i is the initial soil water pressure (suction)
ψ the surface soil water pressure during ponding (hydrostatic)
t_p the time of incipient ponding

The physical meaning of this condition is that the rainfall intensity is exceeding the infiltration capacity of the soil and therefore ponding of water at the surface is taking place. In Eq. (10.42c), ψ_f can be taken as zero.

10.11 SOLUTIONS OF THE EQUATIONS

10.11.1 Ponded condition – linearized solution

Eq. (10.36) can be linearized as follows:

$$\frac{\partial \theta}{\partial t} = D_* \frac{\partial^2 \theta}{\partial z^2} - u \frac{\partial \theta}{\partial z}$$

(10.43)

where D_* $(=D(\theta))$ and $u \left(= \dfrac{\partial K}{\partial \theta} \right)$ are assumed to be constants. The θs in Eqs. (10.36) and (10.43) are identical. The relevant boundary and initial conditions are given by Eq. (10.40). Eq. (10.43) is analogous to the convective dispersion equation with constant parameters for which an analytical solution has been proposed by Ogata and Banks (1961). Their solution, with variables changed to reflect the infiltration process, is of the form:

$$\theta = \theta_i + \frac{\theta_0 - \theta_i}{2} \left\{ \operatorname{erfc} \frac{z - ut}{\sqrt{4D_*t}} + \exp\left(\frac{uz}{D_*} \right) \operatorname{erfc} \frac{z + ut}{\sqrt{4D_*t}} \right\} \tag{10.44}$$

where,

$$\operatorname{erf}(x) = \frac{2}{\sqrt{\pi}} \int_0^x e^{-x^2} dx; \quad \operatorname{erfc}(x) = 1 - \operatorname{erf}(x) = \frac{2}{\sqrt{\pi}} \int_x^\infty e^{-x^2} dx$$

and u can be written as $\dfrac{K_0 - K_i}{\theta_0 - \theta_i}$.

(This is also the downward rate of advance of soil wetness for large times and has units of velocity.)

The error function erf(x) varies from 0 to 1 asymptotically. It is the same as the cumulative normal probability function.

10.11.2 Ponded condition – non-linear solution

The first mathematical solution to the vertical infiltration equation for an infinitely deep soil was proposed by Philip (1957). His initial and boundary conditions were

$\theta = \theta_i$ for $t = 0$, $z > 0$

$\theta = \theta_0$ for $t \geq 0$, $z = 0$

that is, the surface value increasing from an initial value of θ_i to a final value of θ_0 instantaneously.

His solution is based on a power series expansion and takes the form

$$z(\theta,t) = f_i(\theta) + f_i(\theta) \, t^{1/2} + f_2(\theta) \, t + f_3(\theta) \, t^{3/2} \tag{10.45}$$

where z is the depth to any particular value of θ and $f_i(\theta)$ are calculated successively from the $K(\theta)$ and $D(\theta)$ functions.

This shows that $\theta \alpha \sqrt{t}$ for small t (same as horizontal infiltration). For large t, vertical movement of moisture approaches the constant rate u, defined in Eq. 10.44.

Similarly, Philip also showed that the cumulative infiltration, I, can be expressed as

$$I(t) = st^{1/2} + (A_2 + K_0)t + A_3 t^{3/2} + A_4 t^2 + \qquad (10.46)$$

where the As are calculated from $K(\theta)$, $D(\theta)$ functions and K_0 is the conductivity at $\theta = \theta_0$.

Differentiating Eq. (10.46),

$$i(t) = \frac{dI}{dt} = \frac{1}{2} st^{-1/2} + (A_2 + K_0) + \frac{3}{2} A_3 t^{1/2} + \qquad (10.47)$$

Representing Eq. (10.47) by a 2-parameter approximation: (for t not too large - This is done in practice)

$$i(t) = \frac{1}{2} st^{-1/2} + A \qquad (10.48a)$$

or

$$I(t) = st^{1/2} + At \qquad (10.48b)$$

As $t \rightarrow \infty$, i decreases monotonically to its asymptotic value $i(\infty)$. This does not imply $A = K_0$ for small and intermediate values of time.

For large times the series is divergent and it is possible to write Eq. 10.48 as (with A replaced by K)

$$I = st^{1/2} + Kt, \text{or } i = \frac{1}{2} st^{-\frac{1}{2}} + K \qquad (10.49)$$

where K is the hydraulic conductivity of upper layer of soil.

10.11.3 Low rain condition

For rain infiltration (flux boundary condition), there are no known analytical solutions. Numerical solutions are available for specific boundary and initial conditions. Both finite difference and finite element methods have been used (Jayawardena, 1985; Jayawardena and Kaluarachchi, 1986).

10.11.4 High rain condition

Ponding will occur when the rainfall intensity exceeds the surface-saturated conductivity value and overland flow will follow. This state is described by Eqs. (10.24) and (10.26). In the case for high rainfall intensities, solutions are only of numerical type and a method commonly used for verifying the validity of a numerical procedure consists of comparing the depth to the air entry pressure position of the suction profile at ponding (depth of saturated zone at ponding). The analytical value of this depth, obtained by consideration of Darcy's law, is given by the following equation:

$$R = K(\theta_{sat})\left[\frac{\psi_A - \psi_0}{B} + 1\right] \tag{10.50}$$

where B is the depth of saturation at ponding, ψ_A is the air entry pressure and ψ_0 is the surface pressure head. At $t = t_p$, $\psi_0 = 0$, and therefore

$$B(t_p) = \frac{\psi_A}{\dfrac{R}{K(\theta_{sat})} - 1} \tag{10.51}$$

REFERENCES

Boltzmann, L. (1894). *Ann. Phys. Chem.* 53, 959–964.

Brooks, R. H. and Corey, A. T. (1964). Hydraulic properties of porous media, Hydrology Paper No. 3, Colorado State University, Fort Collins.

Brutsaert, W. (1966). Probability for pore size distributions, *Soil Sci.* 101(2), 85–92.

Campbell, G. S. (1974). A simple method for determining unsaturated conductivity from moisture retention data, *Soil Sci.* 117, 311–314.

Clapp, R. B. and Hornberger, G. M. (1978). Empirical equations for some soil hydraulic properties, *Water Resour.*, 14 (4), 601–604.

Fondjo, A. A., Theron, Elizabeth, and Ray, Richard (2020). Assessment of various methods to measure the soil suction, *Int. J. Innovative Tech. Exploring Eng.* 9, 171–184. 10.35940/ijitee.L7958.1091220.

Geotechnical Engineering Office, Civil Engineering and Development Department, The Government of the Hong Kong Special Administrative Region, March 2008: Geoguide 7, Guide to soil nail design and construction, pp 97.

Heath, R. C. (1983). Basic ground-water hydrology, U.S. Geological Survey Water-Supply Paper 2220, 86p. [pdf]

Horton, R., Horn, R., Bachmann, J. and Peth, S. (Editors) (2016). *Essential Soil Physics*, CSIRO, pp. 391.

Jayawardena, A. W. and Kaluarachchi, J. J. (1986). Infiltration into decomposed granite soils: Numerical modelling, application and some laboratory observations, *J. Hydrol.*, 84 (3–4), 1986, 231–260.

Jayawardena, A. W. (1985). Moisture movement through unsaturated porous media: Numerical modelling, calibration and application, Proc. 21st Congress, IAHR, August 19–23, Melbourne, Australia, vol. 1, pp. 12–16.

Jones, J. Benton, Jr. (2012). *Plant Nutrition and Soil Fertility Manual*, 2nd ed., CRC Press, Taylor and Francis Group, Boca Raton, London, New York, pp. 282.

Johnson, A. I. (1967). Specific yield – compilation of specific yields for various materials, Hydrologic properties of earth materials, Geological Survey Water-Supply Paper 1662-D, Prepared in cooperation with the California Department of Water Resources.

Kwan, J. S. H, Chan, M. H. C. and Shum, W. W. L. (2015). A review of slope-specific early warning systems for rain induced landslides, Geotechnical Engineering Office, Civil Engineering and Development Department, The Government of the Hong Kong Special Administrative Region, p. 48.

Ogata, A. and Banks, R. B. (1961). A solution of the differential equation of longitudinal dispersion in porous media, fluid movement in earth materials. Geological Survey Professional Paper 411-a, United States Government Printing Office, Washington, pp. A1–A7.

Philip, J. R. (1957). The theory of infiltration: 4. Sorptivity and algebraic infiltration equations. *Soil Sci.*, 84, 257–264.

Philip, J. R. (1969). The theory of infiltration, *Advances in Hydroscience*, 5, 215–296.

van Genuchten, M. Th. (1980). A closed-form equation for predicting the hydraulic conductivity of unsaturated soils, *J. Soil Sci. Soc. Am.*. 44, 892–898.

van Genuchten, M. T. (1978). Calculating the unsaturated hydraulic conductivity with a new closed form analytical model, 78-WR-08, Water Resources programme, Department of Civil Engineering, Princeton University, New Jersey.

Vauclin, M., Haverkamp, R. and Vachaud, G. (1979). Resolution numerique d'une equation de diffusion non lineaire, Presses Universitaires de Grenoble, 183 pp.

Chapter 11

Water and energy

11.1 INTRODUCTION

Water and energy are closely linked. Energy is required for providing water services and water resources are required for the production of energy. They have a symbiotic relationship. All services in the water sector such as pumping and distribution of water (including lift irrigation), water supply, wastewater treatment and desalination require energy. The energy sector also requires water to cool thermal power plants, generate hydropower and grow biofuels. Energy production comes from fossil fuel, nuclear fuel, hydropower, geothermal, wind, tidal and solar power. Of these, hydropower is the largest renewable source for power generation in the world meeting about 16% of global electricity needs. Approximately 90% of global energy production is water-intensive. In addition to hydropower, other forms of energy production excepting geothermal and photovoltaic require water for steam production and cooling purposes. With increasing demand for energy, there is an indirect increasing demand for water.

The quantities of water required for energy production depend upon the type of energy and sometimes the demand for water for energy production can conflict with other demands for water. Because of the interdependencies and linkages, the optimal allocation of water for different uses becomes very important. Usually, people who are deprived of water services are also deprived of energy services (electricity).

The amount of energy needed to provide potable water varies from source to source. It is reported (WWDR2014, p 24) that if the source is from lakes and rivers, the energy required to produce 1 m^3 of potable water is 0.37 kWh, groundwater 0.48 kWh, wastewater treatment 0.62–0.87 kWh and from sea water 2.58–8.5 kWh.

The provision of water for agriculture, which needs no treatment, requires energy for delivering the water from the source to the demand area, which depends upon the quantity of water delivered. In the case of surface waters, it is usually by gravity or by pumping. The former needs no external energy if the source is at a higher elevation than the demand area. On the other hand, if the demand area is at a higher elevation than the source, the energy required depends upon the elevation difference between the source and the demand area and the distance between them. In the case of groundwater, the energy requirements depend upon the depth of the aquifer as well as the hydraulic properties of the water-bearing formation.

Energy requirements for water treatment depend upon the quality of the source water, the level of treatment and the treatment process. For drinking water, there are many stages and levels of treatment some of which are energy-intensive. For example, reverse osmosis requires energy of the order of 1.5–3.5 kWh per m³ whereas ultraviolet treatment requires much less energy (of the order of 0.01–0.04 kWh per m³).

Desalination is the most expensive method of providing potable water. It is highly energy intensive and therefore does not appear to be a viable option for developing countries.

It is also important to note that the wastewater produced needs treatments before they can be discharged into receiving waters. The processes involved are equally energy intensive. Looking at the reverse process of the symbiotic relationship between water and energy, water is used in many industries that produce fuels such as coal, uranium, oil and gas. Water is an input for growing biofuels such as corn and sugarcane. Water is also crucial for cooling purposes in thermal power plants and is the driving force for hydro-electric and steam turbines.

Water is also a medium of conveyance for the transport of fuel through barges and sometimes via pipelines. A relatively new type of demand for water is in the area of hydraulic fracturing (or fracking) for extracting oil and/or gas from subterranean rocky formations. The process involves injection of fracking fluid (water containing sand or other proppants suspended with the aid of thickening agents) at high pressure into a bore hole to create cracks in the subterranean rocky formations to allow the natural gas and other petroleum products to flow more easily. Fracking fluid typically consists of 98% sand and water and 2% chemicals (acids, surfactants, biocides and scaling inhibitors). The quantities of water injected are of the order of 8–30 million litres per well. Some of the injected water comes back as wastewater.

The water–food–energy nexus is central to sustainable development. Demand for all three is increasing, driven by a rising global population, rapid urbanization, changing dietary habits and economic growth. Agriculture is the largest consumer of the world's freshwater resources, and more than one-quarter of the energy used globally is expended on food production and supply. An integrated approach is therefore required for the optimum management of these three critical domains, which are interlinked. As a universal call to end poverty, protect the planet and improve the lives and prosperity of everyone, everywhere, the member states of the United Nations in 2015 have adopted 17 goals as part of the 2030 Agenda for sustainable development setting out a 15-year plan to achieve the goals. Of these goals, SDG 6 and SDG 7 respectively aim to ensure availability and sustainable management of water and sanitation for all and access to affordable, reliable, sustainable and modern energy for all. These targets are expected to be achieved by 2030.

11.2 WATER FOR ENERGY

Water is essential for all phases of energy production. It is needed for fossil-fuel extraction, transport and processing, power production and irrigation of feedstock for biofuels. Water can also be produced as a by-product of fossil-fuel production. According to World Energy Outlook (WEO), the energy sector is responsible for 10% of global water withdrawals, mainly for power plant operation as well as for the production of fossil fuels and biofuels. In 2014, some 4% of global electricity consumption was used to extract, distribute and treat water and wastewater, along with 50 million tonnes of oil equivalent of thermal energy, mostly diesel used for irrigation pumps and gas in desalination plants. These requirements are expected to grow in the future and by 2040, 16% of electricity consumption in the Middle East is expected to be related to water supply (WEO-2016), especially for desalination (www.iea.org/reports/world-energy-outlook-2016). Because of the strong linkage between water and energy, it is important to follow an integrated approach for the successful attainment of the United Nations Sustainable Development Goals for clean water and sanitation (SDG 6) and for affordable and clean energy (SDG 7).

Energy in water can be in the form of potential, kinetic or pressure. Potential energy is due to the elevation of the location of water with reference to a datum. The theoretical potential energy that can be extracted is given by the simple relationship

$$E = mgh \tag{11.1a}$$

where E is the energy (ML^2T^{-2}), m is the mass (M), g is the acceleration due to gravity (LT^{-2}), h is the elevation difference between the water surface and the reference location (L). Kinetic energy can be expressed as

$$E = \frac{1}{2}mv^2 \tag{11.1b}$$

where v is the velocity (LT^{-1}), and pressure energy can be expressed as

$$E = \rho g h \Delta V \tag{11.1c}$$

where ρ (ML^{-3}) is the density of fluid and ΔV(L^3) is the change in volume.

When the energy quantities are expressed as energy per unit volume, the Bernoulli equation can be obtained.

$$\frac{mgh}{\Delta V} + \frac{1}{2}\frac{m}{\Delta V}v^2 + \frac{\rho gh\Delta V}{\Delta V} = \text{Constant} \qquad (11.2a)$$

All three terms on the left-hand side of the Eq. (11.2a) have the dimension ML^{-1}T^{-2}, which is energy per unit volume. Pressure energy is due to the effects of thermal and kinetic motions of the atoms and the attractive/repulsive forces of fluid molecules on each other (van der Waals effect). Pressure may be considered as energy density. Bernoulli equation (Eq. 11.2a) is also written as

$$P + \frac{1}{2}\rho v^2 + \rho gh = \text{Constant} \qquad (11.2b)$$

Since density of water and acceleration due to gravity are constants, the amount of energy that can be harnessed depends on the quantity of flow and elevation difference between the surface of the reservoir and the location of the turbine, which is normally referred to as the 'head'.

11.2.1 Hydro-electricity

11.2.1.1 Components of a hydro-electricity generating project

The components of a hydro-electricity generating project consist of a reservoir at a high elevation, which is constructed by building a dam across a river or other water course, an outlet structure in the dam to allow the water to flow down in a controlled manner, a penstock/tunnel to carry the water from the outlet to the power generating station, which is at a lower elevation, a nozzle attached to the end of the penstock to discharge the water towards the hydraulic machine, which is a turbine and an electricity generator coupled to the turbine, which rotates together with the turbine to generate electricity. Two types of machines, impulse turbines and reaction turbines, are used to convert the energy of water to mechanical energy. When the head is high, the kinetic energy of the water when it flows down to a lower elevation is high, and in such situations impulse turbines are used whereas in situations where the rate of water flow is high, reaction turbines are used. In both types, fluid passes through a runner having blades. The momentum of the fluid in the tangential direction is changed and so a tangential force on the runner is produced. The runner rotates and does work. The energy in the fluid is reduced.

(i) Impulse turbines

An impulse turbine consists of buckets arranged along the circumference of a wheel that intercepts a high-speed jet of water at the end of a penstock. The potential energy of the water, which flows out from the reservoir, is first converted to kinetic energy as it flows down and transferred to the wheel as mechanical energy, which in turn is converted to electrical energy by the generator that is coupled to the turbine wheel. The widely used impulse turbine in hydro-electric systems is the Pelton wheel named after its inventor Lester Allan Pelton in the 1870s. It is a tangential flow machine used in high head hydro-electric schemes in mountain areas. The rotor consists of a number of buckets attached to its periphery. One or more nozzles are mounted so that they discharge jets along a tangent to the circle through the centre line of the buckets (Figure 11.1).

The buckets are split in half at the centre and the incoming jet divides into two equal halves upon impinging. The notch in the outer rim of each bucket prevents the jet to the preceding bucket being intercepted too soon.

The maximum change of momentum – and hence the maximum force – would be obtained if the fluid is deflected through 180°. But in practice, this deflection is around 165° to prevent the fluid leaving one bucket striking the back of the next.

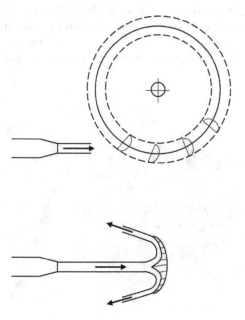

Figure 11.1 Jet and buckets in a Pelton wheel.

The pressure of the fluid before and after striking the buckets is constant at atmospheric pressure. In impulse turbines, there is no change in static pressure across the runner, which is open and at atmospheric pressure. The energy change is from kinetic to mechanical. Normally, the Pelton wheel rotates in a vertical plane, that is, in a horizontal shaft, having only two jets placed symmetrically. In horizontal wheels, a greater number of jets (up to six) can be used.

(ii) Reaction turbines

(a) *Francis turbines*: The basic difference between reaction and impulse turbines is that in reaction turbines there is a change in pressure when the fluid goes through the runner (Figure 11.2). The static pressure decreases as the fluid passes through the runner, which is enclosed.

Francis turbine is one of the common types of reaction turbines and has been developed by the US engineer J. B. Francis (1815–1892). It is also called the radial flow or inward flow turbine. Radial flow is turned into axial flow from the centre of the runner. There is no whirl at exit. It can be thought of as a reverse centrifugal pump. It has a spiral casing through which fluid enters and its cross-sectional area decreases along the fluid path in such a way to keep the fluid velocity constant in magnitude. The flow rate decreases along the spiral casing as fluid enters the runner. As the fluid flows through the turbine, some of the pressure energy is converted into kinetic energy, thereby maintaining a constant velocity despite the decrease in cross-sectional area of flow. From the spiral volute, the fluid passes between stationary guide vanes to the

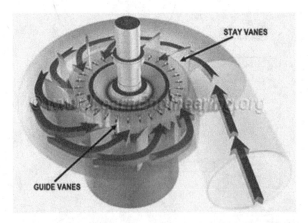

Figure 11.2 Francis turbine (Source: www.learnengineering.org/2014/01/how-does-francis-turbine-work.html).

runner. The stationary guide vanes deflect the fluid by the desired angle according to design.

Blades of Francis turbines resemble air foils. As water flows over a blade, there will be high pressure on one side and low pressure on the other side, giving rise to a lift force. The blades have a bucket-type shape towards the outlet, thereby producing an impulse force before leaving the runner. Both impulse force and lift force make the runner rotate. Guide vanes at the entrance to the runner convert part of the pressure energy to kinetic energy. They also reduce the swirl of inflow into the runner.

When the fluid passes through the runner, the angular momentum is changed by the blades. At exit, it has very little or no whirl velocity (compare with a centrifugal pump, which has the same conditions at inlet).

(b) *Kaplan turbines*: Developed by Austrian engineer V. Kaplan (1876–1934), the Kaplan turbine is equivalent to an axial flow pump in reverse. The flow enters through a spiral casing with decreasing cross-sectional area to ensure uniform velocity as the fluid flows along the periphery. Before the runner, the radial flow is turned into axial flow between the guide vanes and the runner.

(c) *Mixed flow turbines*: In mixed flow turbines, the flow enters radially and leaves with a substantial axial flow component of velocity. All reaction turbines run full of fluid. Usually, they are mounted on vertical shafts, and the motion is in a horizontal direction. They are used on low-head, high-flow situations, with heads of the order of about 15–30 m.

For any reaction turbine, the total head is given by (Figure 11.3)

H = Total head at inlet – Total head at exit

$$= \frac{P_C}{g} + \frac{V_C^2}{2g} + z_C - \frac{V_E^2}{2g} \qquad (11.3)$$

where V_E is the velocity at exit. Effective head across any machine is the difference of head between inlet and exit. Therefore, z_c is included. This does not occur in the case of a Pelton wheel as the fluid enters and leaves at atmospheric pressure.

11.2.1.2 Major hydro-electric projects in the world

Hydro-electricity is the most abundant form of renewable energy in the world. In terms of hydro-electricity generating capacity, the Three Gorges Dam, a concrete gravity dam across Yangtze River in China takes the top place in the world with an installed generating capacity of 22,500 MW. Hoover dam in the United States was the highest and had the largest

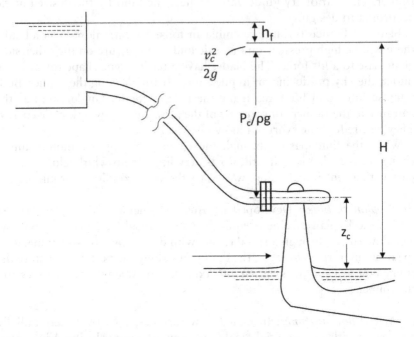

Figure 11.3 Definition sketch for heads in a reaction turbine.

hydropower-generating capacity in the world at the time it was built in 1936. A dam is an essential component of almost all types of water resources development projects including those for hydro-electricity generation.

There are an estimated 845,000 dams in the world. China has about 23,000 large dams followed by United States having about 9,200 large dams. These two countries are followed by India, Japan and Brazil, in that order. In this context, a large dam is a dam with a height between 5 and 15 m and impounding more than 3×10^6 m³ as adopted by the International Commission on Large Dams (ICOLD). They are built to control the flow of water in rivers and other water courses. The structure of a dam can be of concrete gravity type, concrete arch type, buttress type, rockfill embankment type and earth-fill embankment type. The purposes of a dam include generation of hydro-electricity, storage of water for irrigation, domestic water supply and water for industry needs, aquaculture, flood control and recreation. The size of a dam can be expressed in terms of a number of parameters including the hydro-electricity generating capacity, height, capacity and surface area of the impounding reservoir, length of dam, spillway capacity, cost of building the dam and the number of people relocated as a result of building the dam. A dam can be for a single purpose or for multiple purposes. All large dams are multi-purpose type. Based on information available in websites in public domains, Table 11.1 gives some details of the ten largest dams in the world based on the hydro-electricity generating capacity. In terms of the height, the Jinping dam, a concrete arch dam across Yalong River, also in China, with a height of 305 m takes the top place while the Kariba dam, an arch dam in Zimbabwe, with a storage capacity of 180.5×10^9 m^3 takes the top place in terms of storage capacity. Bhakra dam in India was considered as the 'temple of modern India' when it was built. It can be seen that there are more large dams in China than in any other country.

In recent years, the rate of construction of new large dams has gone down mainly as a result of opposition from environmentalists. Construction of large dams also leads to mass relocation of people that lead to many social and economic problems. The Three Gorges Dam in China has displaced over 1.2 million people. Dams are also not without controversies, especially when built across trans-boundary rivers. Construction of large dams, which cost large sums of money, is also prone to abuse of power and corruption by authorities. Very recently (December 31, 2019), it has been reported that the Ethiopian Federal Attorney General charged 50 people with corruption and abuse of office in relation with the construction of the Grand Ethiopian Renaissance Dam, which is being built across the Blue Nile River (www.occrp.org/en/daily/11369-ethiopia-50-charged-with-graft-in-nile-dam-project). The Grand Ethiopian Renaissance Dam, which is expected to be completed in 2022, will be Africa's largest hydro-electricity generating dam.

11.2.2 Pumped storage hydropower

Pumped storage hydropower was first harnessed in 1907, at the Engeweiher pumped storage facility near Schaffhausen in Switzerland. Known as the world's 'water battery', pumped storage hydropower is the cleanest and most cost-effective form of energy storage at the present time. It takes surplus electricity in the system and stores it for use when needed.

Pumped storage hydropower generation makes use of two reservoirs, one at a higher elevation and the other at a lower elevation, which are connected via a penstock/tunnel to a pump-turbine machine at the downstream end. The reservoirs can be open-loop in which there is a hydrologic connection to an outside body of water or closed-loop in which the reservoirs are not connected to an outside body of water. A pump-turbine machine operates as a turbine when its runner is rotating in one direction and as a pump when its impeller is rotating in the opposite direction. This is possible when there is an off-peak surplus of energy generated from other sources such as from thermal or nuclear plants and when there is a significant difference in the power consumption during day time and night time. When the upstream reservoir is full, the flow of water drives the machine as a turbine that generates electricity. When the upstream reservoir is empty or near empty, the off-peak surplus energy is used to pump water from the downstream

Table 11.1 Ten largest dams in the world in terms of hydropower-generating capacity

Name of dam	Country	River	Type	Power (MW)	Height (m)	Length (m)	Capacity (10^9 m^3)	Construction period
Three Gorges	China	Yangtze	Concrete gravity	22,500	185	2,335	42	1994–2009
Itaipu (4 dams)	Brazil	Parana	Earthfill, rockfill, concrete gravity, concrete wing	14,000	225	7,200	29	1960–1984
Xiluodu	China	Jinsha	Double curvature arch	13,800	285.5	700	6.5	2005–2013
Guri	Venezuela	Caroni	Arch + embankment	10,300	162	11,409	138	1963–1978
Tucurui	Brazil	Tocantins	Concrete gravity	8,370	78	12,500	45.5	1975–2012
Grand Coulee	USA	Columbia	Concrete gravity	6,809	168	1,592	11.6	1933–1942
Xinagjiaba	China	Jinsha	Concrete gravity	6,448	161	909	5.1	2006–2012
Longtan	China	Hongshui	Concrete gravity	6,426	216	894	27.3	2001–2009
Krasnoyarsk	Russia	Yenisey	Concrete gravity	6,000	124	1,065	73.3	1956–1972
Robert-Bourassa	Canada	La Grande	Embankment	5,616	162	2,835	61.7	1974–1981
Hoover	USA	Colorado	Arch	2,080	221	379	35	1931–1936

reservoir to the upstream reservoir via the same penstock/tunnel. This cycle is repeated to generate electricity, which usually can be used to provide for fluctuating demands.

There are many pumped storage power plants in the world. The basic requirement is that there should be sufficient surplus energy when the demand is low. In industrial countries, the peak demand is during the day-time when factories are in full swing. In less developed countries, the peak demand is in early night when domestic demand is high. Switching between modes of pumping and generating in seconds, pumped storage hydropower enables harmonizing supply and demand.

The International Hydropower Association (IHA) estimates current total installed pumped storage capacity at 150 GW and store up to 9,000 giga-watt hours (GWh) of electricity globally. As of 2017, China, Japan, the United States, Spain and Italy had the highest pumped storage generating capacities in that order, but Switzerland, Austria, Portugal, Japan and Spain in that order had the highest pumped storage generating capacities as a percentage of total generating capacity (International Renewable Energy Agency, 2017). IHA estimates that pumped storage hydropower capacity is expected to increase by almost 50% to about 240 GW by 2030. With the growing need to de-carbonize the power generation system and to increase the share of renewable energy, many countries are embarking on building pumped storage projects, some with financial incentives from international lending institutions.

11.2.3 Tidal energy

Tidal energy is a form of renewable energy that can be harnessed from water in oceans. Due to the astronomical forces between the sun and the moon and the rotation of the earth, massive amounts of ocean water are moved continuously, creating kinetic energy, which can be harnessed in the same way as in hydro-electricity using turbines. When the moon is directly above a region of the earth, the gravitational pull is strongest due to the shorter distance of separation, causing the sea level to rise. This condition is referred to as the high tide or flood tide. When the moon is furthest away from a region, the gravitational pull is weakest, and the water reaches its lowest level. This condition is known as low tide or ebb tide. High tide and low tide form the tidal cycle, which is continuous and can differ from place to place. During high tide, the shoreline gets extended landwards whereas the opposite effect occurs during low tide.

There are three types of tides identified as diurnal, which has one high and one low each lunar day (e.g. Gulf of Mexico), semi-diurnal, which has two high tides and two low tides each day with the heights of the highs and lows about the same (e.g. US East coast), and mixed semi-diurnal when the highs and lows differ in height (e.g. US West coast). The average tidal ranges are slightly larger when the sun, moon and earth are nearly in alignment, which occurs twice each month. This corresponds to the full moon and new moon conditions. The moon appears dark (new) when it is directly between the earth and the sun. When the earth is between the moon and the sun the moon appears bright (full). In both cases, the gravitational pull of the sun supplements the pull of the moon, causing oceans to bulge a bit more than usual leading to a slightly higher high tide and a slightly lower low tide. This condition is known as the 'spring tide', which has nothing to do with the season. On the other hand, when the sun and moon are at right angles to the earth and opposing each other (first and third quarter moons), the tidal ranges are less than normal and are referred to as 'neap tide'. The world's highest tide is in Bay of Fundy in Canada with a tide of about 16 m. The Severn estuary in UK also has a high tidal range of about 15 m.

There are currently three different ways to harness tidal energy: tidal streams, barrages and tidal lagoons. Tidal stream is a fast-moving body of water generated by tides. A turbine placed in a tidal stream converts the flow energy to mechanical energy, which in turn is converted to electric energy. This type is used in most tidal energy generators. Turbines are large in size, rotate slowly and can disrupt the tide, which they try to harness. Barrages are shallow dams with turbines placed inside them that can harness the tidal energy in the same way as a turbine placed in a river. The generation of energy with a tidal lagoon, which is a body of sea water that is partly enclosed by a natural or artificial barrier, is much like that of a barrage. The advantage is that it can generate electricity when the tide is coming into the lagoon as well as when it is going out, thereby ensuring a continuous operation.

Tidal energy is more reliable than wind energy since tides are regular and predictable. However, much of its potential is yet to be exploited. Currently there are only two major tidal energy plants in operation in the world. They are the barrage type Sihwa Lake Tidal Power Station in South Korea, which generates one-way power twice a day at high tide with a generating capacity of 254 MW, built in 2011 and the Rance River estuary in Brittany, France, that also uses a barrage, built in 1966, with a generating capacity of 240 MW. The tidal stream type MeyGen Tidal Energy Project in Scotland, currently under construction, with a projected generating capacity of 400 MW when completed will be the largest in the world. There are plans to build the world's first tidal lagoon power station in Swansea Bay, Wales, with an estimated generating capacity of 320 MW.

11.2.4 Wave energy

The history of wave energy goes back as far as 1799 when Monsieur Girard and his son proposed using direct mechanical action to drive heavy machinery, including mills, saws and pumps using wave energy in Paris. An early application, also in France was in 1910, where Bochaux Praceique developed a device to light and power his house. From 1855 to 1973, there have been 340 patents related to wave energy in the UK alone. The modern wave energy application is due to Yoshio Masuda who in 1940 tested multiple different wave-energy devices at sea, who is also credited for different wave-energy inventions, such as the 'Kaimei', a large barge used as a testing platform, and the oscillating water column, which was initially used for small-scale navigation. An increased interest in renewable energy appeared in 1973 as a result of the oil crisis during which the Organization of Arab Petroleum Exporting Countries (OAPEC) decided to put a prohibition on oil exports. The new pioneers included Stephen Salter, for inventing the Salter's Duck, or nodding duck, which converted wave power into electricity in 1977 in Scotland. Subsequently, an experimental wave farm has been established in 2008 in Portugal.

Waves derive their energy from wind, which is derived from solar energy due to differential warming of the earth that causes pressure differences. They pick up, store and transmit the energy across oceans until the energy is dissipated at the shore. Wave energy is renewable, clean and available always although the intensity may vary with space and time. Wave energy output is measured by wave speed, wave height, wavelength and water density. Energy is transferred from wind to the waves as wind glides over the surface of oceans. Wave energy is generated by harnessing the up-and-down motion of ocean waves using floating turbine platforms or buoys that rise and fall with the swells as well as by exploiting the changes in air pressure occurring in wave capture chambers that face the sea or changes in wave pressure on the ocean floor.

The wave energy depends on both the amplitude and frequency of the wave and can be expressed as

$$E = \frac{1}{2}\mu A^2 \omega^2 \lambda \tag{11.4}$$

where $\mu = \dfrac{\Delta m}{\Delta x}$; Δm and Δx are the mass and length of an element, and μ is a linear density, A is the amplitude, ω is the angular frequency and λ is the wavelength.

There are five main types of technology used to harness wave energy. They are absorbers, attenuators, oscillation water columns, overtopping and inverted pendulum devices. Absorbers extract energy from the rise and fall of the waves with a buoy. Attenuators do it by being normal to the length of the wave. Oscillation water column (OWC) is a partially submerged enclosed structure with the upper part above the water and filled with air. Incoming waves are directed into the lower part of the structure. When the waves go through the structure, the water column rises and falls with the wave causing the air in the top structure to pressurize and depressurize, which in turn moves a turbine. Overtopping allows the wave to lift over a barrier, which fills a reservoir with water, which is then drained through a hydroturbine. Inverted pendulum device uses the motion of waves to move a hinged paddle back and forth, which drives hydraulic pumps.

Wave energy is renewable, environmentally friendly, leaves only a small footprint, reliable, predictable and has enormous potential. Despite such advantages, wave energy has not yet been fully or optimally exploited. The major reason is the cost of installing wave energy converters (WEC) and their maintenance.

11.2.5 Blue (saline) energy

One of the new technologies that has the potential to be a source of clean and renewable energy, known as blue energy (or salinity gradient energy or osmotic energy), makes use of gradients of salinity in fresh and salt waters. The principle is running the desalination process in the reverse direction. Energy is required to desalinate water and running the process in reverse can generate energy. When a river meets the sea, a large volume of freshwater meets sea water and mixes to reach equilibrium. This mixing process releases energy. The higher salinity of the saltwater draws freshwater through a water permeable membrane, thereby increasing the pressure on the saltwater side. The pressurized water can then be used to drive a turbine and generate electricity.

The energy is captured by placing a water-permeable membrane across the two sources of water. Currently, there are two approaches to harnessing this energy, pressure-retarded osmosis (PRO) and reverse electrodialysis (RED). In the former type, with fresh water on one side of the membrane and seawater on the other, the flow through the water permeable membrane builds up pressure on the seawater side, which can be used to drive a turbine. It is reported that in 2009, Norwegian company Statkraft opened a pilot PRO plant in the Oslofjord with a 2,000 m^2 membrane, achieving a modest proof-of-concept output of 5 kW (https://medium.com/@tjsmed ley/blue-energy-can-we-get-all-our-future-energy-from-salt-water-d1e4f 910d763; http://advantage-environment.com/future/blue-energy/). In the latter method, an ionic current is created by diverting the ions in water, which is converted to electricity by the electrodes. This requires two or more membranes, effectively creating a bank of seawater and freshwater sandwiches. A membrane pair is made up of an anion exchange membrane and a cation exchange membrane with spaces in between through which fresh and saltwater flow. Although the salt in the seawater wants to diffuse into freshwater, only the positive Na$^+$ ions will be able to pass through the cation exchange membrane and the negative Cl$^-$ ions will only be able to pass through the anion exchange membrane thereby making the ions move in opposite directions. Electricity is generated in the same way as in a saline battery by sending the sodium ions (Na^+) and chloride ions (Cl^-) in opposite directions. The membrane used in RED allows the salt ions – rather than the water molecules – to pass through. It is also known that high salinity generates more power and therefore would be ideal to combine RED with desalination plants, which generate highly concentrated brine as a waste product. The plant in Oslofjord as well as a similar plant in Marsala, Sicily, have been subsequently (2013) shut down as they have not been economically competitive but reappeared in the Netherlands (by Redstack) using the same technology in 2014. The location was Afsluitdijk dam with seawater on one side of the dam, and freshwater on the other. The plant is capable of producing 50 kW of power (https://medium.com/@tjsmedley/blue-energy-can-we-get-all-our-future-energy-from-salt-water-d1e4f910d763).

It is estimated that the amount of energy released is equivalent to the potential energy of the same volume of water falling from a height of 250 m. (https://medium.com/@tjsmedley/blue-energy-can-we-get-all-our-future-energy-from-salt-water-d1e4f910d763; http://advantage-environment.com/future/blue-energy/). The amount of electricity that can be produced from a river flowing into the sea depends on a number of factors, including river and sea salinity and temperature, river water quantity and quality, river to sea water volume ratios, sea salt composition, salinity gradient steepness, infrastructure and local energy demand and environmental impacts. The type of membranes used in this type of energy harvesting plays an important role. Current research is focused on replacing polymer membranes with carbon nanotube membranes, which are more efficient than polymers. Although the technology is not yet implemented in full scale, it has improved and pilot plants are currently operational in Oslo, Norway (by Statkraft) and in Harlingen, The Netherlands (by Redstack).

The worldwide theoretical potential for salinity-gradient power is about 1,724 GW. For most regions, the potential is about 300 GW and the continent with largest potential is Asia with 374 GW. Only Europe (94 GW) and Australia (30 GW) have significantly lower potentials. Among the rivers with high technical potentials for salinity power are the Zaire (57 GW), Orinoco (36 GW), Ganges (25 GW), Nile (21 GW), Mississippi (18 GW), St. Lawrence (16 GW), Parana (16 GW), Zambezi (15 GW), and the Mekong (15 GW) (Kuleszo, 2010).

11.2.6 Shale energy – hydraulic fracturing

Hydraulic fracturing, or 'fracking', is an oil and gas well development process that involves injecting high-pressure water, sand and chemicals into a bedrock formation via the well. This process creates new fractures in the rock as well as increases the size, extent and connectivity of existing fractures, thereby improving the permeability of underground formations, which in turn facilitate the movement of gas and oil from petroleum-bearing formations. The wells are drilled vertically to depths far below the water-bearing aquifers and then horizontally and directionally to distances more than 1 km. Fracturing is accomplished by injecting fracturing fluid under high pressure (480–850 bar). Typically, the fracturing fluid consists of about 90% water, 9.5% sand (natural occurring sand grains, resin-coated sand, high-strength ceramic materials and resin-coated ceramic materials), which serves as a proppant that helps to keep the artificially created fractures open after the injection process and 0.5% additives, which help to reduce friction, thereby reducing the amount of pumping pressure from diesel-powered sources, which in turn reduces greenhouse gas emissions and prevent pipe corrosion, which in turn help protect the environment and boost well efficiency. The fracking fluid opens up new or enlarges existing rock fractures that facilitate the migration of natural gas towards the surface. Hydraulic fracturing creates a contact area that is thousand times that achieved by vertical drilling, thereby significantly increasing the production of natural gas from a single well. Modern hydraulic fracturing with horizontal drilling allows multiple wells to be drilled from one spot on the ground.

Hydraulic fracturing requires large volumes of water for a single operation. Fracturing shale gas typically requires around 15×10^3 m^3 of water per well. The actual amount depends on the type of well drilled and the geologic formation of the location. In general, deeper wells in stronger rock formations need more water. The withdrawal of large volumes of water for fracturing processes significantly limits water availability for human consumption, crop irrigation and livestock use.

Once the injection process is completed, the internal pressure of the rock formation causes waste fluid to return to the surface through the wellbore. This fluid, which is known as 'flowback' or 'produced water', may contain the injected chemicals plus naturally occurring materials such as brines, metals, radionuclides and hydrocarbons. The waste fluid is usually stored in onsite tanks and later treated during the waste management process to reduce the toxicity of the fluid and minimize its environmental impacts.

According to US Energy Information Agency (EIA) estimates, in 2016 the United States was the world's largest producer of petroleum and natural gas hydrocarbons. Since 1947, more than 1.7 million wells have been completed using the fracking process, producing more than 7 billion barrels of oil and 17 trillion cubic meters of natural gas. Shale rock formations have become an important source of natural gas in the United States. Hydraulic fracturing is also used in Canada, South Africa, Germany, United Kingdom, Russia and China to increase their natural gas production.

Although the fossil fuels extracted by hydraulic fracturing produce much less carbon dioxide emissions, the process is not without problems. Firstly, the excessive use of water deprives other water users their needs. Secondly, the flowback water that may contain undesirable substances contributes to surface and possible groundwater pollution. Thirdly, well drilling can induce seismic activity. As the world is moving towards the use of renewable energy, the use of more fossil fuels is not favoured by climate activists as demonstrated in the recent COP26 meeting in Glasgow.

11.3 ENERGY FOR WATER

Energy needs water and water needs energy and these linkages have enormous significance for economic growth, life and wellbeing. The interdependencies between energy and water are set to intensify in the coming years, as the energy needs of the water sector rise. Energy is required for a range of water-related processes, such as for provision of freshwater from surface and groundwater sources, water transport, wastewater treatment and desalination. Conversely, energy can be produced as a by-product from wastewater. Desalination and water reuse, which require energy, can help countries that have limited freshwater resources, thereby contributing to the rise in the water sector's energy demand. Although desalination and water reuse meet less than 1% of global water needs today, these processes are projected to increase significantly, particularly in the Middle East where renewable water resources are limited.

Globally, water extraction is estimated to consume over 310 TWh of electricity per year and about 0.5 million barrels of diesel fuel per day. Almost half of global electricity for extraction is consumed in Asia, as this is the continent with the largest water use. India is the world's largest water user by far, partly due to inefficient irrigation in agriculture, and accounts for around a quarter of global water withdrawals, although the per-capita use is well below that of the United States (www.iea.org/reports/world-energy-outlook-2019).

It is not only the amount of energy associated with water that matters but also the emission of carbon dioxide (CO_2) that contributes to climate change. It has been estimated that in 2005, water-related CO_2 emissions were equivalent to the annual greenhouse gas emissions of 53 million passenger vehicles. By sector, water heating was responsible for 70% of the water-related carbon emissions, wastewater treatment for 18%, water supply for 8% and agricultural activities for 6%. Thus, saving water saves energy and also reduces carbon emissions. (Energy–Water Nexus: The Water Sector's Energy Use (fas.org)).

Energy use for water is a function of many variables, including water source (surface water pumping typically requires less energy than groundwater pumping), treatment (high ambient quality raw water requires less treatment than brackish or seawater), intended end-use, distribution (water pumped long distances requires more energy), amount of water loss in the system through leakage and evaporation and the level of wastewater treatment. Likewise, the intensity of energy use for water varies depending on characteristics such as topography, climate, seasonal temperature and rainfall.

In the research and development front, much can be achieved if the water–energy nexus is treated holistically. The present practice in the two industries is to make policy decisions on how to manage each industry independent of the other. Because of the strong inter-dependence of water and energy, it would be best to make sustainable policy decisions considering the combined system. Other areas where research needs to be focused include technologies for saving energy and water as well as energy recovery from wastewater generation, integrated resource management, educating the consumers to use water and energy without unnecessary wastage, a tariff structure that promotes efficient use of water and energy that minimizes wastage.

11.3.1 Energy for agriculture

Energy used in agriculture consists of solar energy used in photosynthesis, direct energy used up by machines used in agriculture and indirect energy used up in producing agricultural machinery, fertilizers, pesticides and chemicals used in agriculture. Part of the energy used is stored in agricultural products as chemical energy for human and animal consumption while the remainder is wasted. Direct energy, including diesel fuel, gasoline, natural gas and liquid petroleum (LP) gas, is used largely for planting, tillage, harvesting, drying and transportation. Part of the solar energy, which is used to heat the soil, water, air and canopy, is either re-radiated or wasted as heat. Direct energy used up by agricultural machinery is wasted as heat and sometimes as noise. Indirect energy that does not contribute to the storage of agricultural chemical energy is also wasted. Chemical manufacturers use natural gas, electricity, fuel oil and other fossil fuels to produce fertilizers and pesticides. Electrical energy, also classified as direct energy, is primarily used for irrigation, climate control in livestock facilities and dairy operations. Indirect energy is used off-farm in producing inputs that are ultimately consumed on the farm. Some indirect energy-intensive farm inputs include pesticides and commercial fertilizer nutrients, which can use large amounts of energy in their production processes. Despite the increase in the amount of fertilizers since mid-1980s, the amount of energy embodied in fertilizers have declined significantly over the past 20 years (The Fertilizer Institute, 2004a; Shapouri et al., 2002).

According to FAO (Home | Energy | Food and Agriculture Organization of the United Nations (fao.org)), the global food system consumes about 30% of the world's available energy and that 70% of that energy is consumed after the food leaves the farm in transportation, processing, packaging, shipping, storage and marketing. It is also estimated that one-third of food produced is lost or wasted with a corresponding loss or wastage of energy.

Agriculture is the largest user of freshwater. Provision of irrigation water for agriculture is mainly by gravity but in regions where the source of water is at a lower elevation than the irrigated area, the water has to be pumped, which requires energy. Since irrigation is a major water user, energy efficiency is highly correlated with irrigation efficiency. Of the different types of irrigation systems practiced worldwide, drip irrigation systems, which operate at low pressures, would be more efficient. More about irrigation can be seen in Chapter 9. Energy efficiency in irrigation can be improved by reducing the quantity of water pumped, by increasing the efficiency of the pumping system or by reducing the system pressure.

FAO has launched the Energy-Smart Food for People and Climate (ESF) Programme, a multi-partner initiative, to assist member countries make the shift to energy-smart agrifood systems. The programme focuses on three thematic areas: energy efficiency, energy diversification through renewable energy and energy access and food security through integrated food and energy production. The ESF programme follows an interdisciplinary 'nexus' approach to ensure that food, energy, water and climate issues are jointly addressed, trade-offs considered and appropriate safeguards are put in place.

11.3.2 Energy for domestic water

Domestic water goes through several stages before it appears at the consumer's tap. It has to be collected, stored, treated, stored again and distributed to consumers. After the water is delivered to the consumers' tap, more energy is needed within the household including the energy to heat the water for showers and baths, to run kitchen appliances such as washing machines, dish washers, kettles, blenders etc. and for cooking. In temperate and cold climates, energy is also needed for central heating of living space, which may be accomplished by a wet system of hot water circulating through pipes to radiators. A boiler heats the water using fossil fuels or electricity and feeds the network of pipes connected to the radiators.

Collection of water for domestic consumption from a source depends upon the location of the source. If it is at a lower elevation than the service area, the water needs to be pumped and transported to the service area, which requires energy. The next stage of public water supply involves treatment, temporary storage preferably at a higher elevation and distribution, all of which require energy. More details of the treatment and distribution are given in Chapter 6 (Sections 6.2 and 6.3). After the consumers have used the water, the wastewater has to be appropriately treated before discharging into receiving waters to prevent pollution of the hydro-environment, which also is energy intensive. Aeration, pumping and solids waste disposal account for most of the electricity used in wastewater treatment plants. In all these stages of providing domestic water and discharge of wastewater, energy efficiency plays a crucial role. Energy efficiency and water efficiency are interrelated and this interrelationship is often referred to as energy–water nexus. Saving in energy also reduces carbon emissions. The energy cost of providing water and wastewater facilities constitute a major fraction of the total energy bill of municipalities.

Solar heating is one of the cleanest means of providing household hot water and is practiced in many places. Although the initial cost of installing solar heaters is high, the running cost is negligible since solar energy is free. One drawback in solar heating systems is that it cannot be guaranteed to provide hot water all the time as it is weather dependent. Therefore, a backup system to supplement the solar energy with conventional energy is necessary.

Globally, it is estimated that water treatment requires about 65 TWh of electricity, of which pumping accounts for 80–85%. For water distribution, the corresponding estimate is about 180 TWh with the energy intensity varying from place to place depending on elevation differences and pressure requirements (WEO 2019). The extraction of groundwater is much more energy intensive than that of surface water, but the energy needs for its treatment are usually less than those for surface water, as groundwater is typically less contaminated. Water losses due to bursts, leaks, thefts and inaccurate meter readings also account for a sizeable loss of energy.

An area where energy for water is likely to increase in future is for desalination, particularly in countries with insufficient freshwater resources to meet the projected demand in regions such as the Middle East. Despite the present high energy cost, harnessing solar energy using solar panels could propel desalination as a major source of freshwater in the future. United Arab Emirates, Saudi Arabia, Qatar and Kuwait have taken the lead in energy demand for desalination and re-use in 2014 (WEO 2016). It is projected that by 2040, desalination projects will account for about 20% of the water-related electricity demand (WEO 2016).

Methods of increasing efficiency of energy–water systems include optimizing system processes, such as modifying pumping and aeration operations and implementing monitoring and control systems through SCADA (Supervisory Control And Data Acquisition). The Electric Power Research Institute (EPRI) estimates that drinking water facilities can achieve energy savings of 5–20% through high-efficiency motors and drives and 10–20% through process optimization and SCADA systems (https://Energy-Water Nexus: The Water Sector's Energy Use (fas.org)).

Some water utilities are generating energy on-site to offset external energy by recovering energy from municipal waste and using the resulting biogas to generate electricity, heat and in some cases sell electricity back to the grid.

11.3.3 Energy for water transfer

Although there is plenty of water on earth, it is not available in sufficient quantities to meet the demand in some places. Therefore, water has to be conveyed from sources to demand areas, which in most cases is by gravity but requires energy in places where the source is at a lower elevation than the demand area. According to World Energy Outlook (WEO 2019), approximately 70 TWh of electricity is used for long-distance water transfer. The largest undertaking is China's South-to-North Water Transfer Project, with capacity projected to increase to 45×10^9 m^3 per year by 2050. Another project of similar magnitude is the State Water Project in California, the United States, which is a multi-purpose water resources development project to transfer water from northern rivers in California to water scarce southern regions that include storage, delivery, aqueducts, power plants and pumping stations. It is 1,100 km long, transferring about 3 km^3 of water annually, and is the single largest energy user in California consuming about 2–3% of all electricity consumed in the state (WEO 2019). Pumping energy is the single largest power load in California (https://Energy-Water Nexus: The Water Sector's Energy Use (fas.org)). Pumps that move water from the San Joaquin Valley to southern California for domestic and irrigation water uses are the single largest power load in the state.

Research targeted at improvements to energy and water efficiency needs to consider the water–energy nexus holistically rather than as separate issues. The current practices of managing water by water authorities (usually managed by public sector) and energy by power authorities (usually private) who make decisions on their respective management policies irrespective of the other do not lead to sustainable management of water–energy nexus.

11.3.3.1 South–North Water Transfer Project

The South-to-North Water Diversion Project, aimed at transferring 44.8×10^9 m^3 of water annually from the Yangtze River Basin in southern China to the Yellow River Basin in arid northern China, via three canal systems is the world's largest and most expensive water transfer project in the world. The three canal systems are known as the Eastern Route, the Central Route and the Western Route.

(i) Eastern route

The Eastern Route generally follows the course of the Grand Canal. The quantity of water transferred annually will increase from 8.9×10^9 m^3 to 10.6×10^9 m^3 to 14.8×10^9 m^3 as the project progresses. These quantities are only fractions of the annual flow in Yangtze River. The completed Eastern Route is 1,152 km long and is equipped with 23 pumping stations with a power capacity of 454 MW. Pumping stations are needed because of the topography of the region. Eastern route will transfer 1×10^9 m³ annually to Tianjin and is not expected to supply Beijing, which is to be transferred by the Central Route. An important component of the Eastern Route is a tunnel under Yellow River in Shandong Province consisting of two 9.3 m diameter horizontal tunnels 70 m below the bed of Yellow River.

(ii) Central route

The central, or middle route, runs a distance of 1,264 km starting from Danjiangkou Reservoir on the Han River, a tributary of Yangtze River, to Beijing and Tianjin. It was completed in December 2014, initially providing 9.5×10^9 m³ of water annually and is expected to increase to about 12–13 $\times 10^9$ m³ by 2030. In dry years, the annual water transfer will be at least 6.2 $\times 10^9$ m³, with a 95% guarantee.

In order to let the water flow by gravity, this route required raising the crest elevation of Danjiangkou dam from 162 to 176.6 m above mean sea level, thereby raising the reservoir water level from 157 to 170 m above mean sea level. A major challenge in this route was the construction of two tunnels under Yellow River.

Construction on the central route began in 2004. By 2008, the 307-km-long northern stretch of the central route has been completed. Water in that stretch of the canal came from reservoirs in Hebei Province and not from Danjiangkou Reservoir. As a result, farmers and industries in Hebei Province had to cut back on their consumption to allow transfer of water to Beijing. Danjiangkou Reservoir is the source of two-thirds of Beijing's tap water and a third of its total supply.

(iii) Western route

The western route was intended to transfer water from three tributaries of Yangtze River near the Bayankala Mountains to Qinghai, Gansu, Shaanxi, Shanxi, Inner Mongolia and Ningxia provinces. There are also plans to divert about 200×10^9 m³ of water annually from the upstream sections of Mekong (Lancang River), Yarlung Zangbo (called Brahmaputra further downstream) and Salween (Nu River) to Yangtze River and Yellow River and to the dry areas of northern China. If implemented, this route is likely to affect downstream water users in India, Bangladesh, Myanmar, Lao, Cambodia, Thailand and Vietnam.

The western route aims to divert water from the headwaters of Yangtze River to the headwaters of Yellow River. This requires dams and tunnels to cross the Tibetan Plateau and Western Yunnan Plateau. Although designed to transfer 3.8×10^9 m³ of water annually from the three tributaries (Tongtian, Yalong and Dadu Rivers) of Yangtze to northwest China, the feasibility of this route is still being studied and a firm decision to go ahead is yet to be made.

11.3.3.2 California State Water Project (SWP)

The California State Water Project (SWP), the largest state-owned water and power development user-financed water system in United States, is a multi-purpose water storage and delivery system that extends more than 1,100 km. It consists of a collection of canals, pipelines, reservoirs and hydro-electric power facilities, delivering clean water to 27 million Californians, 3,035 km² of farmland and businesses throughout the state. Planned, built and operated by the California Department of Water Resources as a user-financed water system designed to deliver about 5.2 × 10⁹ m³ annually, with 30% of the water used for irrigation mostly in San Joaquin Valley and the remaining 70% for domestic, municipal and industrial use. The project includes some 22 dams and reservoirs, a Delta pumping plant, a 714-km-long aqueduct that carries water from the Delta through the San Joaquin Valley to southern California. The project begins at Oroville Dam, the tallest dam in the United States with a height of 230 m on the Feather River and ends at Lake Perris near Riverside. At the Tehachapi Mountains, giant pumps lift the water from the California Aqueduct, some 878 m over the mountains with 587 m at the Edmonston Pumping Plant alone, the highest single water lift in the world. The project provides water to Los Angeles, Riverside, San Bernardino, San Diego and other parts of southern California as well as cities in Napa and Solano counties through the North Bay Aqueduct, Santa Barbara and San Luis Obispo counties through the Coastal Aqueduct and communities in Alameda and Santa Clara counties through the South Bay Aqueduct. In addition to delivering water and flood control, the project is also a major source of power, generating 6,500 GWh annually using a combination of conventional hydro-electric plants and pumped storage power plants, which together constitute the Oroville–Thermalito Complex with a generating capacity of some 819 MW.

11.3.3.3 Other major water transfer projects

Two other major water transfer projects are the 'Great Man-Made River' project in Libya and the Kaleshwaram multi-purpose project built across Godavari River in Telangana state in India. In the former case, fossil water trapped in the Nubian Sandstone Aquifer system with an estimated 373 × 10¹² m³ of groundwater is pumped to provide irrigation and other water needs for most of Libya's urban centres. The aquifer system, which covers some 2 × 10⁶ km², is the largest fossil freshwater reservoir in the world. This man-made river covers a distance of up to 1,600 km and provides 70% of all freshwater used in Libya (Great Man-Made River – Wikipedia). In the latter case, water from Godavari River in Telangana state is lifted to a height of some 500 m and delivered by a complex system consisting of 1,832 km of water supply routes, 1,531 km of gravity canals, 203 km of tunnel routes, 20 lifts, 19 pump houses and 19 reservoirs with a total capacity of 141×10^6 m^3. More details of these two projects can be found in Chapter 9.

11.3.4 Energy for bottled water

The present-day practice of using bottled water for convenience and quality also drains a substantial amount of energy. The energy used in bottled water is used up in manufacturing plastic bottles (polyethylene terephthalate (PET)), extracting water from sources and treatment, transport and distribution.

It has been reported that bottled water consumes 5.6–10.2 Joules of energy per litre of bottled water depending on transport factors. This is almost 2,000 times more than the energy used for producing the same volume of tap water. In the United States, the energy used to produce bottled water accounts for about one-third of one percent of total US energy consumption (Gleick and Cooley, 2009).

In the United States, the water in PET bottles comes from two sources – spring water from underground springs and purified tap water, which respectively account for roughly 56% and 44%. Bottled water from natural springs is marketed under different trade names such as Fijian (Fiji), Evian (France), Hildon (UK), St. Geron (France), Voss (Norway), Volvic (France), Mountain Valley Spring Water (United States), Perrier (France), Farrarelle (Italy) and Gerolsteiner (Germany).

Energy is also required to cool the bottled water prior to sale or consumption. This has two components: the energy needed to cool the water from room temperature to temperature of the refrigerator and the energy needed to maintain the cold water, which depends on the length of time the water is kept in the refrigerator until sold or consumed. It has been estimated that the energy needed to cool from a room temperature of 20°C to a typical refrigerator temperature of about 3.3°C is about 220 kJ per litre (Gleick and Cooley, 2009). An additional energy requirement to maintain the refrigerator temperature for one week is estimated to be 220 kJ per litre. In addition to the contributions to greenhouse gas emissions from the energy used up in bottled water, the problem of disposal of PET bottles is a challenging environmental issue.

11.3.5 Energy for wastewater

Globally, wastewater treatment consumes about 200 TWh or 1% of total energy consumption. In developed countries, wastewater treatment is the largest energy consumer in the water sector. It is projected that global electricity consumption for wastewater collection and treatment will require over

60% more electricity in 2040 than in 2014, as the amount of wastewater in need of treatment increases (WEO 2019).

Wastewater treatment consists of three stages known as primary, secondary and tertiary. Primary treatment, which is physical, involves removal of solids via filters, screens, sedimentation tanks, aeration and UV radiation. Secondary treatment, which is biological, involves removal of dissolved organic matter via the use of aeration tanks, trickling filters, intermittent sand filters, waste stabilization ponds and activated sludge process. Tertiary treatment involves removal of nutrients such as nitrogen and phosphorous via sand filtration and membrane technology. These stages are followed by disinfection by chlorination or ozonation before discharging the treated water to receiving waters. The levels of treatment vary from country to country and sometimes even from region to region within the same country. Primary treatment is the dominant process in developing countries whereas secondary and tertiary treatments are carried out in developed countries. Much of the energy used in wastewater treatment is spent for secondary treatment. Other energy uses include those for pumping and sludge treatment, which involves thickening, digestion, drying, incineration and centrifuging. It is also possible to recover some of the energy used in sludge treatment in the form of heat or electricity from biogas, which is a by-product.

The factors that influence the energy consumption for wastewater treatment include the share of wastewater collected and treated, the level of groundwater infiltration and rainfall into the sewage in combined systems, the treatment level, the contamination level and the energy efficiency of the operations. Aeration systems and pumping are the most energy-intensive processes in wastewater treatment plants. A report by EPRI suggests that 52% of wastewater treatment plants energy consumption is used for aeration, 12% for pumping and 30% for bio-solids processing (Electric Power Research Institute and Water Research Foundation, 2013). The largest component of energy consumed is electrical energy. Other important energy consumers in wastewater treatment plants are pumps (18.9%) (Daw et al., 2012).

The biological processes normally account for 50–80% of the total electrical energy consumption in wastewater treatment plants (Jonasson, 2007; WERF, 2016; Awe et al., 2016). Energy consumption for aeration process is generally between 0.18 and 0.8 kWh per m^3 (Gandiglio et al., 2017). Biological processes consume a high share of energy in wastewater treatment plants (Awe et al., 2016).

A unique property in wastewater treatment is that the influent wastewater carries some amount of recoverable energy. Chemical energy contained in wastewater treatment plants can be partially recovered by means of sludge anaerobic digestion during which biogas is produced, which can be converted to electrical or thermal energy. The kinetic energy of influent wastewater can also be utilized to drive turbines that generate electricity. It is reported that about 25–50% of energy consumed in wastewater plants can be recovered by internal energy generation (Mccarty et al., 2011). In particular, 90% of this energy is thermal (due to the water heat capacity), 20% chemical (expressed by the COD) and less than 1% hydraulic (potential and kinetic) (WERF, 2014b, 2016). Energy consumption could also be reduced by means of online control systems such as, for example, intermittent mixing by turning off the mixers for a period of time without affecting the treatment process (Jonasson, 2007). More information about energy savings in secondary treatment can be found in Panepinto et al. (2016).

Co-digestion of sludge with other organic wastes has received increasing attention in recent years as it has the potential to increase biogas production, thereby increasing the recoverable energy (WERF, 2014a; Shen et al., 2015). Sludge pre-thickening by means of a centrifugal system can achieve the goal of 100% self-sufficiency from the thermal point of view. Sludge pre-treatments and anaerobic digestion management are key drivers to increase biogas yield and reduce the energy consumption of the sludge line whereas the use of co-digestion with other organic wastes is a potential way to increase onsite energy generation.

REFERENCES

Awe, O. W., Liu, R., and Zhao, Y. (2016). Analysis of energy consumption and saving in wastewater treatment plants: case study from Ireland. *Journal of Water Sustainability* 6, 63–76. doi: 10.11912/jws.2016.6.2.63-76

Daw, J., Hallett, K., Dewolfe, J., and Venner, I. (2012). Energy Efficiency Strategies for Municipal Wastewater Treatment Facilities, National Renewable Energy Laboratory.

Electric Power Research Institute and Water Research Foundation (2013). *Electricity Use and Management in the Municipal Water Supply and Wastewater Industries*, 1–194. Available online at: www.epri.com/search/Pages/results. aspx?k=3002001433

Gandiglio, M., Lanzini, A, Soto, A. Leone, P. and Santarelli, M. (2017). Enhancing the Energy Efficiency of Wastewater Treatment Plants through Co-digestion and Fuel Cell Systems, *Frontiers in Environmental Science*, October 30, 2017 https://doi.org/10.3389/fenvs.2017.00070. 21pp

Gleick, P. H. and Cooley, H. S. (2009). Energy implications of bottled water, *Environmental Research Letters* 4, 014009 (6pp).

International Renewable Energy Agency (2017): Renewable power generation costs in 2017. 16 pp. www.irena.org/-/media/Files/IRENA/Agency/Publication/2018/ Jan/IRENA_2017_Power_Costs_2018_summary.pdf? la=en&hash=6A74B8D 3F7931DEF00AB88BD3B339CAE180D11C3)

Jonasson, M. (2007). *Energy Benchmark for Wastewater Treatment Processes – A Comparison between Sweden and Austria*. Lund: Department of Electrical Engineering and Automation Master.

Kuleszo, J., Carolien Kroeze, C., Jan Post, J. and Fekete, B. M. (2010). The potential of blue energy for reducing emissions of CO_2 and non-CO_2 greenhouse gases,

Journal of Integrative Environmental Sciences, 7:S1, 89–96, doi: 10.1080/19438151003680850

Mccarty, P. L., Bae, J., and Kim, J. (2011). Domestic wastewater treatment as a net energy producer – can this be achieved? *Environmental Science Technology* 45, 7100–7106. doi: 10.1021/es2014264

Panepinto, D., Fiore, S., Zappone, M., Genon, G., and Meucci, L. (2016). Evaluation of the energy efficiency of a large wastewater treatment plant in Italy. *Applied Energy* 161, 404–411, doi: 10.1016/j.apenergy.2015.10.027

Shapouri, H., Duffield, J. and Wang, M. (2002). The energy balance of corn ethanol: An update, Agricultural Economic Report 813, US Department of Agriculture, Office of the Chief Economist, Office of Energy Policy and New Uses, Washington DC.

Shen, Y., Linville, J. L., Urgun-Demirtas, M., Mintz, M. M. and Snyder, S. W. (2015). An overview of biomass production and utilization at full-scale wastewater treatment plants, (WWTPs) in the United States: Challenges and opportunities towards energy-neutral WWTPs. *Renewable Sustainable Energy Review*, 50, 346–362, doi: 10.1016/j.rser.2015.04.129

The Fertilizer Institute (2004). Production Cost Surveys. Various issues from 1990–2004. Compiled by International Fertilizer Development Center, Muscle Shoals, Alabama.

World Water Development Report (2014). Water and Energy, World Water Assessment Programme, UNESCO.

World Energy Outlook (WEO) (2016, 2019). (www.iea.org/reports/world-energy-outlook-2016, 2019).

Water Environment Research Foundation (WERF) (2016). WERF Energy factsheet, Alexandria: Water Environment and Reuse Foundation.

Water Environment Research Foundation (WERF) (2014a). Co-Digestion of Organic Waste Products with Wastewater Solids. Alexandria: Water Environment and Reuse Foundation.

Water Environment Research Foundation (WERF) (2014b). Energy from wastewater, Wastewater, Alexandria: Water Environment and Reuse Foundation.

Chapter 12

Water for transport

12.1 INTRODUCTION

Water, apart from being the most important ingredient of life, is also a medium of transport. It carries nutrients necessary for the metabolism of all living things and also carries away the waste generated in living beings. Plants manufacture the food that all other living beings need. The main ingredient in this process is water, which is available in the soil. Plants have devised an ingenious hydraulic system of transporting the water and nutrients available in the soil to the leaves in the canopy, which in tall trees can exceed 100 m in height. This transport system, which is against gravity, can be considered as equivalent to the blood circulation in humans and animals. The difference is that the blood circulation is powered by the heart, which acts like a pump whereas the water and nutrient transport in plants is driven by the hydraulic and chemical potential gradients.

The second type of transport where water is the medium is for the movement of people, goods and services over oceans, lakes, canals, rivers and other types of waterways. It is an important mode of transport that can be via inland transport and/or ocean transport. World trade depends heavily on ocean transport.

Inland transport exists where there are natural rivers or artificial canals of sufficient depth as it is one of the cheapest modes of transporting goods. In landlocked countries such as in Europe, the inland transport system is well developed. Ocean transport can be coastal or across oceans. The former is mainly in countries that have long coastal lines such as in India and mainly for internal trade.

12.2 WATER TRANSPORT IN PLANTS

Plants manufacture the food that all other living beings need. The main ingredient in this process is water. Photosynthesis, which is the process of manufacturing the food, takes place in leaves of plants in the presence of sunlight together with carbon dioxide absorbed from the atmosphere via the stomata of the leaves. The water and nutrients needed for plant development come from the soil. Unlike in humans, where the heart that acts as a pump to keep the blood in circulation, there is no pumping system to keep the water circulated within the plant. Instead, plants have an ingenious hydraulic system with no external energy to absorb, circulate and exhaust water. This system acts against gravity with no external energy. For example, trees that are very tall such as the Redwood trees in California, which are over 100 m tall, the hydraulic system in plants convey water to the canopy against gravity. This is possible due to the presence of two types of vascular tissues in plants known as xylem and phloem.

Xylem is a kind of tough plant tissue found in vascular plants that is made out of dead cells, which are connected end to end, forming a conduit-like structure that allow water to travel with little resistance. These cells, known as tracheids, are made of cellulose and lignin. Water and nutrient transport in the xylem is unidirectional upwards. Phloem, on the other hand, are living cells arranged end to end and their function is to transport food (mainly sucrose and amino acids) from where they are produced to where they are needed for plant growth and reproduction. The transport in phloem is bidirectional. This process of transporting food and nutrients from one part of the plant to other parts of the plant is known as translocation.

The transport process begins at the root system, which consists of a main root, lateral or secondary roots and root hairs. Root hairs are very thin but have a large surface area and can penetrate into the soil water system. The transport process starts at the root hairs assisted by osmosis, which is a spontaneous movement of a solvent from a region of lower concentration to a region of higher concentration across a semipermeable membrane. From the roots, water moves from cell to cell assisted by osmosis until it reaches the xylem. The water is then transported through the xylem conduits to the leaves and evaporated to the atmosphere by transpiration. Part of the water is used up in the photosynthesis process. The food produced by the photosynthesis process is then translocated to areas of the plant that are metabolically active via the phloem consisting of living cells.

The limiting factor for the growth of plants is water. Plants transport water and nutrients from the roots to the tips of their leaves, making transportation a vital process in plants in the same way as blood circulation in humans and animals. Xylem transports water and nutrients from the roots to other parts of the plants whereas phloem transports the food produced by photosynthesis to other parts. Transport by xylem is unidirectional upwards whereas that in phloem is bidirectional upwards and downwards.

Most of the water that is transported via the xylem is transpired through the stomata in the leaves. It is a continuous process that creates a negative pressure on the xylem tubes, which together with the cohesion and surface tension of water molecules and hydrogen bonding helps to pull water against gravity. The bulk of water transported is driven by negative pressure generated by the transpiration process. This transport system enables nutrient transport, provides turgidity to cells in order to support the plant, provides water for photosynthesis and helps to keep leaves cool by evaporation.

12.3 INLAND WATER TRANSPORT

Inland waterways are well developed in Europe and the United States. In Europe, France, Germany, Belgium, the Netherlands and Russia have extensive inland waterways including rivers and canals. The major French rivers have developed to the extent that travelling from Mediterranean Sea to English Channel or from Rhine to Atlantic Ocean can be entirely by rivers and canals.

The Rhine River that connects the oceans to the heart of Europe is the busiest navigable river of the world. It is navigable by small size ocean-going vessels. It is kept busy by the presence of heavy industries on both banks that utilize the benefit from cheap water transport. Starting from the Swiss Alps, the river flows through Switzerland, Liechtenstein, Austria, Germany, France and the Netherlands and empties into North Sea via the Netherlands. It is 1,230 km long and navigable up to 870 km. The river has many tributaries in the countries through which it flows. There are also many canals connected to Rhine to take advantage of its unique position as an inland waterway. Due to the discharge of industrial effluents from the well-developed factories en route, environmental pollution of the river has become an issue of concern in recent years.

The United States has over 36,000 km of navigable inland waterways, of which the two rivers Mississippi and Missouri alone are navigable up to about 8,000 and 1,200 km, respectively. In addition, there are a number of navigable canals including the St. Lawrence canal, which connects Ontario and St. Lawrence, the Sault Sainte Marie canal between lakes Superior and Huron, the canal which links the Chesapeake to Ohio, the New York canal and the canals between North Allegheny and Lake Erie. Some of these waterways cannot be used during winter season due to ice formation.

In South America, the main river Amazon has not been heavily utilized for transport since it runs mainly through relatively less developed areas. The Parana River system, which flows through Argentina, Paraguay, Uruguay and South Brazil, and Orinoco River, which flows through Venezuela, are the main inland waterways in South America.

In Asia, China has the largest inland waterways using the two main rivers Yellow and Yangtze, which run from west to east. In addition, the Grand Canal (see Section 12.3.1), which connects the cities Hangzhou and Beijing, is also a major inland waterway with several ship locks.

In Africa, inland waterways are not well developed as many rivers have rapids and falls.

India has three large navigable rivers, the Ganges (Ganga), the Brahmaputra and the Jamuna. Ganga can be navigated as far as Kanpur, and Brahmaputra is navigable as far as Dibrugarh. One of its tributaries, the Surma, is navigable as far as Sylhet and Cachar. Indus River in Pakistan is navigable as far as Dera Ismail Khan in the North-Western Frontier Province. The Irrawaddy, the most important and the largest river in Myanmar, is navigable up to about 800 km.

12.3.1 Grand Canal

The Grand Canal (also known as the Beijing–Hangzhou Grand Canal, Da Yunhe, Jing-Hang Yunhe) is the oldest and longest canal or artificial river in the world. It is also a UNESCO World Heritage Site since June 2014, as well as a marvel of hydraulic engineering. It starts from Beijing, links the Yellow River, Qiantang River, Huai River, Wei River, Hai River and the Yangtze River and passes through the city of Tianjin and the provinces of Hebei, Shandong, Jiangsu and Zhejiang, and ends up in the city of Hangzhou. Running in a north–south direction when most natural rivers in China run from west to east, this artificial river connects two great rivers in the world. The earliest parts of the canal dates back to the 5th century BC, but the various sections were combined during the Sui dynasty (581–618 AD). Millions of labourers and soldiers were employed in the construction of the canal.

The canal is 1,776 km long, 30–61 m wide, 0.6–4.6 m deep with its highest elevation of 42 m in the mountains of Shandong. The elevation of the canal bed varies from 1 m below sea level at Hangzhou to 27 m at Beijing. It is not a single continuous man-made canal but a series of canals linking several existing artificial or natural canals, lakes, rivers and ship locks. The main purpose of constructing the Grand Canal was to transport grains from south to north using barges. It also served as an internal communication route in China. It was considered as the economic lifeline for internal trade within China. Currently, the canal is navigable only from Hangzhou to Jining, a city located in the southwest of Shandong Province. The route has seven sections – Jiangnan Canal (Hangzhou to Zhejiang), Inner Canal (Yangtze River to Huaian), Middle Canal (Huaian to Weishan Lake), Lu Canal, also called Shandong Canal (Weishan Lake to Linqing), South Canal (Linqing to Tianjin), North Canal and Tonghui River (Tianjin to Beijing). The Grand Canal is currently used as the Eastern Route of the South–North Water Transfer Project. Some details of the Grand Canal are given in Table 12.1 and an approximate profile in Figure 12.1.

Because of the elevation difference along the route of the canal, locks have been built to facilitate navigation from one water level to another water

Table 12.1 Details of Grand Canal

Section	Distance (km)	Intervals (km)	Ground level (m)	Water level (m)	Depth of canal (m)
Beijing	0		36	34	
Tonghui River		29			3
Tongzhou (Bai River)		124			3–8
Tianjin	153		8	7	
Yu River		386			3–10
Linqing	539		36	35	
Huitong River		113			3
Yellow River (North side)	652		38	35	
Yellow River (South side)	652		38	42	
Hui Tong River, Qing Ji Du and Ji zhou River		69			4–4.3
Nan wang Zhen summit	721		52	42	
Jizhou River		21			4–7
Jining	742		40	35	
Huan Gong Gou		143			3
Linjia Ba (South end of lakes today)	885		36	35	
Huan Gong Gou		250			3–10
Huaiyin	1,135		18	16	
Shanyang Yundao		186			4–8
Yangtze River (North side)	1,321		5	0	12–15
Yangtze River (South side)	1,361		5	0	12–15
Jiangnan River		47			4
Danyang	1,408		17	2	
Jiannan River		298			4
Hangzhou	1,706		4	0	

Source: Ronan (1995).

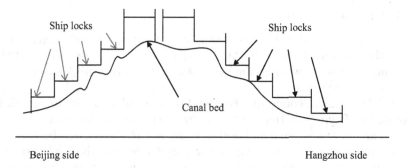

Figure 12.1 Approximate profile of Grand Canal (not to scale).

level. The type of locks used, known as 'pound locks', were invented during the Sung dynasty in 984 AD to help raise or lower the water level in a canal. A pound lock has a chamber with gates at both ends that control the level of water in the pound. A 'pound' in this context is the level stretch of water between two locks (also known as a reach). Nowadays, 'pound' locks are used in many canal/river systems in the world replacing the earlier type of gates with a single gate known as 'flash locks' (also known as staunch locks). At each lock, the water level differs by about 1.2 m and to raise the water level by 42 m required 24 locks. The canal also has 60 bridges across it.

12.4 OCEAN TRANSPORT

The major ocean routes for trade are the North Atlantic Ocean Route that connects the ports of Western Europe with those on the east coast of North America and has the greatest traffic of all ocean routes, the Suez Canal or Mediterranean Asiatic Route that connects the markets of Africa, Asia, Australia, New Zealand with those in Europe and is a vital link between the oil-producing countries with the oil-consuming countries, the Cape of Good Hope Route, which is an alternative to the Suez Canal route for super tankers but is much longer, the Panama Canal, the West Indian Central American Route that connects the Atlantic and Pacific Oceans, the South Atlantic Route that maintains trade connections between Europe and West Indies, Caribbean seaboard, Brazil, Uruguay and Argentina, the Trans-Pacific Route consisting of the North Pacific route linking ports on the western side of the United States and Canada with Asian ports in Japan, China, Philippines and Singapore, and the South Pacific Route that connects Western Europe and North America with Australia and New Zealand. Two of the canals that connect oceans, the Suez and the Panama are briefly described below:

12.4.1 Suez Canal

Two of the well-known waterways that connect seas are the Suez Canal and the Panama Canal. The Suez Canal connects the Red Sea and the Mediterranean Sea whereas the Panama Canal connects the Atlantic Ocean and the Pacific Ocean.

The Suez Canal, connecting Port Said in the north and Suez in the south, located in Egypt and 163 km (193.3 km including the southern and northern access canals) long and 300 m wide, is a sea-level canal because the water levels at the two ends (Gulf of Suez in Red Sea and the Mediterranean Sea) are approximately the same. It can accommodate ships up to a draft of about 19 m. The construction of Suez Canal began officially on April 25, 1859, and opened ten years later on November 7, 1869 at a cost of some $100 million. Initially, construction was carried out using Egyptian manual labour (some forced) but towards the end, many mechanical devices were used for digging and shaping the canal. It is one of the world's busiest shipping lanes. A large percentage of Europe's energy needs are transported from Middle East oil fields via the Suez Canal. It shortens the sea voyage by about 7,000 km. Until recently, most of the Suez Canal was not wide enough for two ships to pass side by side. Passing bays at the Ballah Bypass and the Great Bitter Lake help to alleviate this problem. Typically, it would take 12–16 hours to pass through the canal. The low speed helps to prevent bank erosion by waves.

The canal was closed for shipping twice due to wars. On July 26, 1956, Gamal Abdel Nassar, the President of Egypt at that time nationalized the Suez Canal as a result of United States and UK withdrawing their support for the construction of Aswan dam in Egypt. On October 29, 1956, Israel invaded Egypt and Britain and France followed on grounds that the passage through the canal was to be free. Egypt retaliated by sinking 40 ships, leading to what was known as the Suez Crisis. The Suez Crisis ended in November 1956 as a result of UN intervention by arranging a truce among the United States, UK, Egypt and Israel. Again, in 1967, the canal was closed for commercial shipping during the Six-Day War between Israel and its Arab neighbours. The Suez Canal is currently owned and operated by the Suez Canal Authority of Egypt. It has also been closed several more times due to conflicts between Egypt and Israel.

Suez Canal is not only of strategic importance and national pride for Egypt but also a substantial source of revenue. In recent years, with about 50 ships passing through the canal every day, the annual toll charges run into several billions of dollars. In August 2015, a new 35-km expansion running parallel to the main channel was opened, enabling two-way transit through the canal. It is now called the 'New Suez Canal'. The new depth and width of the canal including the parallel sections range from 23 to 24 m and 205 to 225, m respectively. A railway also runs parallel to the canal on the west bank.

12.4.2 Panama Canal

Panama Canal is an artificial waterway connecting the Atlantic Ocean and the Pacific Ocean. It was opened in August 1914 after more than 30 years of construction and is now managed by the Panama Canal Authority. The long construction time has been due to technical and health issues. Many workers have perished during the construction period.

Unlike the Suez Canal, Panama Canal is not a sea-level canal. It cuts across the Isthmus of Panama and there are locks at the Atlantic side to lift ships up to Gatun Lake, which is an artificial lake 26 m above sea level and to lower them at the Pacific end. The lake has been created to reduce excavation. The canal is 77 km long.

France began the construction of the canal in 1881 but stopped due to technical and worker health problems. The mortality rate was very high due to tropical diseases such as malaria and yellow fever. The United States took over the project in 1904 and completed and opened the canal on August 5, 1914. The total casualty cost has been over 25,000. The canal reduced the sea voyage by 12,875 km and takes about eleven and half hours to pass through. Following a treaty between the United States and Panama, the Panama Canal Authority, an institution answerable to the Panamanian government, took full control of the canal on December 31, 1999. With over 14,000 vessels passing through annually, the Panama Canal is one of the main sources of revenue for Panama. The toll charge for the passage through Panama Canal is of the order of $450,000 for large ships. The lowest toll paid has been only 36 cents by an adventurer who swam along the canal from one end to the other in 1928.

The profile of the Panama Canal (Figure 12.2) consists of an approach channel in Limon Bay on the Atlantic side, which connects to the Gatun locks at a distance of 11 km. At Gatun, ships are lifted by a series of three locks by a height of 26 m to Gatun Lake. The lake is formed by Gatun dam across Chagres River. The channel through the lake varies in depth from about 14–26 m and extends for about 37 km to Gamboa from where the Gaillard Cut (also known as Culebra Cut or Corte de Culebra in Spanish) begins. The channel through the cut extends to a length of about 13 km with an average depth of 13 m up to Pedro Miguel Locks. The locks lower vessels by 9 m to Miraflores Lake, at an elevation of 16 m above sea level. After passing through the channel for about 2 km, vessels are lowered to sea level by a system of two-stepped locks at Miraflores. The final segment consists of a dredged approach channel about 11 km long to join the Pacific Ocean. The original lock chambers were 320 m long, 33.5 m wide and 12.56 m deep constraining the size of the ships that could pass through. The new lock

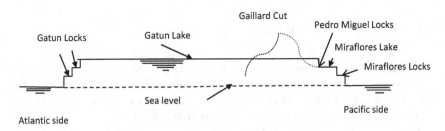

Figure 12.2 Sketch of the profile of Panama Canal (not to scale).

chambers, opened for commercial traffic on June 26, 2016, are 426.72 m long, 54.86 m wide and 18.29 m deep enabling larger ships to pass.

REFERENCE

Ronan, Colin A. (1995). *The Shorter Science and Civilization in China*, An abridgement of Joseph Needham's original text, Cambridge University Press. 364 pp.

Chapter 13

Water for industry

13.1 INTRODUCTION

Industrial demand for water is relatively new compared with agricultural and domestic demands. Perhaps it started in the 20th century when the industrial revolution began to change our lifestyles. It is not a need. It is not necessary for survival but serves humans with improved living standards. The main industrial use of water is for power generation, air conditioning, cooling, manufacturing, among other uses. The quality of cooling water can be low and the water can be recycled. Other industrial uses include water in boilers for steam generation for thermal power stations where high-quality water is needed, in manufacturing processes and for fire-fighting, which is essential in urban living where the rates of flow need to be very high but lasting for a short time.

Industrial water usage is the second largest water user in the world. In industrialized countries, it is the major water user accounting for approximately 55% of total water withdrawals whereas in less developed countries, it is about 9%. Based on information available in different sources (which sometimes differ from source to source), Table 13.1 summarizes the industrial water use as well as the industrial water use as a percentage of total freshwater withdrawals for the year 2015. The world average is about 16%. High-income countries tend to use more water for energy generation and industry. In the United States, in 2010, thermoelectric power used 45% of water (mostly for cooling), irrigation about 32% and public water supply about 12% (Which Industries Use the Most Water – Better Meets Reality). Every manufactured product uses water during some part of the production process including water used for fabricating, processing, washing and diluting, cooling, transporting, incorporating water into a product or for sanitation needs within the manufacturing facility. Industries such as food, textile, pulp and paper, iron and steel, chemicals, petroleum etc. use large quantities of water.

Industries such as mining, chemical, paper manufacturing, textile manufacturing, dyeing, printing and cooling use large quantities of water mainly

DOI: 10.1201/9781003329206-13

Table 13.1 Industrial water withdrawals by region

Region	Industrial water use (km³/year)	Total freshwater withdrawal (km³/year)	Industrial water as a percentage of total freshwater withdrawal
Africa	15	236	5
Americas	271.3	823	34
Asia	225.3	2,637	10
Europe	131	292	57
Oceania	4.2	22	15

from dedicated sources and at the same time they also contribute to chemical and thermal water pollution.

Industries that use significant quantities of water include hydropower generation (see also Chapter 11), hydraulic fracturing (see also Chapter 11), water jet cutting that use pressurized water, cooling and many other industrial processes such as in steam turbines and as a chemical solvent.

13.2 TYPES OF INDUSTRIES

13.2.1 Manufacturing processes

Manufacturing processes use water for creating their products or cooling equipment used in creating their products. According to the United States Geological Survey (USGS), industrial water is used for fabricating, processing, washing, diluting, cooling or transporting a product. Water is also used by smelting facilities, petroleum refineries and industries producing chemical products, food and paper products. Large amounts of water are used mostly to produce food, paper and chemicals. In the United States, according to their Census Bureau estimates, the manufacturing industry requires 18 billion gallons (68 million m³) of water per day in their production operations and is expected to rise significantly by mid-century. Table 13.2 gives typical water requirements for some selected industries and Table 13.3 gives the average water use for some selected products.

13.2.2 Electricity generation industry

Electricity is a basic requirement for all other industries. It can be generated in hydroelectric power plants, thermal power plants, solar power plants, wind power plants, wave energy power plants, tidal energy power plants and blue energy (saline) power plants. Of these, the widely used ones are the first four whereas the last three have not yet been developed on a commercial scale. Details of hydroelectric generation are discussed in Chapter 11. Thermal power plants can be divided into a category that creates electricity by burning fossil fuel and a category that uses nuclear reactions. In both

Table 13.2 Typical requirements for selected industries

Product	Typical water requirement (m^3/ton)
Nitrogenous fertilizer	600
Steel	150
Paper	250
Oil	180
Sugar	100
Artificial silk	1,000
Brick	2
Plastics	750–2,000

Source: *Water for a Starving World* by M. Falkenmark and G. Lindh (1976).

Table 13.3 Average water use for some products

Steel	62,000 US gallons (234.7 m^3) per ton
Average motor car	39,090 US gallons (147.8 m^3)
Pair of blue jeans	1,800 US gallons (6.81 m^3)
Cement	1,300 US gallons (4.92 m^3) per ton
Pair of shoes	2,257 US gallons (8.54 m^3)
Wool	101 US gallons /pound (0.842 m^3/kg)
Cotton	101 US gallons /pound (0.842 m^3/kg)
Plastic	24 US gallons/pound (0.198 m^3/kg)

Source: How Many Gallons of Water Does It Take to Make … (treehugger.com).

categories, the energy of the fuel or nuclear reactions is first converted into mechanical energy via a turbine and then to electrical energy. Solar power plants convert the sunlight into electrical energy using photovoltaics whereas wind power plants convert the kinetic energy of the wind to mechanical energy and then to electrical energy. Worldwide electricity generation has increased from 11,897 TWh in 1990 to 26,730 TWh in 2018 (Global electricity generation I Statista). The United States has been the largest producer and consumer of electricity followed by China, Japan, Russia and India until recently, but in 2011 China has overtaken United States to become the largest producer of electricity. Energy generation has been steadily increasing over the years. Worldwide, fossil fuel remains the major source of energy.

Most of the water used in electricity generation (apart from hydro-electricity) is in thermal power plants as cooling water. Other uses include fuel extraction that consist of drilling and mining of natural gas, coal, oil and uranium, fuel refining before they can be used, fuel transportation and emission control, which require thermal pollution control technologies that need water. In thermal power plants, ultrapure water is used as a source to make steam to drive turbines and for other uses to prevent corrosion, cracking in equipment such as turbine blades, stainless-steel lines, steam circuits and cooling systems.

Water use in power plants consists of withdrawal and consumption. Water withdrawal is removing water from a water source. Some power plants located near the sea use seawater for cooling, but a vast majority of power plants use fresh water for cooling. The withdrawn water may or may not be returned to the source or made available for use elsewhere. Water consumption is the amount of water lost by evaporation during the cooling process.

The global water consumption of energy production (WCEP), based on data from 158 countries, is estimated to be about 52 billion cubic metres of freshwater, of which about 40% accounts for the oil and gas production (Spang et al., 2014).

Although cooling accounts for the greatest use of water in thermal power generation, the actual water use depends upon the type of cooling system used. The cheapest among them is the once-through cooling system, but it has the heaviest water use since there is no re-use. Once it has gone through the cooling system, the effluent water is returned to the water source or to another water body. This practice is not desirable from an environmental point of view since the effluent water is at a higher temperature and may also contain some pollutants that can affect aquatic life in receiving waters. An alternative is to recycle water that has been used for other processes, which may require some form of treatment to prevent scaling and to meet the specifications of cooling equipment. Other less widely used cooling systems include evaporative cooling systems where hot water is brought into contact with cool air, causing heat energy to dissipate through evaporation and dry cooling systems using fans and heat exchangers, which consume little or no water. Other options include using degraded water such as effluents from sewage treatment plants, high salinity water, water from mining operations, agricultural runoff and storm water, but the cost of transporting such water is a factor to be considered.

The outlook for future appears to be an attempt to shift the energy generation from fossil fuels to renewable sources, which is in line with the policies to combat climate change.

13.2.3 Iron and steel industry

Steel is one of the most popular and widely used construction material because of its unique strength, durability, workability and cost. According to the World Steel Association, the largest steel-producing countries are China, India, Japan and the United States, with China accounting for roughly 50% of the production. In modern steel production, the raw materials include iron ore, coal, limestone as well as recycled materials such as scrap iron.

The production of steel in an integrated plant consists of many processes including coke production, sinter production, iron production, iron preparation, steel production, semi-finished product preparation, finished product preparation, heat and electricity supply and handling and transport of raw, intermediate and waste materials.

The steps in the steel industry consists of first converting coal to coke, adding iron ore, coke and limestone to a blast furnace from which pig iron comes out as the output. The pig iron contains impurities such as carbon, silicon, manganese, phosphorus and sulphur. The presence of carbon makes the iron brittle. It is then purified into steel in basic oxygen furnaces (BOF) or electric arc furnaces (EAF). The second stage consists of hot forming and cold finishing operations to convert steel into different shapes and sizes. Water is used in appreciable quantities in almost all operations of the steel-making process. The quantities of water used depend upon the manufacturing processes, which vary from plant to plant, but on average it is estimated that about 40,000 gallons (151 m^3) is used in producing 1 ton of finished steel. Steel manufacturing is one of the highest water users in industry. The sources of water are mainly surface (about 97%), sewage effluents (about 2.2 %) and groundwater (about 1.2%). About 95% of the water used in steel plants is for cooling, which can be recycled. Only a small percentage of water is consumed by evaporation. Pressurized water is also used in the raw material (iron ore) extraction process.

13.2.4 Textile industry

The textile industry is one of the most water-intensive industries in the world, consuming around 190 × 10^6 m^3 of water annually, second only to steel industry. It is reported that India, which is one of the major textile manufacturing countries in the world, consume about 425,000 gallons (1,932 m^3) per day in the textile industry (Textiles and water use I OEcotextiles). It takes about 2.7 m^3 of unseen water to produce the cotton that goes into making one T-shirt (The Impact of a Cotton T-Shirt I Stories I WWF (worldwildlife.org). Table 13.4 gives some typical values of water consumption for different fabric types.

The amount of water used depends upon the type of textile fibre processed, the type of product (woven, knit etc.), and the specific processes and equipment. It is also one of the biggest polluting industries in the world with textile mills generating about one-fifth of the world's industrial water pollution. Textile manufacturing is a very complex process. It starts from fibre, which is made into yarn by the spinning process. The yarn is then made into grey cloth by the weaving process followed by dyeing, printing and finishing and finally to garment manufacturing. Spinning and weaving, which are basically mechanical processes, do not require large amounts of water, but the dyeing, printing and finishing, which together is called wet processing, require large amounts of water. Bulk of the water is used in washing at the end of each process. Textile wet processing includes using water as a solvent for processing chemicals, as a medium for transferring

Table 13.4 Typical water consumption for different fabric materials

Material	Water consumption (kg/kg of fabric)
Cotton	250–350
Wool	200–300
Nylon	125–150
Rayon	125–150
Polyester	100–200
Acrylic	100–200

Source: Water Consumption in Textile Processing Industry – Textile Learner.

Table 13.5 Water consumption pattern in textile mills

Process	Water consumption (% of total)
Bleaching, finishing	38
Dyeing	16
Printing	8
Boiler house	18
Humidification (spinning)	6
Humidification (weaving)	6
Sanitary, Domestic etc.	9

Source: Water Consumption in Textile Processing Industry – Textile Learner.

dyes and chemicals to fabric and for washing and rinsing. Of different types of fibres, cotton requires the largest amount of water for its preparation (Table 13.4). A considerable amount of water is also used in ion exchange, boilers, cooling water, steam drying and the cleaning part of the process. Table 13.5 gives the water consumption patterns in the textile industry.

The wastewater discharged from textile mills contains a cocktail of chemicals including lead, phthalates (a group of chemicals used to make plastics more durable and are often called plasticizers), organochlorines (organic compounds having multiple chlorine atoms) and other chemicals, which can be toxic, and is an issue of concern from consideration of environmental pollution. Globally, it is estimated that up to 20,000 chemicals are used in textile manufacturing.

13.2.5 Pulp and paper manufacturing

The present trend in the world is to move away from printed products and move into digital media, which has its pros and cons. Younger generations dislike reading books and instead focus their eyes on computers or mobile phone screens. The long-term effect of this trend is yet to be understood. Despite the digitalization, pulp and paper industry remains one of the largest users of water in their processes in China, United States, Japan, Europe and several other countries. It appears to be increasing with the increase of gross domestic product.

The stages of paper manufacturing industry consist of pulp making, pulp processing and paper/paper board making with the associated processes such as cooking, bleaching and washing. The large quantities of water used in these processes generate equally large quantities of contaminated wastewater. The water in the industry is used as steam, as process water in the digesters, in transporting and washing the digested pulp and in the preparation and recovery of the chemicals employed in removing lignin and resins from the wood. Water is also used in boilers and for cooling power plants as well as in wood preparation, and clean-up and sanitary requirements in the mills. The water consumption in wood-based mills producing printing paper, packaging paper, newsprint and rayon grade pulp is reported to vary between 96 and 200 m^3 per ton whereas the corresponding range in agro-based mills is 55–225 m^3 per ton (www.dcpulppaper.org/gifs/report20.pdf).

Pulp and paper production is a complex industry with many different kinds of mills, products and processes. Most mills are located in countries with natural forests. Due to dwindling forest resources, bamboo, eucalyptus wood from plantations and mixed hardwood, residues from agricultural wastes and wastepaper are used in India. The raw materials for the industry include chemical pulp, mechanical pulp, wastepaper, filler, pigments and water. Energy is another input. Manufactured products include newsprint, fine paper, magazine paper, tissue, packaging paper and multiply board. Globally, about 422 million metric tons of paper are produced annually with over 60% for packaging, tissue, sanitary and newsprint. Most paper products are manufactured in the United States, Canada, China, Europe, Korea and Brazil. The production volumes of paper and cardboard in the year 2018 in major manufacturing countries are shown in Table 13.6.

The water requirements for the pulp and paper industry depend upon the end product and pulp and are highly variable. In the United States, it ranges from about 10,000 gallons (37.8 m^3) per ton for groundwood pulp to about 90,000 gallons (340.7 m^3) per ton for bleached sulphite paper from

Table 13.6 Production volumes of paper and cardboard

Country of production	Quantity in 1,000 metric tons
China	109,962
USA	72,062
Japan	26,070
Germany	22,682

Source: Paper and cardboard production top countries 2018 | Statista.

pulpwood (report.pdf (usgs.gov)). Water savings can be made by increasing recirculation with the added benefit of energy savings.

The effluents from pulp and paper manufacturing plants contain high values of biochemical oxygen demand (BOD), chemical oxygen demand (COD) and chlorinated chemicals that are collectively termed as absorbable organic halides (AOX). They are treated using conventional treatment methods such as primary clarification, activated sludge process as well as by supplementary treatment methods such as anaerobic biological processes, oxidation procedures and membrane filtration technologies. Future efficiency of pulp and paper industry depends on water usage and the way forward should be to reduce consumption and treat the effluents at their sources rather than lumping all effluents together and treat them. Different methods of treatment at different effluent sources need to be promoted.

13.2.6 Beverage industry

Water is the basic ingredient of all beverages but in addition, water also is used in several processes in the beverage industry of which cleaning and rinsing bottles is a major one. The beverage industry also requires agricultural inputs such as sugar, barley, coffee, chocolate, lemon, vanilla and other plant-derived ingredients. Beverages can be broadly classified as alcoholic and non-alcoholic. Alcoholic beverages include beer, wine, whiskey, cider, gin and rum, whereas non-alcoholic beverages include lemonade, fruit drinks, tea, coffee, milk, sports drinks and bottled water. There are also some country specific alcoholic drinks such as 'sake' made from rice in Japan, vodka made from grains or potatoes in Russia and 'ginseng' made from ginseng roots in Korea with limited global use. Water is also a significant cost factor in the beverage industry. Quantitatively, it takes about 140 litres of water to make a cup of coffee, 170 litres to make a pint of beer, 120 litres to make a glass of wine, 35 litres to make a cup of tea and 200 litres to make a glass of milk. These amounts of water are referred to as 'embedded water' or 'virtual water' in the product (Chapagain and Hoekstra, 2004).

The processes in the beverage industry include distilling and blending of spirits, making wine from grape, making cider and other fruit wines, making non-distilled fermented beverages, brewing beer, making malt, soft drinks and production of mineral waters and other bottled waters. Table 13.7 gives a breakdown of beverage consumption in United States for 2020. Table 13.8 gives a breakdown of consumption of non-alcoholic beverages among adults in the United States during the period 2015–2018.

Consumption of beverages varies from country to country. There is not much alcohol consumption in Asian countries compared to that in Western countries. Europe is perhaps the largest beer brewing region in the world,

Table 13.7 Consumption of beverages in the United States in 2020

Type of beverage	Share of consumption (%)
Bottled water	23.7
Carbonated soft drinks	18.4
Tap/other	11.2
Coffee	11
Beer	10
Milk	8.2
Tea	5.5
Fruit beverages	4.6
Sports drink	2.5
Wine and spirits	2.3
Value added water	1.3

Source: US beverage consumption share by segment 2020 | Statista.

Table 13.8 Non-alcoholic beverage consumption among adults: United States, 2015–2018

Beverage type	Share of consumption (%) (standard error)
Water	51.2 (0.7)
Coffee	14.9 (0.4)
Sweetened beverages	10.2 (0.4)
Tea	8.7 (0.4)
Fruit beverages	5.6 (0.2)
Milk	5.5 (0.2)
Diet beverages	3.8 (0.2)

Source: National Center for Health Statistics, National Health and Nutrition Examination Survey, 2015–2018.

Note: Percentages are based on total grams of reported non-alcoholic beverage intake and may not add up to 100 due to rounding.

with Germany having the largest number of breweries. In Asia, China has a higher per capita consumption of beer than the global average but is behind Japan and Korea. China also has had an upward trend in wine consumption, but very recently it is dampened by the trade dispute between Australia and China. Beer, which is a fermented drink with about 4–6% alcohol, is produced from malted barley, hops, yeast and water, while in some cases, other ingredients such as fruits, wheat and spices are also added. In beer brewing, 20% of influent potable water goes into the product beer while 70% goes as effluent wastewater and 9% evaporation losses and 1% in water losses in spent grains and trub (sediment formed in the brewing process) (Water-food-industry.pdf (water2return.eu). Effluents from beer industry consist of high levels of BOD, COD, total suspended solids as shown in Table 13.9.

Table 13.9 Environmental parameters of wastewater from beer industry

Parameter	Value
COD	2,000–6,000 mg/litre
BOD	1,200–3,600 mg/litre
TSS	200–1,000 mg/litre
pH	4.5–12
Nitrogen	25–80 mg/litre
Phosphorous	10–50 mg/litre

Source: Brito et al., 2007; Water-food-industry.pdf (water2return.eu).

The potential future market for beverages is unlikely to be based purely on the volume. With increasing affluence and disposable income in high-income countries, young customers are moving away from traditional beverages towards mild alcoholic beverages like craft beer, non-alcoholic beverages and spirits as well as healthy alternatives such as 100% natural juices.

13.2.7 Automotive industry

In 2015, the global water use by the major auto manufacturing companies is reported to be 29.3×10^6 m³ by Toyota motor company, 24.9×10^6 m³ by Fiat Chrysler motor company, 33.8×10^6 m³ by Honda motor company and 16.1×10^6 m³ by Ford motor company. General Motors company does not report a corresponding water use in 2015, but they report using 4.31 m³ per car. (Water Use in Car Manufacturing: Over 100 Million Cubic Metres Used by Major Companies (watertoday.ca)). Among the different water use in the automotive industry such as surface treatment and coating, paint spray, washing/rinsing, cooling, boilers and air conditioning systems, the most extensive use of water is the paint application process, including cleaning processing equipment.

The average direct water use in manufacturing is reported to be 5.20 m³ per vehicle while the average direct water consumption is 1.25 m³ per vehicle (Semmens, et al., 2014). Estimates vary depending on the source as well as on the manufacturing processes involved and it has also been reported that the average water consumption of water to produce the average domestic vehicle, including the tyres, is about 39,000 gallons (1,476 m³) (www.automotiveworld.com/articles/water-water-everywhere-vehicle-manufacturing/).

Another associated industry, which uses large quantities of water, is the car wash industry, which uses about 20 gallons on a basic wash for a small car. The quantities vary heavily depending upon the washing process and the size of the vehicle.

The wastewater in the automobile industry contains metals, oils and grease and harmful chemicals from paint residue, which can be harmful to the environment.

Other industries that consume large quantities of water include the petroleum refining industry and the semiconductor manufacturing industry. Oil refining involves a number of processes including separation and blending of petroleum products. Petroleum industry is also a major contributor to water pollution. In addition to discharge of wastewater into surface water bodies, some refineries inject them into wells that can end up in groundwater aquifers. Petroleum industry also pollutes the oceans during accidents.

In the semiconductor industry, the amount of water required depends upon the size and number of layers in chips and electronic components. Industry statistics indicate that creating an integrated circuit on a 300 mm wafer needs about 2,200 gallons (8.33 m³) of water of which 1,500 gallons (5.68 m³) is ultrapure water. Producing ultrapure water involves processes such as filtration, microflocculation, activated carbon, reverse osmosis, degasifiers, electro-deionization and ultraviolet radiation. The total organic constituents in the end product must be less than 1 ppb and the resistivity must be under 18.2 microOhm-cm. A typical fabrication plant may need water equivalent to that used in a city of about 40,000–50,000.

REFERENCES

Chapagain, A. K. and Hoekstra, A. Y. (2004). Water Footprints of Nations, vols. 1 and 2. UNESCO-IHE Value of Water Research Report Series No. 16. Available online at www.waterfootprint.org/Publications.htm, accessed 24 January 2007.

Falkenmark, M. and Lindh, G. (1976). *Water for a Starving World*, Boulder, Westview Press, [Google Scholar].

Spang, E. S., Moomaw, W. R., Gallagher, K. S., Kirshen, P. H. and Marks, D. H. (2014). The water consumption of energy production: an international comparison, *Environmental Research Letters*, 9, no. 10, 14 pp.

Semmens, J., Bras, B. and Guldberg, T. (2014). Vehicle manufacturing water use and consumption: an analysis based on data in automotive manufacturers' sustainability reports. *International Journal of Life Cycle Assessment* 19, 246–256. https://doi.org/10.1007/s11367-013-0612-2

CITED WEBSITES

• Which Industries Use the Most Water – Better Meets Reality (https://betterm eetsreality.com/which-industries-use-the-most-water/)
• How Many Gallons of Water Does It Take to Make . . . (treehugger.com) (www. treehugger.com/how-many-gallons-of-water-does-it-take-to-make-4858491)
• Textiles and water use | OEcotextiles (https://oecotextiles.blog/2010/02/24/ textiles-and-water-use/)

- The Impact of a Cotton T-Shirt | Stories | WWF (worldwildlife.org) (https:// bettercotton.org/cottons-water-footprint-one-t-shirt-makes-huge-impact-envi ronment/)
- Water Consumption in Textile Processing Industry – Textile Learner (https:// textilelearner.net/water-consumption-in-textile-processing-industry/)
- www.dcpulppaper.org/gifs/report20.pdf
- Paper and cardboard production top countries 2018 | Statista (https://hk.sea rch.yahoo.com/search?fr=mcafee&type=E210HK105G91698&p=Paper+ and+cardboard+production+top+countries+2018+%7C+Statista)
- report.pdf (usgs.gov) (chrome-extension://efaidnbmnnnibpcajpcglclefindmkaj/ https://pubs.usgs.gov/wsp/1330a/report.pdf)
- National Center for Health Statistics, National Health and Nutrition Examination Survey, 2015–2018 (www.cdc.gov/nchs/products/databriefs/ db376.htm)
- Water-food-industry.pdf (water2return.eu)
  US beverage consumption share by segment 2020 (www.statista.com/statistics/ 387199/us-consumption-share-of-beverages-by-segment/)
- Water Use in Car Manufacturing: Over 100 Million Cubic Metres Used by Major Companies (watertoday.ca). (www.watertoday.ca/ts-water-use-in-car-manufacturing.asp)
- Water Consumption in Auto Industry (www.automotiveworld.com/articles/ water-water-everywhere-vehicle-manufacturing/)

Chapter 14

Water and recreation

14.1 INTRODUCTION

Recreation is a popular activity among people of all ages. Humans use water for many recreational purposes as well as for exercising and for sports. It can be for pure enjoyment but is also an important activity from a health point of view as most recreation activities involve some degree of physical exercise. Recreation in water can be with active participation such as swimming and diving or passive participation such as observing a recreational activity by others. Observing a calm water body or a waterfall can also be mentally relaxing. Beaches and water parks are popular places for people to relax and enjoy recreation. Some people even keep fish tanks and ponds at home for companionship and enjoyment. There are also adventurous activities such as fishing and hunting in water particularly, among the young generation.

14.2 TYPES OF WATER RECREATION

Types of water recreation can be broadly classified into activities in the water, activities on the water, activities under water, activities on frozen water and activities on snow. Those in the water include swimming, which is the main water recreation activity in the world, diving that can be individual or synchronized, water aerobics, surfing and various games such as water polo, water basketball and water volleyball. Those on the water include boating, canoeing, dragon boat racing, kayaking, white water rafting, jet skiing, sailing, parasailing and yachting. Underwater recreational activities include free diving, scuba diving, snorkelling and various under-water games. Activities on frozen water include ice skating and ice hockey and activities on snow include various forms of skiing. There are also motorized water sports. Most competitive water recreation activities have found their way to the Olympics.

DOI: 10.1201/9781003329206-14

14.3 WATER-RELATED TOURISM

Tourism is one of the largest industries in the world and in some countries and regions, it is the major economic driver. Water-based tourism relates to any activity in water bodies such as lakes, canals, creeks, streams, rivers, waterways, marine coastal zones, ice, snow, seas and oceans. It can be a sports activity or just a means of relaxation. In temperate climates, it is usually a seasonal activity but in tropical climates, it can be all year round. Every year, millions of tourists are attracted to coastal areas. It makes beneficial contributions to the economy, but at the same time leads to some adverse consequences because of the competition for water, which is dwindling in quantities and quality. In many countries with natural tourist attractions, priority for water is allocated to the tourists who contribute to the local economy at the expense of the basic water needs of the local community. It is not only the direct use of water by the tourists that matter, but more importantly the embedded water in the goods and services provided to the tourists, which can be much more than the direct use. It is therefore important to ensure optimal management of the available freshwater resources taking into consideration the supply and the demand.

14.3.1 Types of water-related tourism

14.3.1.1 Ocean tourism

Sailing and boating are active recreations on the water that can also be competitive. Cruises are one of the most popular forms of ocean tourism. They provide a luxurious way to travel and to relax for a short period of time. One of the ill-fated cruises was the *Titanic*, which sank killing over 1,500 people after hitting an iceberg on April 14, 1912. Recreation on cruise ships is a luxury and thus limited to tourists with high income. In the last two years, the demand for cruises has gone down because of the pandemic. A negative effect of cruises is the ocean pollution caused by untreated effluents and solid waste generated by those on board.

Other types of tourism in the oceans include ecotourism where the focus is on the natural environment without harming it. Scuba diving to enjoy coral reefs and other marine features including marine creatures and cruises to view rare sea animals are examples of ecotourism.

14.3.1.2 Canal tourism

Canal tourism is popular in Venice in Italy, Amsterdam in the Netherlands, Suzhou in China and Bangkok in Thailand. Brief descriptions of these four places are given below.

(i) Venice

Venice in Italy consists of 118 small islands separated by some 150 canals and about 400 bridges of various sizes and shapes. It is the water that makes Venice unique. With an aquatic thoroughfare known as the Grand Canal with churches, cathedrals and palaces along its banks reflecting ancient Venetian architecture, the city is famous for its cuisine located in garden islands and lagoons with strips of sand around Venice. Tourists from all over the world travel to Venice to experience the canal system and the Venetian architecture along the banks of the canals.

(ii) Amsterdam

Amsterdam, the capital city of the Netherlands, which is dubbed as the 'Venice of the North', has some 165 canals, about 90 islands and some 1,500 bridges. Amsterdam has more bridges than any other place in the world. The three main canals – Herengracht (Gentleman's canal where rich people used to live on the banks), Keizersgracht (Emperor's canal) and Prinsengracht (Prince's canal) – were constructed during the 17th century at the time of the Dutch golden era. The canal system is also a UNESCO World Heritage site since 2010 and a canal cruise is the most popular tourist attraction in the Netherlands. The canals are about 3 m deep and the locals jokingly define the 3 m as 1 m of water, 1 m of mud and 1 m of bikes. It seems that 12,000–15,000 bikes are fished out of the canals each year. The canal system is fed by freshwater from Rhine River and sea water from IJsselmeer. The water in the canals freeze during parts of winter and recreation activities are then confined to ice skating and walking on the canals.

(iii) Suzhou

Suzhou near Shanghai, dubbed as the 'Venice of the East', is a city in the Yangtze River Delta with a complex network of canals, tea houses, temples, classical gardens and water towns. Suzhou is also a major silk production and manufacturing city recognized as a UNESCO World Heritage site.

(iv) Bangkok

The canal system (or, klongs) in Bangkok, linked to the Chao Praya River, is a bustling network of transportation of people and goods. Constructed during the 18th century and considered as the lifeblood of Thailand, the system consists of four main canals, Taling Chan, Bangkok Noi, Bangkok Yai, and Bang Ramat. It is also dubbed as the 'Venice of the East'. Many of the tourist attractions in Bangkok such as the grand palace and many

Buddhist temples are located along the banks of the canal system. There are also several 'floating markets' on the canal system that sell food and other groceries, which have become popular tourist attractions. With the introduction of the mass transit system in Bangkok, the importance of the canal system as a mode of transportation has gradually diminished.

14.3.1.3 Beach tourism

Beach tourism can be active or passive and can be enjoyed by people of all ages. Popular beaches attract a lot of tourists as well as locals in the summertime for swimming, sunbathing, relaxation and socializing. According to 'Found the World', the ten best beaches in the world are Santa Monica Beach, California; South Beach, Miami, Florida; Waikiki, Honolulu, Hawaii; Mosquito Bay, Vieques, Puerto Rico; Spiaggia Grande, Positano, Italy; Bondi Beach, Sydney, Australia; Bathsheba Beach, Barbados, The Bahamas; Navagio Beach, Zakynthos, Greece; and The Baths, British Virgin Islands (Top 10 Most Famous Beaches in The World | Found the World). The criterion used to select the above ten is not clear and most likely subjective. There are equally beautiful beaches in many other countries too such as for example, Rio in Brazil, and several beaches in the Philippines and Thailand.

14.3.1.4 River tourism

Whitewater rafting is a recreational water sport where an inflatable raft carries four to eight people down whitewater rapids on a river. Whitewater river rafting is considered an adventure sport and has varying levels of difficulty. Typically, an experienced rafting guide will accompany beginners. Whitewater is actually formed from turbulence in the rapids, resulting from fast-flowing currents. Whitewater rivers are divided into several classes depending on the degree of difficulty in paddling the raft. According to 'MapQuest Travel', the ten best places for whitewater rafting are Futaleufu River, Chile, White Nile, Uganda, Pacuare River, Costa Rica, Zambezi River, Zimbabwe, Ottawa River, Canada, Colorado River, Grand Canyon, Magpie River, Canada, Ganges River, India, Salmon River's Middle Fork, Idaho, North Johnstone River, Australia, and Penobscot River, Maine (The 10 Best Places To Go Whitewater Rafting Around the World – MapQuest Travel). Again, the criteria for selecting the above ten places appear to be subjective. Mae Taeng River and Mae Cham River in Chiang Mai, Thailand, and Kamali River and Sun Kosi River in Nepal are also interesting spots for river rafting.

14.3.1.5 Water park tourism

Water parks can be found in almost all countries. According to CNN Travel, the 12 of the best water parks in the world are Aquatica (Orlando,

Florida), Aquaventure Waterpark (Dubai, UAE), Area 47 (Innsbruck, Austria), Beach Park (Fortaleza, Brazil), Caribbean Bay (Gyeonggi-do, South Korea), Siam Park (Tenerife, Spain), Tropical Islands (Krausnick, Germany), Watercube Waterpark (Beijing), WaterWorld Waterpark (Ayia Napa, Cyprus), Wet 'n Wild (Orlando, Florida), World Waterpark (Alberta, Canada) and Yas Waterworld (Abu Dhabi, UAE) (12 of the best water parks in the world | CNN Travel). Water parks can be indoors or outdoors. Their popularity is generally judged by the size, number and complexity of attractions, number of visitors during a given period of time (e.g. annual) and the ease of transportation to the location. The thrilling attractions include water slides, wave pools, water pools of various sizes, swimming pools with varied depths to suit different age groups, free-fall slides, lazy winding rivers, water coasters, aquariums, diving towers, tornado rides, geysers and beaches.

14.3.1.6 Ski tourism

Ski tourism is confined to places in high latitudes with snowfall during the winter season. One exemption is an artificial ski stadium in Dubai where there is no source of snow or even enough water. Ski tourism and mountaineering are adventurous and thrilling recreation activities that require skill and training. According to 'Planetware', the top 12 ski resorts in the world are Whistler Blackcomb, Canada; Courchevel, Alpine ski resort, France; Courchevel, Switzerland; Vail Mountain Resort, Colorado, the United States; Aspen Snowmass, Colorado, the United States; Val d'Isere, France; Cortina D'Ampezzo, Italy; Telluride, Colorado, the United States; Niseko, Hokkaido, Japan; Chamonix, France; St. Anton, Austria; and Kitzbühel, Austria (16 Top-Rated Ski Resorts in the World, 2022 | PlanetWare).

14.3.1.7 Hot spring tourism

Hot springs are formed by geothermally heated water, usually of volcanic origin, that emerge at the surface of the ground due to underground pressure. They bring along a variety of minerals in high concentrations that are beneficial to general health as they have therapeutic properties. Specific properties of hot springs vary depending on chemical composition, mineral concentration and water temperature. Facilities available in hot spring resorts include hydrotherapy, spring pools, spring saunas, spring massage pools, health bathing houses and spring health centres. Other health benefits include taking advantage of the physical properties of water using hydro jets that splash water onto the body, ultrasonic massage equipment and the water's natural buoyancy. There are only a few naturally occurring phenomena that are as relaxing and rejuvenating as hot springs.

There are hot springs in several countries. Lists of the best hot springs in the world, in Japan, in Taiwan and in New Zealand as compiled by 'Trip Savvy', 'Taiwaneverything' and 'Explore New Zealand Travel Blog', respectively, are given below.

According to 'Trip Savvy', the 20 best hot springs in the world are Takaragawa Onsen, Japan; Challis Hot Springs, Idaho, the United States; The Blue Lagoon, Iceland; The Boiling River, Yellowstone National Park, the United States; The Springs in Pagosa Springs, Colorado, the United States; Pamukkale, Turkey; Chena Hot Springs Resort, Alaska, the United States; Olympic Hot Springs, Washington, the United States; Hot Springs National Park, Arkansas, the United States; Yangpachen Hot Springs, Tibet; The Omni Homestead Resort, Virginia, the United States; Terme di Saturnia, Italy; Ojo Caliente Mineral Springs Resort and Spa, New Mexico, the United States; Uunartoq Island, Greenland; Hot Springs Resort and Spa, North Carolina, the United States; Chico Hot Springs Resort, Montana, the United States; Reykjadalur Valley, Iceland; Khir Ganga, India; Termas de Puritama, Chile; and Deception Island, Southern Ocean.

Japan has over 27,000 hot springs (*onsen* in Japanese). According to 'Trip Savvy', the 10 best hot springs are located in Hakone in Kanagawa prefecture, Kusatsu in Gunma prefecture, Beppu in Oita prefecture, Noboribetsu in Hokkaido, Shibu in Nagano prefecture, Ibusuki in Kagoshima prefecture, Kinosaki in Hyogo prefecture, Minakami in Gunma prefecture (Takaragawa Onsen is located in this resort), Yamanaka in Yamanashi prefecture and Kurokawa in Kumamoto prefecture (The Best Hot Springs Destinations in Japan (tripsavvy.com)).

Because of the geological location of Taiwan, the island has one of the highest concentrations of thermal hot springs in the world with about 100 major hot springs and many smaller ones. According to 'Taiwaneverything', the 10 best hot springs in Taiwan are Guanziling in Tainan City, Beitou in Taipei City, Jiaoxi in Yilan County, Zhiben in Taitung County, Guguan in Taichung City, Tai'an in Miaoli County, Jinshan in New Taipei City, Wulai in New Taipei City, Ruisui, a small town in eastern Taiwan about half-way between Hualien City and Taitung City, and Baolai in Kaohsiung City (https://taiwaneverything.cc/2018/08/16/taiwan-hot-springs/).

New Zealand has over 100 hot springs scattered over South and North Islands. According to 'Explore New Zealand Travel Blog', the 10 best hot springs are Polynesian Spa in Rotorua, Parakai Hot Springs in Auckland, Hot Water Beach in Coromandel, Kerosene Creek in Rotorua, Hanmer Springs in Canterbury, Spa Thermal Park in Taupo, Welcome Flat Hot Pools in West Coast, Tekapo Springs in Tekapo, Glacier Hot Pools in Franz Josef and Kawhia Springs in Waitomo (10 of New Zealand's Best Hot Springs – Explore NZ (gorentals.co.nz)).

14.4 QUALITY REQUIREMENTS FOR WATER RECREATION

There are certain water quality requirements for a water body to be used for recreation. Such requirements are based on human health factors as well as ecological factors. Since most recreation activities involve having contact with water, the risk of transmission of diseases by various pathogens present in the water can be high. Exposure to contaminants can be by swallowing water or through skin contact. Effects vary with the type of exposure and the type of contaminant. Ecological conditions include supporting aquatic life, concentrations of various contaminants and the aesthetic appearance of the water body. Contaminants that may affect the quality of water can be of biological or chemical origin. Sources of such contaminants can be storm water, sewage effluents, industrial discharges and atmospheric deposition such as acid rain, organic contaminants from domestic and commercial effluents and the recreation activities themselves.

Some guidelines for the quality of water for recreational activities include a pH value within the range of 5–11 to avoid potential eye and skin irritation, clarity of water to enable the individuals to see sub-surface hazards, colour should not be intense to impede visibility and the recreational area should be free from floating debris.

14.5 HEALTH HAZARDS OF WATER RECREATION

All water recreational activities whether they are in, on or under involve contact with water. With the increase in popularity of water recreation activities worldwide, the risk of accompanying health hazards also would increase. Exposure to pathogens during such activities in water may result in diseases, particularly in populations with reduced immune functions. Pathogens can enter the body through ingestion, inhalation or penetration of the skin. Water contact time as well as the degree of contact are prime factors that influence the amount of exposure to pathogens in water. The longer the contact time and higher the degree of contact with water, the greater the exposure to pathogens. The common symptoms caused by water-related recreational activities are diarrhoea, skin rashes, cough and eye pain.

Chemical and biological hazards may be caused by the ingestion of polluted waters that can lead to infections such as hepatitis, typhoid, cholera, dysentery and other gastrointestinal diseases. Water-borne pathogens can enter the body through eyes, ears, nose, throat, lung and gastrointestinal tracts. Pollution may be caused by faecal contamination as well as non-faecal contamination. Faecal contamination may be caused by direct animal contamination or by accidental faecal release by swimmers. Non-faecal contamination is caused by vomit, mucus, saliva or skin in swimming pools or recreational waters.

Table 14.1 Common germs and illnesses associated with swimming

Rank	Germ	Predominant illness
1	Cryptosporidium	Acute gastrointestinal illness
2	Legionella	Acute respiratory illness
3	Pseudomonas	Acute respiratory illness, skin, ear
4	Norovirus	Acute gastrointestinal illness
5	Shigella	Acute gastrointestinal illness
6	E.coli	Acute gastrointestinal illness
7	Pool chemicals (chlorine, bromine, hydrochloric acid)	Acute gastrointestinal illness / Acute respiratory illness, skin, ear
8	Giadia	Acute gastrointestinal illness
9	Avian schistosomes (cercarial dermatitis)	Skin
10	Algal toxins	Acute gastrointestinal illness / Acute respiratory illness, skin

Source: Recreational Water Illnesses | Healthy Swimming | Healthy Water | CDC.

Diarrhea is the most common water recreational disease. Table 14.1 gives a list of common germs and illnesses associated with swimming.

Of these, Cryptosporidium, Norovirus, Shigella, E-Coli and Giadia are faecal spread germs whereas Legionella and Pseudomonas non-faecal spread germs.

Physical hazards include drowning and bodily injury but in swimming pools they are very unlikely. In the case of scuba diving, physical hazards can be encountered during the descent and ascent as well as when under water. Compressions that can occur during descent affect the ears and nasal sinuses and may even rupture ear drums. During ascent, the air in the lungs expands due to decompression that may cause damage to the lungs if the diver holds his breath. Drowning can occur at any stage if the diver loses his air supply due to equipment failure or otherwise and cannot reach the surface within a short time. Rapid ascent from a deep dive poses a greater risk. Immersion in cold water can lead to a reduction in body temperature that can cause hyperthermia. Water has a much higher thermal conductivity than air and cools the body much faster than air at the same temperature. Hazards can also occur when encountered with dangerous marine creatures such as sharks, crocodiles and other biting creatures.

Overexposure to ultraviolet radiation (UVR) can cause damage to the skin, eyes and the immune system. The most visible effect is the inflammation of the skin commonly referred to as sunburn, which in rare cases can lead to skin cancer. Other hazards include Photokeratitis that causes a painful eye condition affecting the thin surface layer of cornea and photoconjunctivitis, which is also a painful eye condition due to inflammation of the membrane lining inside the eyelids and eye socket. The chronic effect of eye exposure to UV radiation can lead to cataracts of the eyes.

Table 14.2 Precautions to avoid radiation

UV index	Precautions
Less than 2	None required; safe to stay outside
3–5	Protection required; seek shade during mid-day hours
6–7	High; seek shade
8–10	Very high; seek shade
Greater than 11	Extremely high; seek shade

It is also important to realize that all effects of exposure to UV radiation are not adverse. The beneficial effect of UV radiation is the stimulation of the production of vitamin D in the skin. Table 14.2 shows the precautions needed to be taken at various levels of UV exposure.

There is also evidence indicating immunosuppressive effect of acute high doses and chronic low doses of UV exposure on human immune system (Kripke, 1994).

14.6 GUIDELINES FOR AVOIDING HEALTH HAZARDS

The World Health Organization classifies health hazards in recreational coastal and freshwater environments as drowning and injury, exposure to cold, heat and sunlight, exposure to water contaminated by sewage, exposure to free living pathogenic microorganisms, contamination of beach sand, exposure to algae and their products, exposure to chemical and physical agents and exposure to dangerous aquatic organisms (9241545801.pdf (who.int)). The capacity to contain the health risks in all these cases is under the control of the user who should exercise a degree of responsibility when indulging in recreational activities.

Physical hazards such as drowning and injuries are preventable if sufficient care is taken. The effect of exposure to heat and cold in water is more pronounced than in air as the thermal conductivity of water is much greater than that of air. In water, the surface area of the complete body is exposed to heat and cold. The body cools much faster than in air at the same temperature. Exposure to UV radiation can be minimized or avoided by limiting the time exposed to sun and wearing appropriate clothing and applying certain types of lotions on the skin. Exposure to water contaminated by sewage cannot be controlled by the user unless authorities who manage the water recreational areas inform the users of the potential risks. The best way to prevent recreational water illnesses from spreading is to keep germs out of the water in the first place but this cannot be done by the user. Free living pathogenic microorganisms consist of a large number of species of vibrio, found mostly in salt water that can cause an illness called vibriosis. In the context of water recreation, people get exposed to vibrio through open cuts and wounds, but most illnesses are caused by the consumption of raw or uncooked seafood. The best way to avoid exposure to vibrio pathogens is to stay out of water if there are cuts or wounds in the body. Beach sand can be the home for a number of bacteria, fungi and viruses, some of which can be pathogenic. Cyanobacteria or blue-green algae occur worldwide, especially in calm, nutrient-rich waters. Some species of cyanobacteria produce toxins that affect animals and humans. People may be exposed to cyanobacterial toxins by drinking or bathing in contaminated water. The likelihood and nature of human exposure to dangerous aquatic organisms depend on the type of recreational activity and the location. Risks can be from disease vectors including mosquitoes that transmit malaria, dengue fever, yellow fever parasites as well as encountering physically dangerous creatures such as piranhas, snakes, electric fish, sharks, barracudas etc.

In all these cases, the simplest advice is to stay away from waters if not feeling well, not to swallow water and to observe good hygiene.

REFERENCE

Kripke, M. L. (1994). Ultraviolet Radiation and Immunology: Something New under the Sun –Presidential Address, *Cancer Res* 54(23), 6102–6105.

Chapter 15

Water and disasters

15.1 INTRODUCTION

A hazard becomes a disaster when the people or the affected region are vulnerable and lack the coping capacity. Disasters can be natural or human-induced. The former type is difficult if not impossible to prevent whereas the latter type is preventable. In terms of the cost and damage induced by various types of natural disasters, 'water-related disasters' by far exceed those by any other natural disaster. In this context, water-related disasters include all types of floods, land and mud slides, storm surges, tsunamis, tidal waves, debris flow, avalanches, droughts and all types of cyclones. In addition to such geophysical disasters, water-related biological disasters such as epidemics and endemics also take a significant toll in terms of human lives. Human-induced disasters include various types of pollution, accidents and wildfires, among others. In the modern world, pollution of the water environment is a major environmental disaster in many regions, with some places reaching irreversible conditions.

Natural disasters have taken place from time immemorial. In the past, biotic populations living under natural conditions and in harmony with nature were able to live with disasters by adapting their lifestyles or by changing their habitats. With exponential increase in human population and increasing urbanization, natural conditions no longer exist in many places. With increased population density and high value-added infrastructure, the impacts have increased manifold.

The definition of a disaster depends upon the agency or organization that collects and disseminates data. There is a wide variation in the criteria used for inclusion in databases. One of the comprehensive databases on disaster information is EMDAT, which is located in the University Catholic Louvain, Brussels, Belgium (www.EMDAT.net) and which is updated regularly. They define an event as a disaster if there have been more than ten deaths or more than 100 people displaced, or if the government of the affected country has declared a state of emergency and asked for international assistance.

DOI: 10.1201/9781003329206-15

Table 15.1 Worst weather-related disasters during the 50 years from 1970 to 2019 in terms of loss of human lives

Rank	Disaster type	Year	Country	Death toll
1	Drought	1983	Ethiopia	300,000
2	Storm (Bhola)	1970	Bangladesh	300,000
3	Drought	1983	Sudan	150,000
4	Storm Gorky	1991	Bangladesh	138,866
5	Storm Nargis	2008	Myanmar	138,366
6	Drought	1973	Ethiopia	100,000
7	Drought	1981	Mozambique	100,000
8	Extreme temperature	2010	Russian Federation	55,736
9	Flood	1999	Venezuela	30,000
10	Flood	1974	Bangladesh	28,700

Source: Climate and weather related disasters surge five-fold over 50 years, but early warnings save lives – WMO report | | UN News.

Table 15.2 Ten worst weather-related natural disasters in the world in terms of economic losses

Rank	Disaster type	Year	Country	Economic losses in US$ billions
1	Storm Katrina	2005	United States	163.61
2	Storm Harvey	2017	United States	96.94
3	Storm Maria	2017	United States	69.39
4	Storm Irma	2017	United States	58.16
5	Storm Sandy	2012	United States	54.47
6	Storm Andrew	1992	United States	48.27
7	Flood	1998	China	47.02
8	Flood	2011	Thailand	45.46
9	Storm Ike	2008	United States	35.63
10	Flood	1995	Democratic People's Republic of Korea	25.17

Source: Climate and weather related disasters surge five-fold over 50 years, but early warnings save lives – WMO report | | UN News.

According to data compiled by the United Nations, droughts appear to be the cause of the ten deadliest disasters during the period from 1970 to 2019 causing 650,000 deaths, followed by storms causing 577,232 deaths, floods causing 58,700 deaths and extreme temperature events causing 55,736 deaths (Climate and weather-related disasters surge five-fold over 50 years, but early warnings save lives – WMO report | | UN News). In terms of economic losses, storm disasters appear to have taken the top place with the United States suffering the greatest loss. These statistics are summarized in Table 15.1 and 15.2, respectively.

15.2 TYPES OF WATER-RELATED DISASTERS

15.2.1 Floods

Floods have the greatest damage potential of all natural disasters world-wide and affect the greatest number of people. Over the period 1995–2015, floods accounted for 43% of all documented natural disasters, affecting 2.3 billion people, killing over 157,000 and causing $662 billion in damages. Droughts accounted for 5% of natural disasters, affecting 1.1 billion people, killing over 22,000, and causing $100 billion in damages over the same 20-year period. Over the course of one decade (1995–2004), the number of floods rose from an annual average of 127 in 1995 to 171 in 2004 (CRED/UNISDR, 2015).

Floods are caused by heavy rainfall with no significant time lag. Floods caused by snow melting have significant time lags and the potential damages are not as great as those from heavy rainfall. Disasters can also be caused by storm surges in coastal areas, flash floods resulting from dam breaks and floods caused by thunderstorms. Human interventions such as urbanization, reclamation of coastal areas that change the natural hydrological cycle and drainage mechanisms also contribute to flooding.

According to an ICHARM report (Adikari and Yoshitani, 2009) based on data compiled by EMDAT, there have been 3,050 incidents of flood disasters during the period 1900–2006, causing economic damage to the extent of some $342 billion. During the same period, there have been 2,758 incidences of windstorm disasters causing $536 billion worth of damage. The number of people who lost their lives have been in excess of 6.8 million and 1.2 million, respectively, for flood and windstorm disasters. These two types of disasters alone accounted for over 56% of all natural disasters in that period. Of the 1,000 worst natural disasters in terms of the number of human casualties that occurred during 1900–2006, floods accounted for 345, windstorms for 252 and droughts for 273. A summary of the historical major flood disasters in the world including damages caused for the period 1860–2008 is given in WWDR3 (2009, Table 12.1), which is reproduced in Table 15.3. All these facts and figures illustrate the importance of hydro-meteorological disasters. It is also important to note that not only the numbers of disasters are increasing but also the numbers of people affected too because of migration of people into areas with better economic prospects.

Recent flood disasters include the unprecedented rainfall of more than 800 mm equivalent to an years rainfall falling in four days in July 2021 that caused devastating flooding in Henan province of China, a flood in Germany in July 2021 caused by two months of rain falling in two days across Germany, Belgium, Netherlands and Luxembourg that killed 208 people, devastating floods in January 2020 that destroyed Indonesia's capital Jakarta and some neighbouring areas, causing 66 deaths and displacing

Table 15.3 Examples of major floods worldwide 1860–2008

Date	Location	Meteorological condition	Peak discharge (m³/s)	Impact material damage US$ million	Human losses
January 2008	Zambezi River, Mozambique	Heavy torrential precipitation in neighbouring counties	3,800	2	20 dead 113,000 displaced
April–May 2003	Santa Fe, Argentina	Saturated soil due to heavy precipitation in summer 2002 and April 2003	4,100	na	22 dead 161,500 displaced
February 2000	Limpopo River, Mozambique	Extreme precipitation in tropical depression, enforced with torrential rain of three cyclones	10,000	na	700 dead 1,500,000 displaced
July 1997	Czech Republic	Saturated ground after extreme long-lasting precipitation and extreme precipitation	3,000	1.8	114 dead 40,000 displaced
June 1997	Brahmaputra River, Bangladesh	Torrential monsoon rains during monsoon season	10,200	400	40 dead 100,000 displaced
March–April 1997	Red River, United States	Heavy rains and snowmelt	3,905	16,000	100,000 homes flooded 50,000 displaced
November 1996	Subglacial Lake Grímsvötn, Iceland	Jökulhlaup flood	50,000	12	na
February 1996	West Oregon United States	Extreme spring snowmelt and heavy spring precipitation	na	na	9 dead 25,000 displaced
July 1995	Athens, Greece	Storm of a short duration and extreme intensity	650	na	50,000 displaced
November 1994	Po River Italy	Cold front associated with cyclonic circulation and heavy rainfall	11,300	na	60 dead 16,000 displaced
February 1994	Meuse River Europe	Heavy rain due to low pressure system	3,100	na	na
September 1993	Mississippi River United States	Heavy precipitation in June and July; saturated soil due to extremely high precipitation	na	15,000	50 dead 75,000 displaced

Date	Location	Cause			Impact
November 1988	Hat Yai River Thailand	Brief torrential monsoon rain	172	na	664 dead 301,000 displaced
January 1983	Northern Peru	El Niño situation with heavy rains	3,500	na	380 dead 700,000 displaced
August 1979	Machu River India	Exceptionally heavy rainfall, swollen river, resulting in collapse of the Matchu Dam	16,307	100	1,500 dead 400,000 displaced
June–September 1954	Yangtze River China	Intensive rainfall over months	66,800	na	30,000 dead 18,000 displaced
January 1953	North Sea Netherlands	High spring tide and a severe European windstorm	na	504	1,835 dead 100,000 displaced
January 1910	Seine River France	Very wet period for six months followed by heavy rains in January	460	na	200,000 displaced
May 1989	Johnstown Pennsylvania United States	Extremely heavy rainfall due to storm followed by breach of dike	na	17	2,200 dead
July 1860	Eastern Norway	Frost and heavy snowfall followed by snowmelt and heavy precipitation	3,200	na	12 dead

Source: WWDR3, 2009.

Note: na – not available

400,000 people, and flooding in different parts of India that caused landslides and killed over 40 people.

15.2.2 Droughts

Drought can be defined in several ways. In general, it is an extended period of low or no rainfall for a region. It can also be defined with respect to the drop in reservoir levels as well as on crop losses. Although droughts can occur in any country or region, the most drought prone-countries are Morocco, Uganda, Somalia, Iran, Pakistan, China, Afghanistan, Eritrea, Sudan and Ethiopia (The Most Drought Prone Countries in the World – WorldAtlas). The Sahara Desert, which spans across entire North Africa, is the largest hot desert in the world. In terms of population affected by major droughts, India and China have taken the top places during the period from 1900 to 2016.

Factors that contribute to droughts include lack of sufficient rainfall, dryness in the atmosphere, deforestation, greenhouse gas emissions and El-Nino events in certain parts of the world. Deforestation has a significant impact on the occurrence of droughts. Plants absorb water from the soil to build up the food by photosynthesis and the excess water escapes to the atmosphere via transpiration. With deforestation, the process of returning water back to the atmosphere is reduced, resulting in a dryer atmosphere. Dryness in the atmosphere is one of the factors that cause drought. El-Nino events that cause temperature rises in the surface water along the central South American coast influence weather patterns in the Amazon basin, parts of United States and Central America and parts of Africa and Southeast Asia. Human factors such as over-farming can deplete the soil moisture that can lead to droughts.

Droughts are usually followed by famine. During the period 1984/85, the drought-led famine killed some 750,000 people in the Horn of Africa (Djibouti, Eritrea, Ethiopia and Somalia). Table 15.4 gives some examples of the deaths caused by major droughts worldwide during the period 1900–2016 (CRED, Statista, 2019).

Impacts of droughts can be environmental, cultural and societal, economic and health related. Environmental impacts include loss of agricultural fertility, pest infestation and other adverse ecosystem changes; cultural and societal impacts are linked to the value of water attached in different cultures and societies. Drought in such beliefs may be considered as a punishment by god. Economic impacts can be scarcity of food, water and escalating prices of commodities. Health impacts can be exposure to diseases due to malnutrition and physical exhaustion, particularly in poor communities.

More information about droughts can be found in the website National Integrated Drought Information System (NIDIS).

Table 15.4 Number of deaths caused by major droughts worldwide 1900–2016

Year	Country	Number of deaths
1928	China	3,000,000
1943	Present Bangladesh	1,900,000
1942	India	1,500,000
1965	India	1,500,000
1900	India	1,250,000
1921	Soviet Union	1,200,000
1920	China	500,000
1983 May	Ethiopia	300,000
1983 April	Sudan	150,000
1973 December	Ethiopia	100,000

Source: CRED Statista, 2019; five droughts that changed human history | World Economic Forum (weforum.org).

15.2.3 Storms

Storms are caused by severe weather characterized by disruptions to normal conditions such as strong winds, tornados, hail, squalls, storm surges, thunderstorms, heavy precipitation (rain or snow), cyclones, blizzards etc. In desert areas, they may be atmospheric phenomena such as dust storms and sandstorms. They are all meteorological events such as cyclones, tornados, snow blizzards, hailstorms, thunderstorms and heat waves all of which have some link to water that lead to disasters, which last for relatively short periods of time. Tables 15.1 and 15.2 give some statistics of damages caused by storms and other weather-related disasters. Storms were responsible for 54% and floods, 31% of recorded disasters, with the former linked to 71% of deaths and the latter to 78% of economic losses. The United States accounts for 38% of global economic losses caused by weather, climate and water hazards. The different types of extreme weather that can lead to disasters are briefly described below.

15.2.3.1 Monsoons

The word 'monsoon' has its root in the Arabic word *mausim*, which means season. Monsoons are characterized by reversal of wind direction between January and July by at least 120° (180° by definition). Monsoons consist of two seasonal circulations – a winter outflow from a cold continental anticyclone and a summer inflow into a continental heat low (cyclone), that is, surface winds flowing persistently from oceans to continents in summer and just as persistently from continents to oceans in winter. The summer winds blowing from the oceans are warm and moist whereas the winter winds blowing from the continents are dry and cool. There is a corresponding change in the surface pressure gradient and in prevailing weather. These definitions cover the region in the south and south-east of Asia with the south Asian mountains as a natural boundary, that is, approximately 35°N–25°S and 30°W–170°E. Important features of northern summer monsoons are

- Surface pressure – Low in land; high in oceans
- Pressure in the upper troposphere – High in land; low in oceans
- Zonal wind in the lower troposphere – Westerlies on land; easterlies on oceans
- Zonal wind in the upper troposphere – Easterlies on land; westerlies on oceans
- Meridional wind in the lower troposphere – Southerly on land; northerly on oceans
- Tropospheric mean temperature – Warm on land; cold on oceans
- Total moisture – Humid on land; relatively dry on oceans
- Rainfall – Much larger on land than in the trade wind belt on oceans

Monsoons bring large amounts of rainfall. The world's highest recorded annual rainfall of 26,470 mm, and an average annual rainfall of about 12,000 mm have been in Cherrapunji (25°15'N, 91°44'E) in Northeast India, which also has a monthly record of 9,300 mm. This rainfall is brought about by the south-west monsoon. In India, 70% of the rainfall takes place during the south-west monsoon (June–September). In Sri Lanka, SW monsoon, which comes in summer during the period April–September is called *yala* and the north-east monsoon, which comes in winter during the period October to March is called *maha*.

The driving forces in monsoon winds are the pressure gradient between large land mass and the ocean. It can be thought of as a convective motion generated by differential heating of the land and the oceans. The swirl, introduced to wind by the rotation of the earth, is also a contributing factor. The differential heating is caused by the differences in the specific heats of the oceans and the land masses. The specific heat (energy required to raise the temperature by 1°C) of water is twice that of dry soil. Solar energy received on land heats only a few meters of the earth's sub-surface. Much of the energy goes into heating the air. For the oceans, it is quite the opposite. Less energy is available for heating the air. Therefore, for the same amount of energy, the temperature of dry land would be twice that of water. The effective heat capacity of the ocean is very much larger than that of land.

In summer, the rise in temperature over the oceans is less than the rise in temperature over land. The mean summer temperature over the oceans is about 5–10°C less than on land at the same latitude. In winter, large heat storage in the oceans leads to higher temperatures in the oceans. Westerly winds at the lower levels and easterly winds at the higher levels generate the convective motion. The reversal takes place at about 6 km elevation. Monsoon arrival is gradual and starts in June. They last from two to four months. In the Indian sub-continent, an extensive anti-cyclone dominates above the monsoon winds. In mid-latitudes, the pressure gradient force and Coriolis force balance each other. At low latitudes, Coriolis force weakens and there is no geostrophic balance.

15.2.3.2 Cyclones

A cyclone is any circulation around a low-pressure centre regardless of size and intensity. While rotating about the axis, they also move horizontally. They spin (or appear to spin) clockwise in the southern hemisphere and anti-clockwise in the northern hemisphere, that is, the same direction as the direction of rotation of the earth. The main driving force in a cyclone formation is the pressure gradient force, which acts from the high-pressure to the low-pressure region.

(i) Tropical cyclones

Tropical cyclones occur in the tropics (23° 27'N and S). They originate around 5° to 15° latitudes from the equator and are quite common in the Indian and Pacific Ocean parts of the monsoon region. Wind speeds of up to 250 knots at times have been reached. When travelling across continents, they lose energy and die down. Cyclones are usually accompanied by heavy rain. In different regions of the world, tropical cyclones have different names as shown in Table 15.5.

Any storm is a form of cyclone. Tropical cyclone is the proper generic name whereas tropical storm is a less technical term. In the Atlantic and North Pacific, tropical cyclones are called hurricanes and in the Western North Pacific and South China sea, they are called typhoons. The World Meteorological Organization classifies cyclones according to the maximum sustained wind speeds near the centre of the cyclone. In Hong Kong, the classification based on the wind speeds averaged over a period of ten minutes is given in Table 15.6 (Royal Observatory, 1992).

Table 15.5 Names used for cyclones in different regions

Country or region	Name used
North America	Hurricane
Japan, Northern China, South-East Asia, North-Western Pacific Ocean	Typhoon
Indian Ocean	Cyclone
Philippines	Baguios (or Bagyo)
Australia	Willey-Willys

Table 15.6 Classification of tropical cyclones in Hong Kong

Type of cyclone	Speed near the centre	Beaufort scale
Tropical Depression	Up to 62 km/h	6–7
Tropical storm	63–87 km/h	8–9
Severe tropical storm	88–117 km/h	10–11
Typhoon	> 118 km/h	12

(ii) Structure and development of a tropical cyclone

Cyclones have disc-like shape with a vertical scale of tens of kilometres and a horizontal scale of hundreds of kilometres. A tropical cyclone is a heat engine with energy derived from the latent heat of condensation. The power generated in a typical tropical cyclone can be of the order of 20 million MW, and the energy dissipated in one day is sufficient to meet Hong Kong's requirements for 20 years at the 1990 consumption rate (Royal Observatory, 1992).

The physical processes that lead to the formation of a tropical cyclone are still not well understood. It is known that large-scale momentum surges, which can provide inward eddy vorticity flux, are favourable to the development of a cloud cluster to a tropical storm. After the formation of the tropical cyclone, the heating efficiency increases due to the increased vorticity associated with the system. In the northern hemisphere, cyclones have winds in the anti-clockwise direction.

A favourable environment is necessary for a cyclone to develop to a full-scale phenomenon. It is generally known that three conditions must be satisfied for a tropical cyclone to be formed: the sea surface must be warm with temperature exceeding about 26°C, the air at low levels must converge inwards over a large area and the air flow at very high levels must be divergent so that a three-dimensional circulation can be sustained. The life span may vary from about a few days to a few weeks.

Tropical cyclones originate in the oceans with high temperatures. The air masses are lifted from the lowest layers of the atmosphere, which have about the same temperature as the sea and expanded adiabatically, resulting in condensation. Due to the high initial temperature, the air masses remain considerably warmer than the environment at least up to a level of about 12 km.

For tropical cyclones to develop, the value of the Coriolis force must be larger than a certain minimum, which excludes a latitude belt of about 5°–8° on both sides of the equator. Tropical cyclones are initiated by a pre-existing low-level disturbance (e.g. areas of bad weather and relatively low pressure) supported by the Coriolis acceleration.

A fully developed cyclone is a warm core energy-exporting system, which usually remains intense for many days over the ocean. Essential to this is an extremely large surface pressure gradient near the core of the cyclone, sometimes exceeding 3 hPa per km.

The energy source for a tropical cyclone is mainly the latent heat of condensation of water vapour. Thus, if it remains over warm waters, it has a continuous source of energy and therefore can continue indefinitely.

At a given level and latitude, the pressure gradient force is balanced by the Coriolis force. The wind, which acts normal to the Coriolis force, is called the gradient wind. This force system is prevalent only if other forces such as friction are neglected. The result, in the northern hemisphere, is a system of isobars in concentric circles with the gradient wind blowing counterclockwise for cyclonic systems and clockwise for anti-cyclonic systems. For cyclones in the northern hemisphere, the only way the air could move in a circular path is by having a net centripetal force (directed towards the centre). This can happen when the pressure gradient force is greater than the Coriolis force. Near the ground surface, frictional forces are significant and the winds tend to get deflected inwards, thereby distorting the concentric circular isobar pattern.

(iii) Extra tropical cyclones (wavy cyclones)

Extra tropical cyclones are developed in the mid-latitudes (50°S to 50°N). They tend to develop whenever air masses of different properties converge, such as for example when polar and tropical air masses meet in the mid-latitudes. The principal source of energy is the temperature and density differences in the two air masses. They dissipate the energy by mixing of the air masses. Winds in the region of 30–80 km/h are common. An extra tropical cyclone is essentially frontal type and has different stages of development.

When warm air ascends over a cold wedge, adiabatic cooling takes place, which leads to condensation and precipitation. As the cyclone occludes, cold air replaces warm air at low levels. The centre of gravity of the system is lowered and the potential energy is converted to kinetic energy of the winds. Lee sides of mountains are favoured for cyclone development.

(iv) Hurricanes

Hurricane derives its name from the Carib god 'Hurican', which in turn was derived from the Mayan god 'Hurakan', one of their creator gods, who blew his breath across the chaotic water and brought forth dry land and later destroyed the men of wood with a great storm and flood. In the United States, a cyclone with wind speeds in excess of 32.6 m/s (119 km/h) is called a hurricane. Speeds of up to 90 m/s (324 km/h) have been recorded. It has a calm central area called the eye (common to all cyclones). In most cases, the surface wind speeds do not usually exceed 67 m/s (241 km/h), but they may occur over a large area. The time scale of a hurricane is of the order of a few days.

(v) Thunderstorms

When the atmosphere is unstable and the moisture content is high, convective cloud development once started proceeds at a rapid rate. The cloud air because of its buoyancy continues rising. In a very unstable air mass (i.e. where the lapse rate is very high), the rising parcel of air becomes more and more buoyant with altitude. This is because of the temperature decrease with altitude. In some cases, the cloud air may be warmer than the environmental air up to the lower layers of the stratosphere. A cloud air ascending at the rate of perhaps 1 m/s at 1,500 m may attain speeds of 25 m/s at an altitude of 7,500 m. In this manner, small clouds become bigger and in turn develop into cumulonimbus clouds or better known as thunderstorms. These extend to altitudes of about 10–20 km. The upper limit of the growth of a thunderstorm is determined by the height of the stratosphere. This is so because the lower layers of the stratosphere are very stable, the temperature gradient at the stratosphere is zero or negative. Once it has reached an altitude where the cloud is colder than the environment, it begins to slow down but will continue upward movement a few thousand meters because of its momentum.

When the thunderstorm is mature, the upward movement takes place at its maximum speed. Because of the growth of precipitation particles, which coalesce and move downwards, there is a downward draft of equal magnitude. At this stage, heavy rain, electrical effects and gusts at the surface are common. The lifting of moist low-level air to high troposphere can take place by three mechanisms: (i) convectional lifting – when low level moist air is heated by high surface temperatures caused by solar radiation; (ii) orographic lifting – when moist air is forced up by topographical barriers such as mountain ranges; and (iii) frontal lifting – convergence of low level air in the vicinity of cold fronts. Lightening is another feature of thunderstorms. Sometimes it cannot be seen but can be heard.

Electrons from the water droplets accumulate at the base of the cloud. This negative charge induces a positive charge on the earth's surface below the cloud. A potential gradient of about 1,000 volts per meter occurs between the cloud and the ground. When this is too large, a discharge of electrons takes place. The rapid heating of the air in the lightning path produces a violent expansion of air, which initiates a sound wave propagating outwards at the speed of sound. (Lightning travels at about 10^9 km/h, whereas sound travels at about 960 km/h.) By recording the times between seeing the flash and hearing the sound, it is possible to calculate the approximate distance from the place of lightning. Thunderstorms can affect a large area but will not last more than a day. They bring large amounts of rain. Gustiness and falling temperature are signs of an approaching thunderstorm.

(vi) Tornados

Tornados are quite common in the United States. They last only for a few minutes but with extreme force. Wind speeds are of the order of 130–180 m/s (480–640 km/h). Distances affected are of the order of 100–1,000 m. Because of the extreme low pressure, no man-made structure can survive a direct hit by a tornado. When tornados occur in water, a phenomenon known as waterspout is formed.

(vii) Tropical depressions

These are centres of low pressure, which form in the troughs. They produce deep clouds and much precipitation mainly of the convective type. By classification, wind speeds are less than 17.4 m/s.

(viii) Tropical storms

These are well-developed low-pressure systems surrounded by strong winds and much rain. By convention, a system qualifies as a tropical storm if winds range from 17.4 to 32.6 m/s (40–120 km/h).

15.2.4 Landslides

A landslide (see also Chapter 10), sometimes referred to as a slope failure, can be defined as the uncontrollable downward movement of a solid mass like soil, rock or debris under the effect of gravity. Landslides can cause damages to property, infrastructure and even death. They are triggered by weak geological formations, earthquakes, volcanic activities, soil erosion and heavy rain and their combinations. Mudslides and debris flow are also triggered by similar factors. Based on geotechnical information, it is possible to identify slopes that are vulnerable to failure, but the exact timing of failure is difficult to predict. Since rainfall is a major triggering factor, it is possible to give warnings of impending slope failures based on the intensity and duration of rainfall. A list of the deadliest landslides of the 20th century is given in Table 15.7. In the United States, largest known recorded landslide in history was the one caused by the eruption of Mount St. Helens in 1980 that is estimated to have moved about 2.8 km^3 of debris at a downslope speed of about 112–240 km/h.

According to the International Disaster Database of the Centre for Research on the Epidemiology of Disasters (CRED) (EM-DAT), since 1900 some 130,000 persons have lost their lives because of landslides and flash

Table 15.7 Deadliest landslides of the 21st century

Rank	Date	Place	Name/article	Death toll
1	June 16, 2013	Kedarnath, Uttarakhand, India	2013 North India floods	5,700
2	August 8, 2010	Gansu, China	2010 Gansu mudslide	1,287
3	February 17, 2006	Southern Leyte, Philippines	2006 Southern Leyte mudslide	1,126
4	August 9, 2009	Siaolin Village, Kaohsiung, Taiwan	Siaolin mudslide	439–600
5	May 2, 2014	Argo District, Badakhshan Province, Afghanistan	2014 Badakhshan mudslides	350–500 reported
6	April 2, 2017	Mocoa, Colombia	2017 Mocoa landslide	314
7	October 1, 2015	El Cambray Dos, Guatemala Department, Guatemala	2015 Guatemala landslide	220
8	August 2, 2014	Sunkoshi, Sindhupalchok District, Nepal	2014 Sunkoshi blockage	156+
9	June 12, 2017	Rangamati, Chittagong and Bandarban, Bangladesh	2017 Bangladesh landslides	152
10	July 30, 2014	Malin, Ambegaon taluka, Pune district, Maharashtra, India	2014 Malin landslide	136
11	June 11, 2007	Chittagong, Bangladesh	2007 Chittagong mudslides	123
12	September 6, 2008	Cairo, Egypt	2008 Cairo landslide	119
13	March 1, 2010	Bududa District, Uganda	2010 Ugandan landslide	100–300
14	October 29, 2014	Badulla District, Sri Lanka	2014 Badulla landslide	100+
15	May 18, 2015	Salgar, Antioquia Department Colombia	2015 Colombian landslide	78
16	April 23, 2015	Badakhshan Province, Afghanistan	2015 Badakhshan landslides	52
17	August 20, 2014	Hiroshima Prefecture, Japan	2014 Hiroshima landslides	50+
18	March 22, 2014	Oso, Washington, United States	2014 Oso mudslide	43
19	February 20, 2010	Madeira Island, Portugal	2010 Madeira floods and mudslides	42
20	November 9, 2001	Amboori, Kerala, India		40

Source: The Deadliest Landslides of the 21st Century – WorldAtlas.

floods with economic losses amounting to over $50 billion. In the period from 2000 to 2014, the corresponding figures have been around 26,000 deaths and $40 billion in losses. The damages from the landslides triggered by the Wenchuan Earthquake (Magnitude 7.9), in Sichuan, China, on May 12, 2008, have been in excess of 87,000 casualties. In general, the potential destructiveness of landslides is a function of the volumes of material that are mobilized, and their velocity as well as the slope of the terrain.

15.2.5 Avalanches

An avalanche can be defined as a large mass of sliding or flowing snow. It is similar to a landslide except that the moving material is snow that may sometimes be mixed with other debris. Since snow is the driving force, avalanches are restricted to locations receiving large amounts of snow every year. Typical locations where avalanches occur are Rocky Mountains in North America, the Alps in Europe and the Himalayas in Asia. People who are most likely to face disasters caused by avalanches are skiers and mountaineers. There are also threats for communities in alpine environments as avalanches may contain large slabs of snow and ice that can destroy lodges and even towns downstream of the avalanche sites.

15.2.6 Tsunami

Tsunami is a Japanese word; 'tsu' meaning harbour and 'nami' meaning wave. The phenomenon is usually associated with earthquakes, landslides or volcanic eruptions in or adjacent to oceans and results in sudden movement of the water column. Until recently tsunamis were called tidal waves, even though the event has nothing to do with tides. Tsunamis are long-wavelength, long-period sea waves produced by the sudden or abrupt movement of large volumes of water. In the open ocean, the distance between wave crests can surpass 100 km and the wave periods can vary from five minutes to one hour. Such tsunamis travel 600–800 km/h, depending on water depth. Large waves produced by an earthquake or a submarine landslide can overrun nearby coastal areas in a matter of minutes. Tsunamis can also travel thousands of kilometres across open oceans and cause destruction on far shores hours after the occurrence of the earthquake. About 80% of tsunamis occur in the Pacific Ocean, but they can also occur in other large bodies of water such as lakes.

A tsunami is different from a wind-generated surface wave on the ocean. The passage of a tsunami involves the movement of water from the surface to the seafloor, which means its speed is controlled by water depth according to the relationship $v = \sqrt{gh}$ where g is the acceleration due to gravity and h is the depth of water. In the open ocean, even the largest tsunamis have relatively small wave heights. As the tsunami travels along a long and gradual slope, the wave height increases and this process is called shoaling. Consequently, it slows down as the wave approaches land and reaches increasingly shallow water. However, the water column still in deeper water is moving slightly faster and catches up, resulting in the wave height becoming much higher. A tsunami is often a series of waves and the first may not necessarily be the largest.

Tsunami waves have small amplitudes offshore and a very long wavelength usually of the order of hundreds of kilometres compared to normal ocean waves that have wavelengths of only 30 or 40 m. Because of the small amplitude, tsunamis, which form a swell of about 300 mm, are generally unnoticed at sea. They grow in height as they approach shallower water. A tsunami can occur in any tidal state and even at low tide it can still inundate coastal areas.

The temporary rise in water level when the tsunami's wave peak reaches the shore is called the 'run up'. There can be several waves arriving at the shore with significant time between wave crests. The first wave may not always have the highest run up. When the first part of a tsunami to reach the shore is a trough rather than a crest, the water level along the shoreline recedes, exposing the normally submerged areas significantly. Many of the people who perished during the Indian Ocean tsunami were those who went into the exposed area out of curiosity.

15.2.6.1 Tsunami generated by seismicity

Earthquakes are the main cause of tsunami generation. Tsunami can be generated when the sea floor abruptly deforms and vertically displaces the overlying water. Tectonic earthquakes are a particular kind of earthquake that are associated with the Earth's crustal deformation. During an earthquake, the rock on one side of the fault suddenly slips with respect to the other. The fault surface can be horizontal, vertical or at an angle. A thrust fault is a discontinuity in the earth's crust across which older rocks are pushed above younger rocks. When this happens, abrupt water displacement due to the vertical movement takes place.

Tsunamis can also be generated by giant landslides falling onto a water body. Such an event displaces large volumes of water as the energy from the falling landslide is transferred to the water at a rate faster than it can absorb. Mega-tsunamis that can cross oceans can also be generated from landslides and from volcanic island collapses. Tsunami can travel at speeds up to 950 km per hour in deep water, which is approximately the speed of a passenger jet.

Damage by tsunamis can be due to the smashing force of a wall of water travelling at high speed and the destructive power of a large volume of water draining off the land and carrying everything with it. Some recent tsunamis are briefly described below.

15.2.6.2 Tsunami in Tonga (2022)

The most recent tsunami that occurred on January 15, 2022, as a result of the eruption of the Hunga Tonga-Hunga Ha'apai underwater volcano, located about 65 km north of Tonga's capital, Nuku'alofa, caused a 1.2-metre tsunami and has been felt as faraway places as the Pacific coast, Japan, New Zealand, Fiji, Vanuatu, Peru and Australia. According to reports there have been three casualties, but the major problem from this volcanic eruption has been the water pollution from volcanic ash.

15.2.6.3 Tohoku earthquake in Japan (2011)

The most recent mega-tsunami was caused by an earthquake of magnitude 9 on the Richter scale that occurred on March 11, 2011, off the coast of Japan with its epicentre located approximately 72 km east of the Oshika Peninsula of Tohoku region with the hypocentre at an underwater depth of about 32 km. This earthquake which lasted for about 6 minutes is generally known as the 'Tohoku earthquake'. The earthquake triggered an extremely destructive tsunami with waves heights more than 30 m in several places travelling up to 10 km inland with smaller waves reaching other countries including the entire Pacific coast of North and South America after several hours. This earthquake is estimated to be the most powerful one to hit Japan and one of the five most powerful earthquakes in the world since record keeping began in 1900. It is estimated to have moved Honshu, the main island of Japan, 2.4 m east and shifted the earth on its axis by almost 10 cm. This deviation would have led to some planetary changes such as length of the day and the tilt of the earth. The axis shift would have been caused by the redistribution of the mass on the earth's surface, which would change the planet's moment of inertia, which in turn would cause small changes in the earth's rate of rotation. The quake occurred some 373 km from Tokyo and the nearest major city was Sendai. There have been over seven foreshocks of which six were with magnitude greater than 6 and over 671 aftershocks of which 44 were with magnitude greater than 6.

The damages caused by this mega-tsunami include some 19,747 deaths, 2,556 missing and 6,242 injured (2011 Tōhoku earthquake and tsunami – Wikipedia), damages to houses, buildings and infrastructure, railways and roads, disruption to basic facilities such as electricity and water, fires in some areas and above all the damages to the Fukushima Nuclear Power Plant. The leading cause of death has been drowning. The economic damages resulting from the earthquake and tsunami have run into several hundreds of billions of US dollars.

The energy released by the earthquake and dissipated as shaking and tsunami energy is estimated to be of the order of $1.9 \pm 0.5 \times 10^{17}$ Joules, which is sufficient to power a city of the size of Los Angeles for an entire year, or equivalent to 9,320 gigatons of TNT, or 600 million times the energy of the Hiroshima bomb.

The United Nations, through UN Resolution 70/203 adopted on December 22, 2015, has designated November 5 as the 'World Tsunami Awareness Day'. This date is based on an anecdote and example of good practice known in Japan as 'Inamura no hi' (fire of rice heaves), which took place on November 5, 1854, following a massive Ansei Nankai earthquake in 1854.

(i) *Fukushima Daiichi nuclear accident*

The Fukushima nuclear accident happened after the Tohoku earthquake that was followed by a 15 m high tsunami (the original design tsunami height has been 3.1 m for Daiichi (Number One) plant based on assessment of the 1960 Chile tsunami) on March 11, 2011 in Japan and is considered as the second worst nuclear accident in history after the Chernobyl disaster, which had occurred in the former Soviet Union in 1986. The earthquake and tsunami led to power loss in Fukushima Daiichi plant that caused a cooling system failure in three reactors, which led to partial meltdown of fuel rods. The incident led to explosions in the containment buildings and release of radiation into the atmosphere and the ocean. On detecting the earthquake, the active reactors automatically shut down their normal power-generating fission reactions. The flooding caused by the tsunami caused the failure of the emergency generators and loss of power to the circulating pumps.

The nuclear power plant, operated by the Tokyo Electric Power Company (TEPCO), consisted of six boiling-water reactors constructed between 1971 and 1979. At the time of the accident, only reactors 1–3 were operational, and reactor 4 served as temporary storage for spent fuel rods. After the incident, the radiation levels in food, water and the ocean near Fukushima increased and some 150,000 people living within a 20 km radius from the power plant, which was later extended to a 30 km radius, were forced to be evacuated. By March 2017 all evacuation orders in the areas outside the difficult-to-return zone (which continued to impound some 371 square km) had been lifted. Large amounts of water contaminated with radioactive isotopes were released into the Pacific Ocean during and after the disaster. TEPCO, which operated the plant, built new walls along the coast including a 1.5 km long 'ice wall' of frozen earth to stop the flow of contaminated water into the ocean and to the aquifers. The tsunami inundated about 560 km² and resulted in a human death toll of about 19,500 and much damage to coastal ports and towns, with over a million buildings destroyed or partly damaged.

The main radionuclides that emitted radiation had been volatile iodine-131, which has a half-life of 8 days, and caesium-137, which has a half-life of 30 years, which in its decay produces another radionuclide caesium-134, which has a half-life of two years. Although there were no casualties directly resulting from radiation during the immediate aftermath of the accident, there have been reports of abnormal growth defects in children as observed about a year later.

As of October 2019, a large volume of contaminated water had been stored in the plant area, which was subsequently purified to remove radionuclides except tritium to a level the Japanese authorities consider as safe for releasing to the sea. According to the calculations by the Japanese authorities, the discharge of purified water to the ocean would expose a radiation dose much less than the natural background dose of radiation which the International Atomic Energy Agency (IAEA) considers as appropriate. However, there is opposition to the discharge of such water to the ocean by neighbouring countries, specially Korea and fishing communities.

15.2.6.4 Indian Ocean – Sumatra and Andaman earthquake and tsunami (2004)

On December 26, 2004, a massive earthquake measuring 9.2 on the magnitude scale occurred off the west coast of northern Sumatra in Indonesia, causing a tsunami affecting several countries across the Indian Ocean Basin. The epicentre of the main earthquake was approximately 160 km off the western coast of northern Sumatra in the Indian Ocean with the hypocentre some 30 km below mean sea level. The resulting tsunami affected Australia, Bangladesh, India, Indonesia, Kenya, Malaysia, Myanmar, Thailand, Sri Lanka, Seychelles, Somalia and the Maldives. The tsunami waves caused widespread damage, including death and injuries and devastated coastal areas, destroyed towns, homes, infrastructure and the livelihoods of thousands of displaced people.

The International Federation of Red Cross and Red Crescent report the human toll from the disaster, as of March 23, 2005, as 273,636 dead, including 220,153 in Indonesia, 7,253 missing, more than 1,590,707 displaced and in excess of 507,496 made homeless. With the objective of mitigating risks, the UN launched an International Early Warning Programme (IEWP) at the World Conference on Disaster Reduction held in Kobe, January 2005. Subsequently, the Indian Ocean Tsunami Warning Centre was established in response to the UN initiative for an international early warning system and became operational in 2006.

15.2.6.5 Indonesia – Krakatau (1883)

On August 27, 1883, Krakatau volcano produced four massive explosions with each explosion producing equally strong tsunamis with wave heights of about 30 m. When the volcano cracked and collapsed into the ocean, the resulting tsunami has been over 36 m and is estimated to have caused over 40,000 casualties and destroyed over 200 villages in the Indonesian islands.

Since 1927, small eruptions have taken place and a new island named Anak Krakatau (child of Krakatau) has appeared in the caldera.

15.2.6.6 Europe – Lisbon earthquake and tsunami (1755)

This earthquake occurred on All Saint's Day on November 1, 1755, while many of the 250,000 inhabitants of Lisbon were in church. Human casualties from the earthquake were probably in excess of 30,000 with another 1,000 or so from the resulting tsunami. Stone buildings swayed violently before collapsing and fire ravaged the city. Many who sought safety on the river front were drowned by a large tsunami. In all, one quarter of Lisbon's population perished.

15.2.6.7 Deadliest tsunamis since 1900

Destructive tsunamis have occurred in all of the world's oceans and seas. Table 15.8 gives a list of the deadliest tsunamis since 1900 extracted from 'WorldAtlas' (Here Are The Deadliest Tsunamis In History – WorldAtlas). More information about tsunamis can be found in the international tsunami database.

15.2.7 Water-related biological disasters

Polluted waters can cause a number of water-borne diseases, some of which can be fatal. These were discussed in Chapter 7.

15.3 MITIGATIVE MEASURES

Disaster mitigation measures include those that eliminate or reduce the impacts and risks of hazards through proactive measures taken before an emergency or disaster occurs. They can be structural or non-structural and depend on the type of hazard.

15.3.1 Structural

In the case of flood hazards, building embankments, dikes, storage reservoirs or basins to restrict overflow, retarding basins to lower the flow, levees and floodwalls to confine floodwaters and improvement of channel capacity can reduce or eliminate damages resulting from floods. They are all physical means of mitigation that cost money. An assessment of the cost–benefit analysis should be made prior to making investments on structural means. In the case of hazard posed by storms, mitigation measures include building shelters at carefully chosen places whereas for drought hazards, provision of deep wells as well as provision of alternative irrigation facilities to crops.

Table 15.8 Deadliest tsunamis since 1900

Rank	Name	Location	Year	Casualties (estimated)
1	2004 Indian Ocean earthquake and tsunami	Indian Ocean	2004	230,210
2	1908 Messina earthquake	Messina, Italy	1908	123,000
3	2011 Tōhoku earthquake and tsunami	Japan	2011	18,550
4	1960 Valdivia earthquake	Valdivia, Chile, and Pacific Ocean	1960	6,000
5	1976 Moro Gulf earthquake	Moro Gulf, Mindanao, Philippines	1976	5,000
6	1945 Baluchistan earthquake	Arabian Sea, Indian Ocean	1945	4,000
7	1933 Sanriku earthquake	Sanriku, Japan	1933	3,068
8	1952 Severo-Kurilsk tsunami	Severo-Kurilsk, Kuril Islands, USSR (Russia)	1952	2,336
9	1998 Papua New Guinea earthquake	Papua New Guinea	1998	2,200
10	1946 Nankai earthquake	Nankai, Japan	1946	1,500
11	1944 Tōnankai earthquake	Tōnankai, Japan	1944	1,223
12	2006 Pangandaran earthquake and tsunami	South of Java Island	2006	800
13	2010 Chile earthquake	Chile	2010	525
14	1906 Ecuador–Colombia earthquake	Tumaco-Esmeraldas, Colombia-Ecuador	1906	500
15	2010 Mentawai earthquake and tsunami	Sumatra, Indonesia	2010	408
16	1979 Tumaco earthquake	Tumaco, Colombia	1979	259
17	1994 Java earthquake	Java and Bali, Indonesia	1994	250
18	1993 Hokkaido earthquake	Okushiri, Hokkaido, Japan	1993	197
19	2009 Samoa earthquake and tsunami	Samoa	2009	189
20	1983 Sea of Japan earthquake	Sea of Japan	1983	170
21	1999 Izmit earthquake	Sea of Marmara	1999	150
22	1964 Alaska earthquake	Alaska, USA and Pacific Ocean	1964	121
23	1992 Nicaragua earthquake	Nicaragua	1992	116

Source: Here Are The Deadliest Tsunamis In History – WorldAtlas.

15.3.2 Non-structural

Non-structural measures for mitigating flood disasters include flood zone mapping and zoning, early warning systems that require flood forecasting, evacuation, training, preparedness and awareness and flood insurance. The objective of an early warning system is to empower individuals and communities threatened by natural (or similar) disasters to act in sufficient time and in an appropriate manner so as to reduce the possibility of personal injury, loss of life and damage to property or nearby and fragile environments (IDNDR, August 1997). The components of an early warning system for flood forecasting include a flood forecasting model, data, calibration algorithm to test the forecasting model, translating the technical information to common language comprehensible to the vulnerable community and issuing warnings. An important point in issuing warnings is that it should be from a single authority and unambiguous.

There are several types of mathematical models that can be used in an early warning system. Broadly, they can be classified as physics-based models, conceptual models and data-driven models with each type having its own pros and cons. For example, physics-based models have the potential to understand the underlying mechanisms of rainfall–runoff transformation but require high resolution data as well as complex mathematical formulation. Conceptual models are relatively easy to formulate but require assumptions such as linearity, which sometimes may not be realistic. Data-driven models on the other hand do not require a detailed description of the processes in the hydrological cycle but only considers the variation of a hydrological variable with time as in the case of stochastic models (time series models), or the input–output transformation as in the case of deterministic models. The latter type includes but not limited to various forms of artificial neural networks, fuzzy logic systems, adaptive neuro-fuzzy systems, support vector machines, dynamical systems approach, genetic algorithms and genetic programming, some of which are discussed in an earlier publication by the author (Jayawardena, 2014).

15.4 INTERNATIONAL INITIATIVES

With the involvement of the United Nations, a number of international initiatives to reduce disaster risks have been launched since the late 1980s. Some of the better-known initiatives are listed below.

15.4.1 International Decade for Natural Disaster Reduction (IDNDR)

The International Decade for Natural Disaster Reduction (IDNDR) was founded in the 1990s with the basic objective to reduce through concerted international actions, especially in developing countries, loss of life, property destruction and social and economic disruption caused by natural disasters such as earthquakes, tsunamis, windstorms, floods, landslides, volcanic eruptions, wildfires and other calamities of natural origin such as grasshopper and locust infestations.

15.4.2 Yokohama strategy and related plan of action

The Yokohama Strategy for a Safer World was adopted at the first World Conference on Natural Disaster Reduction, Prevention, Preparedness and Mitigation held in the city of Yokohama, Japan during May 23–27, 1994, in partnership with the scientific community, business, industry and the media, deliberating within the framework of the International Decade for Natural Disaster Reduction (IDNDR).

15.4.3 United Nations International Strategy for Disaster Reduction (UNISDR)

The United Nations International Strategy for Disaster Reduction (UNISDR) has been created as the successor to the secretariat of the International Decade for Natural Disaster Reduction, in December 1999, to ensure the implementation of the International Strategy for Disaster Reduction (ISDR). It is mandated by the United Nations General Assembly to serve as the focal point in the United Nations system for the coordination of disaster reduction and to ensure synergies among the disaster reduction activities of the United Nations system and regional organizations and activities in socioeconomic and humanitarian fields.

In response to disaster trends and the increased expectations and demands by governments and other stakeholders to implement the Hyogo Framework for Action, the International Strategy for Disaster Reduction (ISDR) has evolved into a global system of partnerships consisting of national authorities and platforms, regional, international, intergovernmental and non-governmental organizations, the United Nations system, international financial institutions, and scientific and technical bodies and various specialized networks.

15.4.4 United Nations Office for Disaster Risk Reduction (UNDRR)

The United Nations Office for Disaster Risk Reduction (formerly UNISDR) was established in 1999 and mandated by the United Nations General Assembly to serve as the focal point of the United Nations system for the coordination of disaster reduction and to ensure synergies among the disaster reduction activities of the United Nations system and regional organizations and activities in socio-economic and humanitarian fields. The objectives include bringing governments, partners and communities together to reduce disaster risk and losses and ensuring a safer sustainable future. UNDRR oversees the implementation of the Sendai Framework for Disaster Risk Reduction 2015–2030, supporting countries in its implementation, monitoring and sharing what works in reducing existing risk and preventing the creation of new risk.

15.4.5 Hyogo Framework for Action (HFA) 2005–2015

The Hyogo Framework for Action (HFA) was adopted at the World Conference on Disaster Reduction, held during 18–22 January 2005, in Kobe, Hyogo, Japan as the global blueprint for disaster risk reduction efforts between 2005 and 2015. Its goal was to substantially reduce disaster losses by 2015 – in lives, and in the social, economic and environmental assets of communities and countries. The framework offers guiding principles, priorities for action and practical means for achieving disaster resilience for vulnerable communities.

15.4.6 Sendai Framework for Disaster Risk Reduction 2015–2030

The Sendai Framework for Disaster Risk Reduction 2015–2030 was adopted during the Third UN World Conference on Disaster Risk Reduction held in Sendai, Japan, in March 2015. It is the successor instrument to the Hyogo Framework for Action (HFA) 2005–2015. Its objective is to work hand in hand with other 2030 agenda agreements including the Paris agreement on climate change. It recognizes that the state has the primary responsibility to reduce disaster risk that should be shared with other stakeholders as well.

In addition, other UN agencies such as UNEP has been promoting ecosystem-based approaches such as river basin management, coastal zone management and protected area management as a means to reduce disaster risk and build resilience of vulnerable communities and countries. In Latin America and the Caribbean, a number of projects have been carried out to promote ecosystem-based disaster risk reduction (Eco-DRR).

15.4.7 International Day for Disaster Risk Reduction

After a call by the United Nations General Assembly, the International Day for Disaster Risk Reduction started in 1989. It is on October 13, the day on which people and communities around the world celebrate the reduction of their exposure to disasters and raise awareness of the importance of controlling the risks they face.

15.4.8 UNESCO's contribution

Education, science and culture are the three broad pillars upon which UNESCO has been formed and expected to function. There are many centres and institutes in different parts of the world established to promote these themes. Broadly, there are two types of centres or institutes identified as category 1 and category 2. Category 1 centres or institutes are institutionally and legally part of UNESCO and are considered an integral part of the organization's programme and budget. On the other hand, category 2 centres and institutes under the auspices of UNESCO are institutes of excellence that contribute to the implementation of UNESCO's priorities, programmes through international and regional cooperation. Such centres are scattered around the globe. Category 2 centres and institutes are proposed by member states to contribute to the achievement of UNESCO's approved programme through international and regional cooperation. They are not legally part of UNESCO but are associated with UNESCO through formal arrangements approved by the General Conference and/or the Executive Board. UNESCO is not legally responsible for them and shall bear neither responsibility nor liabilities of any kind, be it managerial, financial or otherwise.

In the context of water-related centres under the auspices of UNESCO, the IHE Delft Institute for Water Education based in Delft, the Netherlands, was a category 1 centre from 2001 to 2016. It is the largest international graduate water education facility in the world. In order to facilitate continuous funding beyond 2016 by the Dutch government, it has been turned into a category 2 centre under the auspices of UNESCO from 2017. As a result, there are no category 1 water-related centres at the present time. There are some 38 water-related category 2 centres worldwide. Of these, the International Center for Water Hazard and Risk Management (ICHARM), established in 2006 within the Public Works Research Institute (PWRI) of the Japanese government and located in Tsukuba, is the only dedicated centre for water-related disaster mitigation and risk management. ICHARM conducts training courses as well as graduate programmes leading to masters and doctoral degrees in collaboration with the National Graduate Institute for Policy Studies (GRIPS) in Tokyo.

15.4.9 International Flood Initiative (IFI)

International Flood Initiative (IFI) is a joint initiative in collaboration with international organizations such as UNESCO-IHP, World Meteorological Organization (WMO), UNISDR, United Nations University (UNU), International Association of Hydrological Sciences (IAHS) and International Association for Hydro-Environment Engineering and Research (IAHR). It focuses on research, information networking, education and training, empowering communities and providing technical assistance and guidance and will address issues identified by the Hyogo Framework of Action (2005–2015), and contribute to the UN International Decade for Action, 'Water for Life' (2005–2015) (4) and UN Decade on Education for Sustainable Development. It promotes an integrated approach to flood management including reducing social, environmental and economic risks and increasing the benefits from floods and the use of flood plains. The initiative is based on the concept of integrated flood management considering positive and negative impacts of floods including prevention, mitigation, preparedness, response and recovery. The IFI aims to foster the mobilization of resources and networks of the UN system, non-governmental organizations (NGOs), donor agencies and the insurance industry in order to assist communities and governments in developing culturally sensitive flood management strategies comprising of optimal structural and non-structural measures thereby targeting sustainable development.

REFERENCES

Adikari, Yoganath and Yoshitani, Junichi (2009). Global trends in water-related disasters: an insight for policymakers, United Nations World Water Assessment Programme side publication, pp. 26.

Jayawardena, A. W. (2014). *Environmental and Hydrological Systems Modelling*, CRC Press, Taylor and Francis Group, Boca Baton, FL, 516 pp.

Royal Observatory (1992). Tropical cyclones in 1992, 96 pp.

UNESCO World Water Assessment Programme (WWAP) (2009). World Water Development Report 3 – Water in a Changing World.

WEBSITES CITED

www.emdat.net

Climate and weather related disasters surge five-fold over 50 years, but early warnings save lives – WMO report | | UN News. https://news.un.org/en/story/2021/09/1098662

CRED/UNISDR, (2015). www.undrr.org/publication/unisdr-annual-report-2015

The Most Drought Prone Countries in the World – WorldAtlas. www.worldatlas.com/articles/the-most-drought-prone-countries-in-the-world.html

CRED Statista, (2019). https://globalimpactnews.com/2020/08/22/cred-publishes-2019-disaster-statistics/

5 droughts that changed human history | World Economic Forum (weforum.org). www.weforum.org/agenda/2019/05/5-droughts-that-changed-human-history/

The Deadliest Landslides of the 21st Century – WorldAtlas. www.worldatlas.com/articles/the-deadliest-landslides-of-the-21st-century.html

2011 Tōhoku earthquake and tsunami – Wikipedia. https://en.wikipedia.org/wiki/2011_T%C5%8Dhoku_earthquake_and_tsunami

International Federation of Red Cross and Red Crescent. www.ifrc.org

Here Are the Deadliest Tsunamis In History – WorldAtlas. www.worldatlas.com/articles/deadliest-tsunamis-since-1900.html

IDNDR, (August 1997). chrome-extension://efaidnbmnnnibpcajpcglclefindmkaj/https://www.unisdr.org/2006/ppew/whats-ew/pdf/guiding-principles-for-effective-ew.pdf

Chapter 16

Water-related conflicts

16.1 INTRODUCTION

Water is a finite resource that is indispensable for all living things that include humans, animals and other biotic species. There is no life without water. Water problems can lead to food shortages, energy shortages and hardships in day-to-day living that can lead to economic and governing instability. With the unabated increase in human population, the competition for water is rapidly increasing, resulting in conflicts related to the use and availability of water. Such conflicts can be regional and/or across countries. Water conflict in this context implies tensions and disputes between regions, countries or groups of people about the utilization, consumption and control of water resources. Conflicts arise when the ownership of a water source such as a river, lake or an underground aquifer is contested.

For a long time, water, unlike other resources, has been a resource owned by communities and governments. In recent times, it has become a tradable commodity, resulting in depriving this precious resource to those who cannot afford it. As the level and rate of poverty increase, this inequality will lead to conflicts between those who control water and those who consume water. More about the disparity in the availability of drinking water and the consequences are discussed in a separate publication by the author (Jayawardena, 2021).

There are some 286 international rivers and 592 transboundary aquifers shared by 153 countries (UN, 2018). In a non-homogeneous world, it is therefore natural to expect some conflicts about the way the water has to be shared. In addition to international conflicts, there are also internal conflicts within countries when there are disparities in the use and accessibility to water from different communities. With the per capita share of water availability decreasing with time, such conflicts could also be expected to increase in the future.

Water conflicts can arise because of territorial disputes, competition over resources and/or political reasons. During the period 2000–2009,

DOI: 10.1201/9781003329206-16

there had been 94 registered conflicts where water played a role (49 as a trigger, 20 as a weapon and 34 as a casualty[1]). The period 2010–2018 (up to May 2018) reported 263 registered conflicts (123 with water as a trigger, 29 as a weapon and 133 as a casualty) (WWDR, 2019). Conflicts are most likely to emerge when the downstream country is militarily stronger than upstream countries, and the downstream country believes its interests in the shared water resource are threatened by the actions of the upstream countries.

Water conflicts have existed in pre-historic times, historic times and now in the present world. The earliest conflict may have been in Mesopotamia over a dispute about depriving water to Girsu, a city in Umma. Since then, there have been conflicts of all three types in ancient Western Asia, Northern Africa, Southern Europe, Eastern Asia, North America and Latin America until the beginning of the 20th century. In the 20th century, conflicts extended to sub-Saharan Africa and Southern Asia. More recent conflicts include the division of Indus River between India and Pakistan, Arab forces cutting water to Israel, which led to the first Arab–Israel War, North Korea releasing flood waters to damage floating bridges operated by UN troops in the Pukhan Valley, several conflicts of all types among Israel, Syria and Jordan, trigger-type conflict between Egypt and Sudan over Nile River water, destroying irrigation facilities in North Vietnam by US bombing, between Paraguay and Brazil over Parana River and Cuba cutting off water supply to US bases in Guantanamo Bay. In the 21st century, a conflict arose in Bolivia as a result of water privatization.

In the recent times, three conflicts of international concern are over the Brahmaputra River, which has its headwaters in Tibet, China and flows through India and Bangladesh, the Grand Ethiopian Renaissance Dam across Blue Nile River, which will control downstream water flow to Egypt and the Ilisu Dam across Tigris River in Turkey, which will control downstream flow to Iraq and Syria. All these three conflicts are about dams been built on the upstream of rivers, which will control/reduce downstream flow. Egypt is sensitive about water from the Nile as it depends almost entirely on Nile water for drinking, farming and industrial water supplies. Very recently, it is reported that Ethiopian Attorney General has filed charges related to the 6,450-MW Grand Ethiopian Renaissance Dam project. (www.hydroreview.com/2020/01/02/ethiopian-attorney-general-files-charges-related-to-the-6450-mw-grand-ethiopian-renaissance-dam-project/?utm_medium=email&utm_campaign=hydro_weekly_newsletter&utm_source=enl&utm_content=2020-01-07)

---

1 The different categories add up to more than the total number, because some conflicts have been listed in more than one category.

16.2 CAUSES OF WATER-RELATED CONFLICTS

The causes of water conflicts can be political, economic and/or diplomatic. Water is a basic human need and politicians always seek a controlling interest in water affairs as it can attract votes and support from the community if their interest leads to positive impacts. On the other hand, if the impact to the community is negative, the politicians will lose the community support and may not get elected again. Water governance plays a crucial role in avoiding water conflicts in the management of water resources. In many countries, all major rivers are managed by the state. In development activities in countries with inadequate resources, financing by international lending organizations tend to dictate the way the projects should be managed that can lead to internal disputes. Hydro hegemony is a practice followed by powerful entities, national or international to control water by strategies such as coercion pressure, treaties and sometimes through corruption.

There are countries with abundant water resources such as Brazil, Canada, China, Iceland, Norway, Russia and the United States. At the same time, there are also water- poor countries such as Kuwait, Egypt, United Arab Emirates, Malta, Jordan, Saudi Arabia, Singapore, Maldovia, Israel and Oman. The water-poor countries often suffer from water stress, which becomes a precursor to conflicts, which may lead to violence in socially and politically fragile countries.

It is usually the downstream countries or regions that suffer from inequitable control from upstream countries or regions. The construction of dams across many rivers worldwide have led to disputes and complaints from downstream countries. Examples include the construction of the Grand Renaissance Dam across Blue Nile in Ethiopia, construction of several dams across the upper reaches of Mekong River in China, control of downstream flow of Ganges River by building dams for energy by India among others. Over-extraction of groundwater from aquifers is another cause for disputes as it can cause land subsidence in addition to mining the aquifers.

Another cause of conflict is the non-sharing of hydrological and other related information by upstream countries with downstream countries. Other factors that drive conflicts include population growth, economic expansion, prolonged drought, climate change, pollution, inefficient water use in agriculture, destruction of nature, poor water resources management and weak institutional governance.

16.3 TYPES OF CONFLICTS

Conflicts are categorized as trigger, weapon and casualty based on the use, impact or effect that water has within the conflict. Trigger refers to the situation when there is a dispute over the control of water or water systems or where economic or physical access to water or scarcity of water triggers violence. Weapon refers to the situation when water resources or water systems themselves are used as tools or weapons in a violent conflict. Casualty refers to the situation where water resources, or water systems, are intentional or incidental casualties or targets of violence.

Prior to the simplification to the above three types, there have been six categories identified as 'control of water resources' (state and non-state actors) where water supplies or access to water is at the root of tensions, 'military tool' (state actors) where water resources, or water systems themselves, are used by a nation or state as a weapon during a military action, 'political tool' (state and non-state actors) where water resources, or water systems themselves, are used by a nation, state or non-state actor for a political goal, 'terrorism' (non-state actors) where water resources, or water systems, are either targets or tools of violence or coercion by non-state actors, 'military target' (state actors) where water resource systems are targets of military actions by nations or states and 'development disputes' (state and non-state actors) where water resources or water systems are a major source of contention and dispute in the context of economic and social development (Pacific Institute, 2019; Definitions, Methods, & Sources – World's Water (worldwater.org)).

16.4 STRATEGIES TO REDUCE WATER-RELATED CONFLICTS

There is no single solution or strategy to reduce or eliminate water-related conflicts since water is a basic ingredient of life but used inequitably by humans. The World Resources Institute and the Pacific Institute have (Ending Conflicts Over Water | World Resources Institute (wri.org)) identified the following four categories of strategies.

16.4.1 Natural resources, science and engineering approaches

Water is a finite resource and should be used in the most efficient way, thereby eliminating wastage. For example, improved irrigation technologies such as drip irrigation, improvements in water use in industrial processes as well as restrictions in water use in water-stressed regions. This requires both supply and demand management.

16.4.2 Political and legal tools

Over-extraction from water bodies can lead to ecological problems, resulting in protests from affected communities. Political and legal tools can then be used to diffuse such tensions.

16.4.3 Economic and financial tools

The true value of water is generally more than the cost. Increasing the cost to reflect the true cost can help improve the service and maintenance of water infrastructure, which can adversely affect the poor and disadvantaged communities. Water pricing should reflect human rights, societal values and aim at an equitable strategy.

16.4.4 Policy and governance strategies

Good governance is a basic tool for efficient and optimal use of any resource. Policies should be formulated with the involvement of professionals, stakeholders and users of the resource.

16.5 MAJOR WATER CONFLICTS IN THE WORLD

There have been water-related conflicts within countries as well as across countries. The countries that have had conflicts within their borders include India, Pakistan, Mali, Iran, Iraq, Nigeria, Yemen, Somalia, Egypt and Bolivia. Prominent cross-country conflicts occurred in Turkey, Syria and Iraq over the Euphrates–Tigris River, Afghanistan and Iran over harnessing of the waters of the Helmand River, which has alarmed Iran as they perceive it as a threat to Iran's water security, China and downstream countries across which Mekong River flows as China has built dams in the upstream reaches of Mekong (Langcang Jiang), thereby controlling the downstream flow, Jordan River Basin with riparian rights shared by Israel, Lebanon, Jordan and the State of Palestine, and Turkey and Armenia on sharing the water of the Arpacay River, which they have resolved in an equitable manner. At times, the conflicts in these regions have led to 'water wars'. A few of the major transboundary river basins where disputes exist are briefly described below.

16.5.1 Nile River basin

River Nile is considered as the longest river (6,695 km) in the world with its headwaters originating in Burundi. With its main tributaries the White Nile, the Blue Nile and Atbara, it flows through Burundi, Congo, Egypt, Eritrea, Ethiopia, Kenya, Rwanda, Sudan, South Sudan, Tanzania and Uganda. Historically, Egypt, being the downstream country, benefitted the most from the resources of Nile. Of all Nile riparian countries, Egypt depends on external sources to the tune of about 97% of her water needs while the other riparian countries are able to somehow manage with resources from within their borders. Since the upstream riparian countries, after years of marginalization, began to harness the resources of the Nile, Egypt is feeling the pinch as it will no longer have the full (or major) benefit of Nile resources. In particular, Ethiopia, which is the source of about 85% of Nile water, is building a major dam to harness the Nile water for irrigation, power and other needs. The dam, being built across Blue Nile, and known as the Ethiopian Renaissance Dam will be the largest dam in Africa capable of producing 6,000 MW of power. Egypt and to some extent Sudan have been opposing this project from the inception as Egypt would lose its control over the Nile water. Historically, an agreement made between the colonial power Britain and Egypt in 1929 gave Egypt exclusive right to use Nile water. This agreement was renegotiated after the independence of Sudan in 1956 by an agreement signed between Egypt and Sudan in 1959 but excluded all other upstream countries. Both these agreements have given Egypt the veto power in all developments of the Nile and exclude the interests of other upstream countries. The present conflict is mainly due to the diminishing authority that Egypt enjoyed over the control of Nile water.

In many international relations, water has become a major political issue. Each country considers access to and control of water is important for national security. Water is also the basic requirement for economic development, including hydropower generation. Instead of becoming the basic ingredient for development in a region, water has become the main cause of conflicts and instability in some regions. Nile River basin is one of them. Although the Nile Basin Initiative of 1999 as a framework for cooperation has achieved some success, diverging interests of the upstream and downstream countries have become obstacles for fair and equitable sharing of Nile water. The Nile Basin is considered as a high-risk area for 'water wars'.

16.5.2 Mekong River basin

Mekong River is an international river that originates in China and flows through Myanmar, Laos, Thailand, Cambodia and Vietnam before emptying into South China Sea. The construction of dams in the Upper Mekong River has already raised serious concerns in the downstream riparian countries. However, the downstream countries themselves have also been planning, in recent years, the construction of a number of dams in the Lower Mekong River to meet their various water demands. Since these downstream countries are only starting to develop now, the demand for Mekong water will continue to grow in the foreseeable future. These only add to the already existing concerns about the sustainability of the Mekong River basin and, consequently, highlight further complications in the management of its water resources.

The upper Mekong River (also known as Langcang Jiang) has a high potential for hydropower development. As of now, there are plans for the construction of 22 hydropower stations with a total installed capacity of

3,200 MW. It is the fourth largest in China in terms of hydropower potential, only after Chang Jiang (Yangtze), Yarlung Tsangpo (upper Brahmaputra) and Nu Jiang (upper Salween). A cascade of 11 dams have already been built across the mainstream of Langcang Jiang in the Yunnan Province in China. Several others, including some in Tibet, are either under construction or being planned. The combined installed capacity of these hydropower projects would run into over 31,665 MW. The construction of dams on the upper reaches of Mekong (Langcang Jiang) has both positives and negatives for the downstream riparian countries. On the positive side, such dams control flood flows during the rainy season and, thus, mitigate damages to life and properties that would otherwise occur; they also augment flows during the dry season. At the same time, however, there are also significant concerns that the construction of dams on the upper reaches will result in significant reduction of flows to the downstream riparian countries and will adversely affect the ecological balance in such countries, including those related to sediment transport and fish migration.

To be fair and equitable, the development of the Mekong River resources needs to be done without limiting the right to the use of the river and its resources by any riparian country. Benefits of any development activity may come in different forms, and so are the associated costs. In this context, a fair and equitable guiding principle is to have the beneficiaries bear the costs. It is also important to realize that the development of the river and its resources should consider not only the human population that benefit from the river but also other biotic populations as well. To avoid conflicts and minimize negative outcomes, not only engineers but also many other professionals from many diverse disciplines, such as biologists, geographers, geologists, social scientists, economists and politicians should have a synergistic role to play to sustain the life and services of the river and its environment. The way forward should be to follow a holistic approach in which the river is considered as belonging to all riparian countries, their people and other biotic populations that depend on the river when development activities are planned. To some extent, the management of the Mekong River basin follows an agreement signed by Cambodia, Lao PDR, Thailand and Vietnam under the auspices of the Mekong River Commission (MRC) established on April 5, 1995, which mandates

> to cooperate in all fields of sustainable development, utilization, management and conservation of the water and related resources of the Mekong River Basin, including, but not limited to irrigation, hydropower, navigation, flood control, fisheries, timber floating, recreation and tourism, in a manner to optimize the multiple-use and mutual benefits of all riparians and to minimize the harmful effects that might result from natural and man-made activities.

However, as China and Myanmar are only 'dialogue partners' of MRC, the effectiveness in resolving tensions and conflicts so far has been limited due to lack of enforcement powers. More about the Mekong River can be found in a separate publication (Jayawardena and Sivakumar, 2017).

16.5.3 Indus River basin

Indus River, also referred to as 'Sindhu', is one of the world's largest rivers draining an area of approximately 1.12 million km² of which 47% is within Pakistan, 39% within India, 8% within China and 6% within Afghanistan and a length of approximately 3,200 km. Originating from the Himalayan mountains and running through the four countries, it discharges into the Arabian Sea via the large Indus Valley Delta near the port city of Karachi. Its main tributaries are the five rivers, Jhelum, Chenab, Ravi, Beas and Sutlej that give the name Punjab ('Five Rivers') to the region divided between India and Pakistan.

The first dispute arose when India and Pakistan were separated by the partition of British India in 1947 and the irrigation system of the Bari Doab and Sutlej Valley Project that was originally designed as one scheme was separated into two schemes. The headworks were in India and the canal ran through Pakistan. This dispute continued for some years and was later resolved through The Indus Water Treaty brokered by the World Bank (then known as the International Bank for Reconstruction and Development) and signed by the then prime minister of India, Jawaharlal Nehru, and the then president of Pakistan, Ayub Khan, on September 19, 1960, which paved the way for an equitable sharing of the waters of Indus River. This treaty is recognized as one of the successful water treaties in the world that provides control of the waters of the three eastern rivers, namely Sutlej, Beas and Ravi to India and the three western rivers, namely West Indus, Jhelum and Chenab to Pakistan. The technical details of the treaty have been a subject of negotiation from time to time, but the treaty has so far been a success. The treaty gives 20% of Indus water to India and the remaining 80% to Pakistan.

In addition to water disputes between India and Pakistan, there are also regional conflicts between the provinces of Sindh and Punjab in Pakistan. One of the main concerns is about the construction of dams such as Kalabagh dam (KBD) and Bhasha dam on the upstream Punjab province that deprives Sindh province its due share of Indus water under the Indus Basin Irrigation System (IBIS). Over-extraction of groundwater is another issue that leads to conflicts.

India too has had several inter-state water conflicts over sharing of water. Such disputes are settled by tribunals on a case-by-case basis. Major disputes include the Godavari Water Disputes Tribunal, Krishna Water

Disputes Tribunal – I, Narmada Water Disputes Tribunal, Ravi and Beas Water Tribunal, Cauvery Water Disputes Tribunal, Krishna Water Disputes Tribunal – II, Vansadhara Water Disputes Tribunal and Mahadayi Water Disputes Tribunal.

16.5.4 Jordan River basin

The Jordan River, which is only 251 km long, with most of its length flowing below sea level, and its waters are central to both the Arab–Israeli conflict (including Israel–Palestine conflict), as well as the more recent Syrian Civil War. Its waters originate from the high precipitation areas in and near the Anti-Lebanon mountains in the north, and flow through the Sea of Galilee and Jordan River Valley ending in the Dead Sea at an elevation of minus 400 metres, in the south.

Israel and Palestine share three main water sources, the Jordan River basin, the Coastal Aquifer with Israel upstream and Gaza downstream and the Mountain aquifer, which starts in the heights of the West Bank and flows to the Jordan Valley. The two nations have a long history of water disputes, which became a major conflict after the Six-Day War in 1967 when Israel nationalized the water resources of the occupied territories. The Mountain Aquifer under the West Bank and Israel is the only remaining water resource in Palestine and one of the most important underground water resources of Israel. The Israeli government limits Palestine to 20% of the annual yield of Mountain Aquifer. Palestine needs permits from the Israeli government to carry out any infrastructure development work in the occupied territories, which are often not given. On average, Palestinians use 70 litres per capita of water a day while Israelis consume 280 litres per capita per day. The World Health Organization threshold for a healthy life is 100 litres per capita per day (Holy water in the Israeli–Palestinian water conflict – LifeGate). The stumbling block for a fair and equitable solution to this conflict is the Israeli government's view that their asymmetric authority over unrestricted access to West Bank and Gaza water as a non-negotiable right is central to its security.

The Israel–Palestine water conflict is dragging on because of distrust between the two parties and the asymmetric political authority of Israel. Israel has made headway in securing water through desalination but not in the occupied territories. Scientists have pointed out ways of resolving the water problem in the region, for example, by installing desalination plants on the Israeli and Palestinian Mediterranean coasts as well as by producing solar energy in northern Jordan that has ideal conditions for producing solar energy (EcoPeace Middle East, 2014). Trade of water with energy would lead to inter-dependence as a means of resolving the conflict.

Two separate incidents involving conflicts and violence over water were reported in 2011 in Israel and Palestine, a region with an especially long history of water disputes. In the first, Israel's military was reported to have destroyed nine water tanks in the Bedouin village of Amniyr in the South Hebron Hills in the West Bank, Palestine. Later, soldiers destroyed pumps and wells in the Jordan Valley villages of Al-Nasaryah, Al-Akrabanyah and Beit Hassan (Abuwara, 2011). In the second incident, Israeli settlers near Qasra, a West Bank village of some 6,000 inhabitants, were reported to have destroyed crops including olive trees and a water well (Bsharat and Ramadan, 2011).

16.5.5 Brahmaputra River basin

Brahmaputra River, known as Yarlung Tsangpo in Tibet, originates in Tibet, China and flows through India and Bangladesh before emptying into Bay of Bengal after joining with the Ganges, Padma and Meghna rivers. It is considered as the 15th longest river and the ninth largest river in the world by discharge volume. It has caused tension between India and China for not sharing information on the status of the river upstream during the run-up to landslides in Tibet in 2000, which caused flooding in north-eastern India and Bangladesh. Between India and Bangladesh that share the waters of Ganges River as a major source of water for their people, an ad-hoc water sharing agreement has been reached in 1983, whereby the two countries were allocated 39% and 36% of the water flow, respectively. The new bilateral treaty expands upon this agreement by proposing an equal allocation of the Teesta River water, which flows through the northern part of Bengal and merges with Brahmaputra River after entering Bangladesh. The conflict has taken a political turn in recent years.

16.5.6 Chronology of water conflicts

There have been conflicts of different type in several other places. Pacific Institute has compiled a chronological list of conflicts spanning from 3000 BC to 2019. Table 16.1 is a partial list of conflicts that occurred in the 21st century, extracted from this list.

Table 16.1 Partial list of the chronology of water conflicts in the 21st century

Year	Headline	Conflict type	Country	Region	Description
1991–2001	US sanctions against Iraq target water systems	Casualty	Iraq	Western Asia	United States deliberately pursues policy of destroying Iraq's water systems through sanctions and withholding contracts
1991–2007	Violence over use of India's Cauvery River	Trigger	India	Southern Asia	Violence erupts when Karnataka, India, rejects an Interim Order handed down by the Cauvery Waters Tribunal, set up by the Indian Supreme Court. The Tribunal was established in 1990 to settle two decades of dispute between Karnataka and Tamil Nadu over irrigation rights to the Cauvery River.
2000	Central Asian nations cut off water to neighbours	Weapon	Kyrgyzstan, Kazakhstan, Uzbekistan	Central Asia	Kyrgyzstan cuts off water to Kazakhstan until coal is delivered; Uzbekistan cuts off water to Kazakhstan for non-payment of debt.
2000	French workers pollute river over labour dispute	Casualty	France	Western Europe	In July, workers at the Cellatex chemical plant in northern France dump 5,000 litres of sulphuric acid into a tributary of the Meuse River after they are denied workers' benefits. A French analyst points out that this is the first time 'the environment and public health were made hostage in order to exert pressure, an unheard-of situation until now'.
2000	Terrorist drill gets out of hand in California	Casualty	United States	North America	A drill simulating a terrorist attack on the Nacimiento Dam in Monterey County, California, got out of hand when two radio stations reported it as a real attack.

(continued)

Table 16.1 Cont.

Year	Headline	Conflict type	Country	Region	Description
2000	Riots in northern China	Trigger, Casualty	China	Eastern Asia	Civil unrest erupts over use and allocation of water from Baiyang Lake, the largest natural lake in northern China. Several people die in riots by villagers in July 2000 in Shandong after officials cut off water supplies. In August 2000, six die when officials in the southern province of Guangdong blow up a water channel to prevent a neighbouring county from diverting water.
2000	Aqueduct in Colombia is attacked	Casualty	Colombia	Latin America	A bombing attempt occurs at an aqueduct in Colombia. The bomb does not detonate. The National Liberation Army is suspected in the attempt.
2000	Water company is bombed to demand release of prisoners	Casualty	France	Western Europe	A water company office in Ustaritz, France, is damaged by a bomb thought to be placed in support of a group called the Basque Fatherland and Freedom (aka ETA). The attackers leave writing on the wall in the building, demanding the release of ETA prisoners.
2000	Settlers attack a village in Palestine, injuring civilians and cutting water supply	Casualty	Palestine	Western Asia	In a village in Palestine, three Palestinians are injured by gunshots from Israeli settlers. The same settlers also cut the water supply to the village by damaging the pipes.
2000	Water storage facility is destroyed	Casualty	Eritrea	Sub-Saharan Africa	Eritrea accuses Ethiopia of bombing the Red Sea port at Assab, saying several bombs are dropped on Assab, destroying a water storage facility on the outskirts of the town.

Year	Name	Type	Country	Region	Description
2000	Moroccan army destroys wells	Casualty	Morocco	Northern Africa	Units of the Moroccan army destroy the wells situated in the municipalities of Guelta and Boujdour. Citizens usually use these wells to provide water for their sheep in the summer.
2001	Palestinians destroy water supply to Israeli settlements	Casualty	Israel, Palestine	Western Asia	Palestinians destroy water supply pipelines to the West Bank settlement of Yitzhar and to Kibbutz Kissufim. The Agbat Jabar refugee camp near Jericho disconnects from its water supply after Palestinians loot and damage local water pumps. Palestinians accuse Israel of destroying a water cistern, blocking water tanker deliveries, and attacking materials for a wastewater treatment project.
2001	Water cut off in Macedonian conflict	Weapon, Casualty	Macedonia	Southern Europe	Water flow to Kumanovo (population 100,000), Macedonia cut off for 12 days in conflict between ethnic Albanians and Macedonian forces. Valves of Glaznja and Lipkovo Lakes damaged.
2001	US bombs Afghan powerhouse	Casualty	Afghanistan	Southern Asia	American forces bomb the hydroelectric facility at Kajaki Dam in Helmand province of Afghanistan, cutting off electricity for the city of Kandahar. The dam itself is not targeted.
2001	Nepali soldiers stop rebels from targeting drinking water supplies	Casualty	Nepal	Southern Asia	According to the Nepal Defence Ministry, soldiers repulse attempts by rebels to blow up the sources of Kathmandu's drinking water supplies at Shivapuri and Sundarijal. No details of these incidents are given.

(continued)

Table 16.1 Cont.

Year	Headline	Conflict type	Country	Region	Description
2001	Bombing disrupts Lebanese water supplies	Casualty	Lebanon	Western Asia	Water supplies to more than 20 towns are disrupted after a bomb explodes near the town of Baalbek in Lebanon.
2001	A water tank car is destroyed by a landmine	Casualty	Russia	Eastern Europe	A water tank car is destroyed by a landmine in Grozny, Chechnya, Russia. A teenage girl is also wounded in the explosion.
2002	Nepal rebels blow up powerhouse	Casualty	Nepal	Southern Asia	The Khumbuwan Liberation Front (KLF) blows up a 250 kW hydroelectric powerhouse in Nepal's Bhojpur District, cutting off power to Bhojpur and surrounding areas. The damages take six months to repair and cost 10 million rupees ($120,000). During 2002, Maoist rebels destroy more than seven micro-hydro projects, a water-supply intake and supply pipelines to Khalanga in western Nepal.
2002	Colombian rebels bomb dam	Casualty	Colombia	Latin America	Colombian rebels in January damage a gate valve in the dam that supplies most of Bogota's drinking water. Revolutionary Armed Forces of Colombia (FARC) detonate an explosive device planted on a German-made gate valve located inside a tunnel in the Chingaza Dam.

2002	Botswana authorities destroy wells of Kalahari Bushmen	Casualty	Botswana	Sub-Saharan Africa	Botswana's president Festus Mogae sends troops to the Kalahari Desert to destroy wells and empty water sources of indigenous Khoisan (also known as Bushmen), ostensibly in an effort to remove them from their ancestral lands and assimilate them into modern society. Critics blame the government of taking away water rights in favour of mining interests and labelled the government's actions a 'siege'; Botswana is condemned by international observers. Against expectations, a band of Bushmen retreat into the desert and survive for years with little outside assistance.
2002	Bolivian irrigators and townsmen clash over water rights	Trigger, Casualty	Bolivia	Latin America	Tarata, Bolivia, experiences violent confrontations between irrigators from the region of Arbieto and residents of the nearby town of Tarata. The dispute centres on the allocation of water among these users and new water infrastructure that took water from the Laka dam. In October, irrigators destroy approximately two kilometres of water supply pipeline. In response, residents destroy part of a primary irrigation channel. After failed negotiations, the water system is damaged again in December in another confrontation between irrigators and village residents.

(continued)

Table 16.1 Cont.

Year	Headline	Conflict type	Country	Region	Description
2002	Indian military protecting reservoir attacked by terrorists	Casualty	India	Southern Asia	A camp of the Assam Rifles of the Indian military at Yairipok in Thoubal district with a mission to protect a water reservoir is attacked by suspected terrorists who fired grenades and other explosives on January 27. Later, a spokesman for the United National Liberation Front (UNLF) claims that the attack was carried out by its armed wing, the Manipur People's Army, which has been fighting for what it called an 'Independent Manipur'.
2002	Youths destroy local water, electric and telecommunication facilities in Cote d'Ivoire	Trigger, Casualty	Cote D'Ivoire	Sub-Saharan Africa	On October 10, youths claiming to be members of the Group of Patriots for Peace (GPP) destroy facilities of the Ivoirian Water Distribution Company, the Ivoirian Electricity Company, and Cote d'Ivoire Telecommunications. Demonstrators say they were protesting the 'free' supply of water, electricity and telephone in rebel-controlled areas.
2002	Four wells found with human remains in Cote d'Ivoire	Casualty	Cote D'Ivoire	Sub-Saharan Africa	Conflict between the government of Laurent Gbagbo and opposing rebel forces leads to extensive violence and regional conflict. In late September 2002, there are reports of several mass graves discovered in the Bangolo. In Zeregbo and Bahably, four water wells are found with human remains. Early reports indicate that western rebel groups who captured the area killed the persons in the mass graves and wells between December 2002 and January 2003.

2003	Water systems damaged in US–Iraq war	Casualty	Iraq	Western Asia	During the US-led invasion of Iraq, water systems are reportedly damaged or destroyed by different parties, and major dams are military objectives of the US forces. Damage directly attributable to the war includes vast segments of the water distribution system and the Baghdad water system, damaged by a missile
2003	Iraqi water treatment plant damaged	Casualty	Iraq	Western Asia	The Abu Nawas pumping station on the Tigris River, which supplies water to Sadr City and other treatment plants in Baghdad, is severely damaged during the 2003 conflict.
2011	NATO attacks water pipe factory in Libya	Casualty	Libya	Northern Africa	NATO forces attack on a factory that produces pipes for the 'great man-made river' water supply pipeline near Brega, reporting that it was used as a base for military operations and the launching of missiles.
2011	Multiple attacks on Libyan water networks disrupt supply.	Casualty	Libya	Northern Africa	NATO bombs a plant that builds pipelines for the Great Manmade River project that carries water from southern groundwater to coastal cities. Major water networks are disrupted in this conflict, which affects access to safe drinking water, sanitation and hygiene throughout Libya. Armed groups attack water wells cutting off water to civilians.

(continued)

Table 16.1 Cont.

Year	Headline	Conflict type	Country	Region	Description
2012	Pakistani militants attack water systems in Kashmir	Trigger, Casualty	India, Pakistan	Southern Asia	Violence erupts in the latest event in the dispute between Pakistan and India over the waters of the Indus Basin. Pakistani militants attack and sabotage water systems, flood protection works and dams in the Wular Lake region of northern Kashmir. They attack engineers and workers and detonate explosives at the unfinished Tulbul Navigation Lock/ Wular Dam. Pakistan claims the new dam violates the Indus Water Treaty by cutting flows to Pakistan.
2012	Violent conflict among Kenyan and Ugandan herders	Trigger	Uganda, Kenya	Sub-Saharan Africa	Tensions lead to violence between Uganda and Kenya after Kenyan Pokot herdsmen cross the border seeking water and pasture. In October, the Ugandan government sends 5,000 soldiers to control violence among pastoralists from the two countries.
2014	Ukraine accused of cutting water supply	Weapon	Ukraine, Crimea	Western Asia	After Russia annexes Crimea from the Ukraine, they accuse the Ukraine of cutting off the water supply in the North Crimea Canal, leading to a water shortage for Crimea's agricultural fields growing grapes, rice, maize and soya. The canal delivers water from the Dnieper River in southern Ukraine and accounts for 80% of Crimea's water supply. The Ukraine government denies any political motive for the cut in supply, describing the reduced flow as a temporary result of building a flow structure to measure water deliveries.

2014	Islamic State contaminates drinking water supplies, Syria	Weapon	Syria, Iraq	Western Asia	Islamic State deliberately contaminates drinking water with crude oil in the Balad district of the Salahaddin Governorate. Poisoned water supplies are also reported from Aleppo, Deir ez Zor, Raqqa, and Baghdad.
2014	Palestinian water supply and wastewater system hit by Israeli war planes	Casualty	Palestine	Western Asia	Israeli warplanes hit water supply wells, pipelines and wastewater treatment plants servicing hundreds of thousands of Palestinians in the Gaza region. Palestinian officials state that the attacks intentionally targeted the infrastructure, mounting a war crime under the Geneva Protocol.
2019	Israel reportedly cuts off water supplies to approximately 1,500 Palestinians	Weapon	Palestine	Western Asia	Palestine News Network reports a water shut off to 17 Palestinian communities in Hebron, West Bank. The report suggests that the shut off was a purposeful act by Israel to force the movement of local Palestinians out of their homes to make room for more Israeli settlements.

Note: This is not a complete list of conflicts. A complete list can be found at the above website.

REFERENCES

Abuwara, A. (2011). Can water end the Arab-Israeli conflict? *Al Jazeera English*, July 29, (http://english.aljazeera.net/indepth/features/2011/07/2011727851 9784574.html)

Bsharat, M. and Ramadan, S. A. (2011). Fears of violence mount in West Bank on eve of Palestinian UN bid, Xinhua News Agency, September 21. (http://news. xinhuanet.com/english2010/world/2011-09/21/c 131150218.htm)

EcoPeace Middle East (2014). https://ecopeaceme.org/water-energy-nexus/

Jayawardena, A. W. (2021). An inconvenient truth about access to safe drinking water, *International Journal of Environment and Climate Change*, 11(10): 158–168, Article no.IJECC.75343, ISSN: 2581-8627

Jayawardena, A. W. and Sivakumar, B. (2017). Mekong River, Chapter 102 in *Handbook of Applied Hydrology*, 2nd ed. (Editor -in-Chief: Vijay P Singh). McGaw Hill Education, pp 102–1 to 102–9.

Pacific Institute (2019). *Water Conflict Chronology*. Pacific Institute, Oakland, CA. www.worldwater.org/water-conflict/.

United Nations (2018). Sustainable development goal 6: Synthesis Report 2018 on Water and Sanitation, United Nations.

World Water Development Report (2019). *Leaving No One Behind*, World Water Assessment Programme, UNESCO.

Appendix

Names in use for water in some languages

Arabic	Maa
Africanos	Water
Bengali	Jal
Burmese	Yei
Chinese	Soi
Czech	Voda
Dutch	Water
Danish	Vand
English	Water
French	Eau
German	Wasser
Greek	Nero
Hebrew	Maim
Hungarian	Viz
Hindi	Paani
Indonesian	Air
Irish	Uisce
Italian	Acqua
Japanese	Mizu
Korean	Mool
Khamer	Thuk
Lao	Nam
Latin	Aqua
Luxembourgish	Waasser
Malayalam	Vellam
Malaysian (Malay)	Air
Maori	Wai
Norwegian	Vaan
Punjabi	Jal
Persian	Ab
Polish	Woda
Portuguese	Agua
Russian	Voda
Rumanian	Apa
Sindhi (Pakistan)	Panhi
Spanish	Agua
Sinhalese	Wathura, Jalaya
Swahili	Maji

Swedish	Vatten
Tagalog	Tubig
Tamil	Thaneer
Thai	Naam
Turkish	Su
Ukrainian	Voda
Urdu	Pani
Vietnamese	Nuoc

Author index

Subject index

Printed in the United States
by Baker & Taylor Publisher Services

Printed in the United States
by Baker & Taylor Publisher Services